Transonic Wind Tunnel Testing

Bernhard H. Goethert

Edited by
Wilbur C. Nelson

DOVER PUBLICATIONS, INC.
Mineola, New York

Bibliographical Note

This Dover edition, first published in 2007, is an unabridged republication of the work published by Pergamon Press, New York, in 1961, for and on behalf of the Advisory Group for Aeronautical Research and Development, North Atlantic Treaty Organization.

International Standard Book Number: 0-486-45881-4

Manufactured in the United States of America
Dover Publications, Inc., 31 East 2nd Street, Mineola, N.Y. 11501

This is one of a series of publications by the NATO-AGARD Fluid Dynamics Panel (formerly Wind Tunnel and Model Testing Panel). Professor Wilbur C. Nelson of the University of Michigan is the editor.

CONTENTS

		page
FOREWORD		xi
ACKNOWLEDGMENTS		xiii
LIST OF SYMBOLS		xv
1. INTRODUCTION		1
2. REVIEW OF THE HISTORY OF TRANSONIC TEST FACILITY DEVELOPMENT		2
1. Low Speed Wind Tunnels		2
2. High Speed Subsonic Wind Tunnels		7
3. Drop Tests		11
4. Rotating Rigs		12
5. Transonic Testing with Wing-Flow Technique		14
6. Transonic Testing in Closed Wind Tunnels with Movable Sidewalls		17
7. Transonic Testing in Closed Wind Tunnels with Thick Boundary Layer		18
8. Wind Tunnels with Slotted Walls		21
9. Wind Tunnels with Perforated Walls		25
10. Concluding Remarks		29
References and Bibliography		30
3. FLOW AROUND AIRPLANES IN FREE FLIGHT		32
1. Subsonic Speed Range		32
2. Mach Number Influence in Subsonic Speed Range		33
3. Sonic Speed		37
4. Supersonic Speed Range		39
References and Bibliography		40
4. WALL INTERFERENCE CORRECTIONS IN CONVENTIONAL WIND TUNNELS		41
1. Open-Jet Wind Tunnels		41
2. Closed Wind Tunnels		44
3. Comparison of Wall Interference Corrections for Various Types of Test Sections (Subsonic Flow)		53
References and Bibliography		56
5. SUBSONIC FLOW IN WIND TUNNELS WITH LONGITUDINAL SLOTS		57
1. Physical Aspects and Wall Interference in Slotted Wind Tunnels (Without Lift)		57
2. Characteristics of Special Slot Configurations		68
3. Distribution of Velocity Corrections along Model		71
4. Mach Number Influence on the Effectiveness of Walls with Longitudinal Slots (Without Lift)		74

CONTENTS

     5. Lifting Wing in a Circular Wind Tunnel with Longitudinal Slots. 75
     References and Bibliography 84

6. SUBSONIC FLOW IN WIND TUNNELS WITH PERFORATED WALLS . 86
     1. Basic Characteristics of Perforated Walls . . . 86
     2. Flow Distortions in Perforated Wall Wind Tunnels . . 88
     3. Velocity-corrections in the Vicinity of One Perforated Wall (Two-dimensional, Incompressible Flow) . . . 92
     4. Flow Disturbances in a Perforated Two-dimensional Wind Tunnel (Two Perforated Walls) 100
     5. Velocity Corrections in Circular Perforated Wind Tunnels . 103
     6. Comparison of the Flow at Large Distance Behind a Lifting Wing in Slotted and Perforated Wind Tunnels . . . 106
     Appendix 107
     References and Bibliography 108

7. EXPERIMENTS IN SLOTTED AND PERFORATED WIND TUNNELS AT SUBSONIC AND SONIC MACH NUMBERS 109
     1. Introduction 109
     2. Two-dimensional Double-wedge Airfoil Models in Conventional and Slotted Wind Tunnels 109
     3. Two-dimensional Airfoil Model in Different Slotted Wind Tunnels 116
     4. Body of Revolution in Slotted Wind Tunnels and in Free Fall . 120
     5. Experiments with Bodies of Revolution in Perforated Test Sections 123
     6. Experiments with Wing-fuselage Models in Perforated Test Sections 125
     7. Bodies of Revolution at Sonic Speed 130
     References and Bibliography 134

8. SUPERSONIC WALL INTERFERENCE IN PARTIALLY OPEN TEST SECTIONS FOR TWO-DIMENSIONAL CONFIGURATIONS . . 136
     1. General Considerations Concerning Wave Reflections in Open, Closed, and Partially Open Test Sections . . 136
     2. Theory of Wave Cancellations in Wind Tunnels with Longitudinal Slots 140
     3. Theory of Wave Cancellation in Perforated Wind Tunnels . 142
     4. Experimental Results for Two-dimensional Shock-wave Cancellation 144
     5. Experimental Results for Two-dimensional Expansion-Wave Cancellation 157
     Appendix 159
     References and Bibliography 165

9. SUPERSONIC WALL INTERFERENCE IN PARTIALLY OPEN TEST SECTIONS FOR MODELS OF ROTATIONAL SYMMETRY AND COMPARISON WITH TWO-DIMENSIONAL WALL INTERFERENCE . . 167
     1. Boundary Conditions along Wind Tunnel Walls: Two-dimensional and Three-dimensional Models 167

2. Model Disturbances Compared with Cross-flow Characteristics of Perforated Test Sections (Conventional and Differential-resistance Type). 171
   3. Model Disturbance Characteristics Compared with the Cross-flow Characteristics of Combined Slotted–perforated Test-section Walls 174
   4. Wind Tunnel Test Results Obtained for Cone-cylinder Models in Various Partially Open Test Sections at a Mach Number of 1.20 178
   5. Results of Wind Tunnel Testing with Three-dimensional Models in Perforated Test Sections at Various Mach Numbers and Blockage Ratios 190
      References and Bibliography 198

10. TYPICAL TEST RESULTS FROM COMPLETE AIRPLANE MODELS IN TRANSONIC WIND TUNNELS 200
    1. Influence of Flow Non-uniformities 200
    2. Comparison of Force Measurements for a Fuselage-wing Model in Slotted and Perforated Test Sections . . . 203
    3. Tests in Slotted Test Sections 209
    4. Tests in Perforated Test Sections 217
    5. Comparison between Wind Tunnel and Flight Test Results . 229
    6. Concluding Remarks 232
       References and Bibliography 234

11. CROSS-FLOW CHARACTERISTICS OF PARTIALLY OPEN WALL ELEMENTS 236
    1. Physical Aspects of the Flow through Partially Open Walls . 236
    2. Theoretical Calculations for the Pressure Drop through Perforated Walls 240
    3. Experiments to Determine the Cross-flow Characteristics of Perforated Walls 248
    4. Experiments to Determine the Influence of Detailed Wall Geometry and Flow Parameters 256
    5. Experiments to Determine Cross-flow Characteristics of Walls with Longitudinal Slots 270
    6. Flow Disturbances of Wall Openings and Their Decay . . 275
    7. Concluding Remarks 279
       References and Bibliography 280

12. FLOW ESTABLISHMENT IN TRANSONIC WIND TUNNELS . . 281
    1. General Discussion of the Problem Areas . . . 281
    2. Establishment of Subsonic Flow in Perforated Test Sections . 283
    3. Establishment of Subsonic Flow in Slotted Test Sections . 296
    4. Supersonic Flow Establishment in Perforated Test Sections . 300
    5. Supersonic Flow Establishment in Slotted Test Sections . . 312

CONTENTS

6. Requirements for Mass Flow Removal from Partially Open Test Sections 319
7. Calibration of Partially Open Test Sections . . . 325
   References and Bibliography 331

13. **BOUNDARY LAYER GROWTH ALONG PARTIALLY OPEN WALLS** 334
    1. Simplified Theoretical Considerations 334
    2. Refined Theory for Boundary Layer Growth along Perforated Walls 339
    3. Experimental Determination of Velocity Profiles and Displacement Thickness of Boundary Layers Adjacent to Perforated Walls 346
    4. Experimental Determination of Boundary Layer Development in Slotted Test Sections 352
    References and Bibliography 355

14. **POWER REQUIREMENTS OF TRANSONIC WIND TUNNELS** . 356
    1. Influence of Auxiliary Plenum Chamber Suction on Power Requirements 356
    2. Power Requirements of Perforated Test Sections for Various Combinations of Auxiliary Plenum Chamber and Diffuser Suction 358
    3. Influence of Wall Setting on Power Requirements (Perforated Test Sections) 364
    4. Influence of Wall Geometry (Perforated Walls) . . 367
    5. Influence of Model Size (Perforated Test Section) . 369
    6. Comparison of Perforated Test Sections Equipped with Sonic and Supersonic Nozzles 370
    7. Power Requirements of Wind Tunnels with Slotted Test Sections 371
    8. Full-scale Wind Tunnels 378
    References and Bibliography 381

15. **LIST OF TRANSONIC WIND TUNNELS AND THEIR MAIN CHARACTERISTIC DATA** 383
    1. Abbreviations Used 383
    2. Table of Transonic Wind Tunnels 384

16. **CONCLUDING REMARKS** 390

INDEX 392

FOREWORD

TRANSONIC aerodynamics and especially transonic wind tunnel testing constituted for a long time one of the critical chapters of aerodynamic science. The theoretical reason can be easily seen: purely subsonic and purely supersonic flows can be described by approximate solutions derived from linear differential equations; therefore scaling rules and fundamental properties of such flows can be described in physical terms, which can be visualized in a relatively easy manner. If we have to deal with a flow in the transonic regime, the similarity laws and scaling rules are not directly apparent and the measurements become highly dependent on the ratio between dimensions of the wind tunnel and the dimensions of the model. The first observations showed, for example, a rapid increase of the drag of wing profile as the speed approached the sound velocity; however, the value of the speed at which this drag increase occurred, and its magnitude, depended more on the characteristics of the wind tunnel than on the characteristics of the wing profile. Some of the quite experienced experimentalists published test results before they realized the enormous influence of wind tunnel effects.

But even today with all the experiences derived from a large number of interesting publications, a comprehensive study of the problems of transonic wind tunnel testing has a great value. I am convinced that active workers in this field, as well as students, should find the present work of considerable interest and value, combining as it does vast knowledge and broad recent experience.

TH. VON KÁRMÁN

ACKNOWLEDGMENTS

THE author wishes to express his appreciation to the organizations and individuals who generously supported the preparation of this AGARDograph with their cooperation and assistance.

In particular, the author is indebted to the Arnold Engineering Development Center, Arnold Air Force Station, Tennessee, U.S.A., for approval of the extensive use of AEDC material and assistance in the literature survey and manuscript preparation. Thanks are also due to the many research and industrial organizations, too numerous to be cited here individually, who approved the use of both published reports and unpublished material.

Grateful acknowledgment is made of the contributions of individuals who assisted during the various phases of preparation of the AGARDograph; especially to Col. B. W. MARSCHNER, Ph.D., for his assistance in the selection of theoretical material on Mach one flow, and to Major J. T. WAREING for preparing the list of wind tunnels in Chapter 15 and for other valuable help.

Personal thanks are extended to Mrs. M. D. WALL and GEORGE W. CHUMBLEY for editorial assistance, to Mrs. M. R. DICKERSON for typing the manuscript, and to W. S. WISEMAN and D. NORTHCUTT for illustration work.

LIST OF SYMBOLS

A Area
ΔA_B Frontal area of bulging streamlines
A_M Cross-sectional area of model
A_T Cross-sectional area of test section
A_W Wing area
A_{wp} Area of perforated wall
A, B Constants (Chapt. 5)
a Velocity of sound
b/a Side ratio of rectangular tunnel test section
c Wing chord
c Open area ratio (Chapt. 5)
C, c Experimental constants (Chapts. 8, 13)
C_{Div} Sum of widths of divergent walls
C_T Circumference of test section at upstream end
C_L Lift coefficient
C_D Drag coefficient
$(C_D)_{\text{sub}}$ Drag coefficient at sub-critical Mach number
C_M Moment coefficient
C_p Pressure coefficient, $\Delta p/q$
D Diameter of circular test section
d Distance between slot centers
F Normalized stream function: $\psi = v_\infty x F$
H Test section height
h Distance between streamlines
K Characteristic wall constant for "no-reflection" condition (Chapt. 8)
K Constant for crossflow pressure drop through perforated wall (Chapt. 6)
K Correction factor (Chapt. 4)
L Length of test section walls
l Mixing length in boundary layer (Chapt. 8)
l Wave length in x-direction (Chapt. 6)
M Mach number
m 1, 2, 3 ... constants (Chapt. 5)
m Intensity of three-dimensional doublet (Chapt. 6)
m Mass flow through test section (Chapt. 13)
Δm Mass flow removal from test section (Chapt. 13)
m' Mass flow ratio, $\Delta m/m$ (Chapt. 13)
m_a Mass flow between two streamlines
n Location of image doublet (Chapt. 6)
n Coordinate normal to flow (Chapt. 5)
$2n$ Number of slots of equal width, equally spaced
p Static pressure

LIST OF SYMBOLS

p_0 Total pressure
Δp_r Pressure rise due to partial shock reflection
Δp_{sh} Pressure rise due to shock wave
Q Source intensity
q Dynamic pressure, $\rho V^2/2$
R Ratio of open to total wall area
R Radius of curvature
R Tunnel radius (Chapts. 5, 6)
R_e Reynolds number
R_c Radius of curvature (Chapt. 5)
r Radial distance (Chapt. 3)
r Radial coordinate (Chapt. 5)
r, x, ω Cylindrical coordinates
r_0 Radius of model cylinder
Δr_{max} Maximum radial bulging of streamlines
s Slot width
t Wing thickness
t Circumferential coordinate (Chapt. 5)
v Velocity
Vol Volume of model
W Complex velocity function
w Width of perforated wall
x, y, z Coordinates
y_1 Thickness of mixing layer
Δy_D Characteristic lateral width of disturbance field around model
Z Complex coordinate, $x + iy = re^{i\delta} = Z$
dZ Element of centrifugal force (Chapt. 4)
α Angle of attack
α Experimental constant (Chapt. 13)
α_0 Constant referring to mixing zone of free jet (Chapt. 13)
θ Flow inclination
θ Direction angle of vector Z (Chapt. 5)
2ν Angle indicating width of slots (Chapt. 5)
λ_v Form coefficient (Chapt. 6)
Φ Complex stream function (Chapt. 6)
Φ Total velocity potential (Chapt. 4)
ϕ Velocity potential
ρ Density of air
γ Ratio of specific heats for air
ψ Stream function
δ^* Boundary layer displacement thickness
η Coordinate ratio, y/x
η_1 Mixing layer coordinate ratio, y_1/x
η_w Wall divergence
ξ Dimensionless coordinate, $\eta/(c\eta_1)^{2/3}$
τ Shear stress
Γ Circulation (Chapt. 6, 11)

SUBSCRIPTS

- 1 Singularity in free flight
- 2 Singularity to satisfy perforated wall condition
- 2c Singularity to satisfy closed wall condition
- c Compressible flow
- D Displacement correction
- n Component normal to flow direction
- p Porous wall
- r Component in radial direction
- s Slot region
- w At or near the wall
- x, y, z Component in x, y or z direction
- ∞ Undisturbed flow

CHAPTER 1

INTRODUCTION

SINCE the turn of this century the art of wind tunnel testing has progressed very rapidly. In the course of the development cycle from small, low-speed wind tunnels to the large, high-speed wind tunnels of the 1930's and 40's, the aerodynamic design problems remained essentially the same, and the mechanical construction of the tunnels was readily accomplished on the basis of the general knowledge in this field. During the 1930's, the design and construction of supersonic wind tunnels was also mastered by several independent groups of investigators, and rapid progress was made in establishing test flows of good quality and steadily increasing supersonic Mach numbers.

The Mach number range around sonic speed, however, exhibits unusual difficulties, both in the design of airplanes as well as in that of wind tunnels. This difficulty is caused by the well-known aerodynamic fact that in the transonic speed range the flow changes its character from a "single-type flow", either subsonic or supersonic, to a "mixed-type flow" with local supersonic fields imbedded in subsonic flow or local subsonic fields imbedded in supersonic flow. Even today, the complexity of transonic flow has made it impossible to establish simple transonic theories with which the characteristics of airplanes or components can be reliably predicted. Consequently, the transonic speed range depends more than either the subsonic or the supersonic speed ranges on experiments to provide the required aerodynamic information for airplane and missile designers.

Obviously the reliability and the limits of wind tunnel testing must be thoroughly understood before the results of such tests can be safely applied to airplane design. This rather general requirement is much more pronounced in the transonic than in either the subsonic or supersonic speed ranges because the transonic wind tunnels developed to date are very sensitive with respect to the arrangement of slots, perforation pattern, model size, etc.

This AGARDograph therefore presents a review of the extensive efforts made during the last 15 years to develop practical transonic wind tunnels and discusses their performance and their limitations. Emphasis is placed on the design and operational characteristics of both types of modern transonic wind tunnels, that is, on wind tunnels with either longitudinally slotted or perforated test sections. Each chapter of the AGARDograph has been made as complete as possible, with its own series of figures and a separate list of references and bibliographical material.

CHAPTER 2

REVIEW OF THE HISTORY OF TRANSONIC TEST FACILITY DEVELOPMENT

1. LOW-SPEED WIND TUNNELS

a. Period before 1930

An appraisal of the advances in wind tunnel test techniques can be readily obtained by comparing the modern wind tunnel facilities with their historical predecessors. For example, test equipment such as the wind tunnel built and used by the Wright Brothers in the early 1900's paved the way for the development of the early flying machines of this century. Since the basic problems were very elementary, relatively crude wind tunnels sufficed to provide the required information. The Wright Brothers' wind tunnel (see Fig. 2.1), for instance, was used to provide basic information on the lifting characteristic of wings, on the drag of the wings and the entire flying article, and particularly on the center of pressure and its location and shifting, which are so vitally important to the stability and control of an airplane (Fig. 2.2).

After they had succeeded in building and flying the first airplanes of this century, the inventors and engineers of the period required more refined wind tunnels. Subsequently Eiffel developed the "open return" type wind tunnel in France and Prandtl the "closed return" type wind tunnel in Germany. Problems investigated in these wind tunnels were concerned with the effectiveness of wing contours, the lift distribution in chord and spanwise directions, wing-planform influence, control, etc. The foundations of our knowledge of aerodynamic drag were established; and its principal categories, that is, pressure drag which is produced mainly by wing tip vortices (induced drag) and by separation of the flow, and surface friction were explored. Wind tunnels of this period had a more uniform flow, less turbulence, and more precise measuring equipment to determine the forces and pressures acting on the wing surfaces than did those of the earlier period.

The engineering efforts before approximately 1930, including extensive use of wind tunnels, had as the end product an airplane from which most unnecessary external structural elements such as wing struts, landing gear, and unshrouded engine installations had been eliminated. Aerodynamically clean aircraft with smooth wings and fuselage, and shrouded engine installations were produced for numerous commercial uses (Fig. 2.3).

b. Period after 1930

After the external drag of the airplanes had been largely reduced, further progress was made by reducing the wing area and by making the airplane surfaces as smooth as possible. The influence of wing roughness, rivets, steps in the contours, etc., was considered in the research for further improvements. It became necessary to conduct reliable investigations of these individually small effects which could, in their entirety, result in large performance improvements. Hence, the period of the 1930's is characterized by the development of large wind tunnels in which full-scale aircraft or full-scale components could be tested to provide reliable information on

FIG. 2.1. *Wright Brothers' wind tunnel (1900)*.

FIG. 2.2. *Wright Brothers' airplane (1903)*.

small-scale disturbances. The wind tunnels of this era are, for instance, the large 30 × 60 ft wind tunnel of the NACA at Langley Field, Virginia, and similar wind tunnels at Berlin, Paris (Fig. 2.4), and other places.

The product of this era of aerodynamic progress was the clean airplane achieved at the end of the 1930's and at the beginning of the 1940's (Fig. 2.5). Wing areas had been reduced to a minimum, compatible with flight performance and take-off characteristics and with full utilization of various lift augmenting devices such as trailing and leading-edge flaps.

Fig. 2.3. *Airplane of the 1930's—period preceding the DC-3.*

During the 1930's it was also recognized, on the basis of theoretical and experimental work, that one predominant part of the airplane drag was the turbulent boundary layer drag. Sizable reductions of this drag could be achieved when the turbulence of the boundary layer could be eliminated, and laminar boundary layer flow could be established. To study this peculiar effect it was necessary to develop wind tunnels with exceptionally smooth flow. The turbulence generally inherent in conventional wind tunnel flow makes it nearly impossible to investigate aerodynamically smooth airplane models with proper simulation of free flight in the turbulence-free atmosphere. As a result, a number of special wind tunnels were designed at the end of the 1930's and during the 1940's to investigate the laminar friction effect. Wind tunnels of this type were characterized mainly by large contraction ratios, the installation of several fine-mesh turbulence screens, and careful contouring of the inlet nozzle as well as the diffuser behind the test section.

At the present time, that is, around 1960, we are still in the midst of attempts to exploit the definitely demonstrated opportunity of significantly reducing the drag of a subsonic airplane (mainly its friction drag) by stabilizing the laminar boundary layer.

a. *Test section of NACA Langley 30 × 60 ft wind tunnel with model installation.*

b. *Open-circuit wind tunnel at Chalais-Meudon, France.*

Fig. 2.4. *Large wind tunnels of the 1930's—period for tests of full-scale airplanes.*

Fig. 2.5. *Airplane He-70 for passenger and transport service (1934)*.

2. HIGH-SPEED SUBSONIC WIND TUNNELS

a. General consideration of compressibility effect

The wind tunnels used to develop aircraft before the early 1940's were not designed for a study of the compressibility effect of the air. At the flight speeds of most airplanes of that time (only a few of them exceeded velocities of 50 or 60 per cent of sonic speed), the air behaves much like an incompressible fluid. This fact can be readily recognized by considering the density of air at various flight Mach numbers in comparison with the stagnation point density:

Density Changes at Various Flight Speeds

Speed at Sea Level, m.p.h.	Mach No.	Density Change, $\Delta\rho/\rho_0$, per cent
304	0.40	7.6
456	0.60	15.9
608	0.80	26.0
760	1.00	36.6

It is readily seen from this table that at flight speeds corresponding to Mach number of 0.50 the compressibility of the air can be taken into consideration by a small correction. Thus reliable results can be obtained in practically incompressible flow, that is, in low-speed wind tunnels. However, the table also shows that the density change grows very rapidly and becomes a major parameter when flight Mach numbers of 0.6 are exceeded. Hence, at the end of 30 years, when the trend of the fast airplanes of that period (fighter and bomber aircraft) was toward maximum flight speeds approaching sonic speed, the need for reliable aerodynamic design information resulted in major efforts to develop high speed wind tunnels.

b. Closed wind tunnels

In the 1930's, experimental research in small high speed subsonic wind tunnels was begun in order to obtain insight into the manner in which the characteristics of airplane components are influenced by the compressibility of the air. Typical small wind tunnels of this period are the wind tunnels of the NACA in the United States and the wind tunnels of the Aerodynamic Research Institute at Goettingen (supported by the theoretical work of Prandtl and Buseman). The first experimental information on the pressure distribution and the characteristics of simple two-dimensional wing sections at high subsonic speeds was obtained in these small wind tunnels. Drastic changes in drag, lift, and center-of-pressure location and its dependence on wing thickness, camber, and thickness distribution were documented in a cursory manner for the first time.

Testing in these small wind tunnels revealed such significant aerodynamic changes that extraordinary efforts were soon made in several countries to investigate the observed effects more reliably in wind tunnels of larger scale.

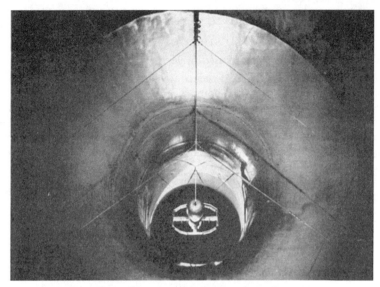

a. *View downstream into test section.*

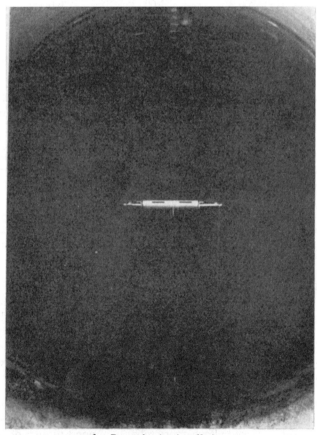

b. *Research wing installation.*

FIG. 2.6. *High-speed wind tunnel (2.7 m diameter) of the DVL, Berlin, Germany (1939).*

In Berlin the DVL, the first large wind tunnel (2.7 m diam) with velocities up to choking, started operation early in 1939 (Fig. 2.6). In the United States, several wind tunnels approximately 8 ft in diameter were put into operation by the NACA in 1940. Wind tunnels of this type provided the information required to develop the high-speed military aircraft used in World War II.

This period of airplane development was characterized by a rapid improvement in performance made possible by the advent of the turbojet

a. *Heinkel 178 (1939)*.

b. *Messerschmitt 262 (1941)*.

Fig. 2.7. *First jet-powered airplanes.*

engine. The new-type engine provided a much more powerful and efficient propulsion system at the high speeds involved than the reciprocating engine–propeller system (Fig. 2.7). Also the aerodynamic shape of the airplanes of this period was so clean that, in slightly descending trajectories (not considering steep dives), it was possible to obtain velocities which approached sonic speed. The flight tests paralleling the tunnel investigations showed a tremendous increase in drag and unpredictable behavior of the airplanes, coupled with ineffectiveness or reversal of the controls, near sonic speeds. This period of development was marked by many total losses of airplanes caused by disintegration of structural components caused by changes in the pressure distribution or by the inability of the airplane controls to maintain level flight.

During these years the pressure on wind tunnel operators was very great to provide test facilities in which the behavior of airplanes could be studied at speeds immediately below and through sonic speed. Normal closed-type wind tunnels could not provide the required aerodynamic information in the speed range close to Mach number one because of the choking phenomenon. The difficult situation facing the wind tunnel operators can best be described by a cursory review of the permissible blockage of a model at high subsonic Mach numbers (blockage calculated for one-dimensional flow):

Blockage for Choking in Closed Wind Tunnels
10 ft Wind Tunnel (Circular)

Mach No.	Blockage, per cent	Wing Thickness, in.	Fuselage Diam, in.
0.8	3.7	3.5	23.1
0.9	0.88	0.83	11.2
0.92	0.56	0.53	8.9
0.94	0.31	0.29	6.7
0.96	0.14	0.13	4.5
0.98	0.03	0.03	2.1
1.00	0.0	0.0	0.0

The table shows that two-dimensional wing tests become practically impossible in closed wind tunnels at speeds above Mach number 0.90. In the case of fuselage models with circular cross sections, the problem is somewhat less acute since, at Mach number 0.90 a fuselage nearly 1 ft in diameter can still be tested in a closed wind tunnel as can a fuselage of 6.7 in. diam at a Mach number of 0.94.

It is obvious, however, that in closed wind tunnels the model size rapidly becomes impracticably small when sonic speed is approached. Consequently, during the 1940's much emphasis was placed on developing other types of wind tunnels or test facilities which could provide the required aerodynamic information.

c. Open-jet wind tunnels

In order to alleviate the choking phenomenon, frequent attempts were made to investigate the transonic aerodynamic problems in open-jet wind tunnels. These attempts, made principally in the late 1930's and early 1940's, did not prove successful. The failure was the result of difficulties connected with the strong tendency of open-jet tunnels to develop pulsations in the test-section flow. These pulsations originated in the turbulence of the mixing layer along the jet boundaries and made the flow around the model and its shock waves very unsteady. Also, tests in open-jet wind tunnels, even when the pressure in the surrounding open chamber was kept at a known static pressure (for example, corresponding to Mach number one), required the application of a Mach number correction of unknown magnitude. Added to this were the large power requirements of open-jet wind tunnels, which again can be traced back to the mixing layer of the jet boundaries and the resulting low diffuser efficiency. It is therefore understandable that serious consideration of this type of wind tunnel was dropped at an early stage. Attention was again concentrated on the closed wind tunnel which, if extremely small models were used, allowed sonic speed to be approached from below and from above; however, a gap was left which was unsuitable for the acquisition of test data.

3. DROP TESTS

In order to obtain interference-free experimental data on the characteristics of airplane components in the critical speed range near Mach number one, efforts were made beginning around 1941 and continuing for several years to deduce such data from tests with unpowered models dropped from high altitudes. The model trajectory was measured by means of ground-based optical observation stations or by built-in telemetry instrumentation, and the velocity, acceleration or deceleration, and aerodynamic forces were determined.

Some typical tests of this period were conducted in the DVL, Berlin, with bodies of revolution.[1] By using the ultra-high-altitude aircraft of this period (maximum altitude close to 50,000 ft) and dropping several models with the same external shape but different weights, it was possible to obtain aerodynamic data in the critical speed range, as depicted in Fig. 2.8. The accuracy of these data is quite satisfactory. Other model shapes were also examined using the same method; one had rectangular and swept-back wings stabilized by means of long cylindrical centerbodies with small thin tail surfaces. Interference-free drag and other aerodynamic data were obtained for such model shapes.

The method of free-flight model testing was subsequently greatly improved, particularly by activities carried on in the United States. Models were equipped with their own power supply, usually rockets, so that they would accelerate to large velocities when launched from the ground.[2] In these tests, the booster, properly equipped with test surfaces, could serve as the test model, or the booster could be used to launch a highly instrumentated test model into the air where it would separate from the model after the boost phase. Test data could then be obtained for the separated model alone.

The advantages of the free-flight model testing technique lie in the fact that there is no possibility of wall interference effects, as in wind tunnels. Large models of airplanes or missiles can be used to provide reliable stability and control data of both the static and the dynamic type. However, such

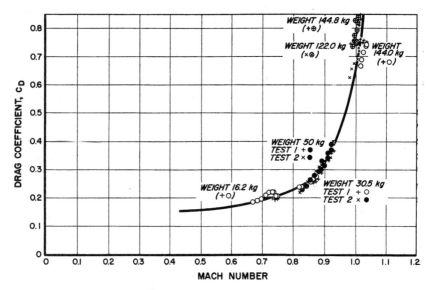

Fig. 2.8. *Drag coefficient of body of revolution with stability fins from drop tests (1944)*.[1]

tests are very expensive, and the production of test data is relatively slow. Also the number of parameters for which information can be obtained (pressures, forces, moments, etc.) is relatively limited in comparison with the abundant possibilities offered in ground test facilities.

The free-flight model testing technique, therefore, even though it is still used today, is used only for special problems. These special problems are mainly concerned with the dynamic stability of airplanes or missiles or with the establishment of reference data to check the reliability of wind tunnel data in cases of sensitive flow conditions.

4. ROTATING RIGS

During the search for experimental methods suitable for exploration of the transonic speed range, it was frequently suggested that rotating rigs be built. With such rigs, the models were mounted on a rotating arm and were thus subjected to high relative velocities. Usually the disturbances by the wake of the model flow were reduced by superposing a small perpendicular flow to remove the wake from the path of the model during the succeeding revolutions.

Successful applications of this test method are known only in cases of small, very simple models. In the test rig of the NACA, for instance, a diamond-shaped two-dimensional wing was investigated through the transonic speed range. The data obtained compared quite well with the values

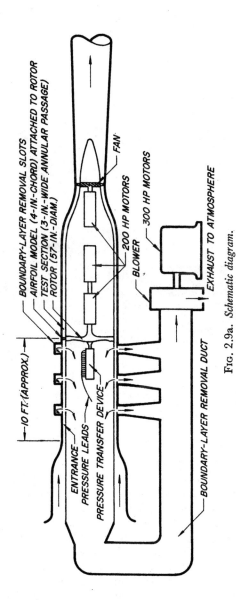

Fig. 2.9a. Schematic diagram.

predicted by theory and supported the validity of these theoretical methods, especially at Mach number one (Fig. 2.9).

However, because of the overall complexity and the influence of the relative curvature of the test flow, the application of the whirling-arm method has not been extensive. It has generally been restricted to use in research investigations of simple two-dimensional airfoil shapes similar to that discussed above.

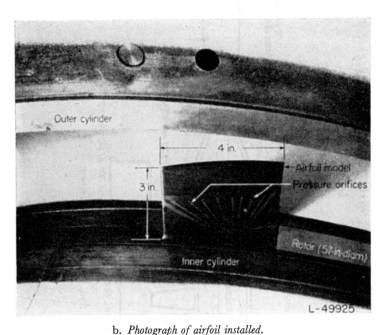

b. *Photograph of airfoil installed.*
FIG. 2.9. The Langley annular transonic tunnel (rotating rig) (1948).[3]

5. TRANSONIC TESTING WITH WING-FLOW TECHNIQUE

a. *Wing-flow technique with airplanes in free flight*

The technique of wing-flow testing stemmed from the urgent need which arose during the 1940's to provide aerodynamic information in the transonic speed range from slightly below Mach number one to low supersonic Mach numbers. It makes use of the fact that a local high-velocity field occurs around the wing with considerably higher Mach numbers than the free-flight Mach number. For instance, a typical airplane wing will have developed local supersonic disturbance fields when the flight Mach number is still as low as 0.7 or slightly above. Thus, by flying an airplane in the speed range of Mach numbers 0.7 or 0.75 and placing a model on the wing in the region of the local high Mach number field, aerodynamic information can be obtained about model characteristics in the critical transonic Mach number range. A picture of such an installation is given in Fig. 2.10. Numerous tests of this type were conducted in the United States during the 1940's by the NACA.[4]

The advantage of this method was that for the first time it was possible to obtain aerodynamic data concerning the overall forces and the pressure distribution for wings and other model components through the critical transonic speed range. On the other hand, the disadvantages of the method are very apparent. It has all the difficulties associated with flight tests of

Fig. 2.10. *Wing-flow method demonstrated by model mounted on wing of P-51D airplane (1947)*.[4]

full-scale airplanes. Also, the low rate of data collection and the high cost of operation weigh heavily against it. Obviously only very small models can be used and the resulting low Reynolds numbers contribute to further uncertainty in the evaluation of such data. In addition, the most serious handicap of the wing-flow method derives from the non-uniformity of the flow field in which the model is placed. In contrast to the uniform flow existing in good wind tunnels, in the wing-flow method the model is exposed to a test flow field which has velocity gradients in the flow direction and perpendicular to it. Since near Mach number one the flow is extremely sensitive with respect to changes in stream density, the reliability of test data obtained by the wing-flow method cannot be readily assessed. As a consequence, this method was completely abandoned as soon as more reliable test methods became available.

b. Wing-flow technique using bumps in wind tunnels

The transonic bump technique in wind tunnels represents an exact parallel to the wing-flow method in flight testing described in the preceding paragraphs. This method, which was employed in the United States in the 1940's, provided a curved surface like a large bump in a subsonic wind tunnel. Along this curved surface or bump, a disturbance field was established similar to that occurring above an airplane wing. Mach numbers extending into the low supersonic range could be established in the local disturbance field. By suitable shaping of the contour of the bump, the

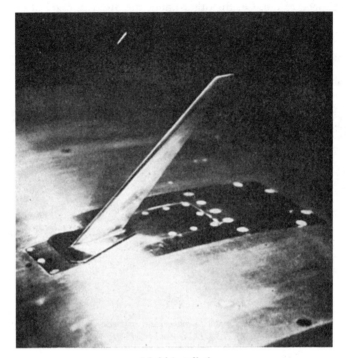

a. *Model installation.*

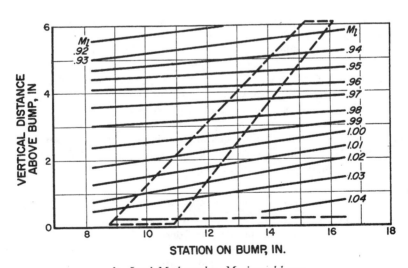

b. *Local Mach number, M_l, in model area.*

FIG. 2.11. *Transonic bump in NACA 7×10 ft wind tunnel (1951).*[5]

velocity gradients in both the horizontal and vertical directions could be greatly reduced.[5,6] A typical installation is shown in Fig. 2.11a.

This method had the advantage that, by a relatively simple modification of an existing subsonic wind tunnel, it was possible to obtain aerodynamic data in the range from high subsonic to low supersonic Mach numbers. Also the rate of data collection and the instrumentation of the model could be more complete than in flight tests so that increased data accumulation could be accomplished.

However, the method did not eliminate the major disadvantage of the wing-flow method, that is, only relatively small models could be installed. For instance in the Ames 16 ft high-speed wind tunnel of the NASA, half-wing models with a half span of no more than 18 in. were typical installations.[7] Also the uniformity of the flow was subject to the same criticism discussed before (Fig. 2.11b).

Thus the bump method of wind tunnel testing was also completely abandoned as soon as more efficient transonic wind tunnels could be established.

6. TRANSONIC TESTING IN CLOSED WIND TUNNELS WITH MOVABLE SIDEWALLS

During the search for a suitable transonic test method, some subsonic wind tunnels were modified in the late 1940's to provide low supersonic velocities. These modifications were made in a crude but simple manner. The purpose was to provide in one wind tunnel the possibility of gathering information in both the subsonic speed range close to Mach number one and the supersonic speed range beginning at a Mach number slightly above one. Only a narrow gap around Mach number one remained in which no data could be collected.

An example of a wind tunnel of this type is the Wright Field 10 ft wind tunnel as it was during its initial phase of operation with solid walls (Fig. 2.12a and Ref. 8). In this tunnel, movable sidewalls were provided to increase the effective flow area in the test section downstream of the throat and to thus establish supersonic flow in the test region. The models naturally had to be very small, first to avoid excessive flow distortion in the high subsonic Mach number range and second, to avoid interference effects due to shock reflection on the solid tunnel walls.

The velocity distribution for the Wright Field 10 ft wind tunnel for various settings of the sidewalls are indicated in Fig. 2.12b. With a model 4 ft long, a Mach number non-uniformity of no more than ± 1.0 per cent occurred at the highest attainable test Mach number of 1.18.

A similar approach with the same objective was selected by a French group of the Institut Aéronautique, Saint-Cyr. This group provided a pair of movable inserts along the tunnel walls. Movement of the inserts changed the size of the throat and, consequently, allowed establishment of supersonic Mach numbers in the test section. The velocity distribution obtained with this device was also found to be satisfactory for some test work, the non-uniformities being mainly a function of the slenderness of the movable inserts.

Because of the limitations of model size and wall interference, these schemes also are no longer used in transonic testing.

TRANSONIC WIND TUNNEL TESTING

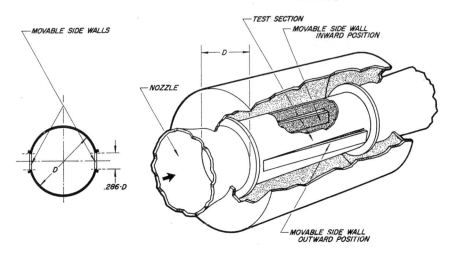

a. *Test section configuration.*

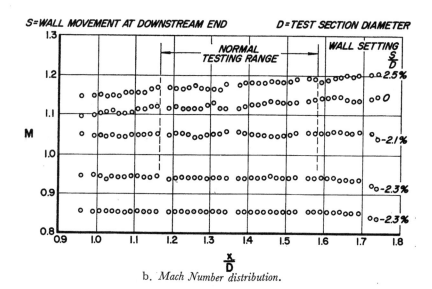

b. *Mach Number distribution.*

Fig. 2.12. *Wright Field 10 ft wind tunnel (closed) with movable sidewalls (1951).*[8]

7. TRANSONIC TESTING IN CLOSED WIND TUNNELS WITH THICK BOUNDARY LAYERS

a. Influence of wall boundary layer on choking Mach number in closed wind tunnels

Accurate experiments concerning the choking Mach number in closed wind tunnels as a function of the model displacement showed that the choking Mach number is frequently higher than predicted by inviscous flow theory. This phenomenon is particularly pronounced at small Reynolds numbers when the boundary layer along closed wind tunnel walls is relatively thick. Calculations on the displacement thickness of the wall boundary layer in the vicinity of the model, based on experimental data, indicated that the increase in the choking Mach number can be attributed to the thinning of the boundary layer as a result of the disturbed pressure field around the model.[9]

Normally, the increase in the choking Mach number, that is, in the maximum Mach number attainable in a closed wind tunnel, has the character of a relatively small correction. However, it can be reasoned that by artificial thickening of the wall boundary layer a sizable increase in test Mach number can be accomplished, even to the extent that supersonic flow in a test section with a constant geometrical cross section can be established. Experiments were conducted in several small model tunnels in which the thick boundary layer was produced either by installing screens in the stagnation section of the test section[10,11] or by recirculating low Mach number air along the test-section walls and through separate channels around the test section (see Fig. 2.13 and Ref. 11). Typical Mach number distributions are presented in Fig. 2.13b. It is apparent that the Mach number distribution in the test section is not very uniform. Furthermore, the center core of the flow is very small, less than $\frac{1}{4}$ the tunnel diameter or $\frac{1}{16}$ the tunnel cross-sectional area.

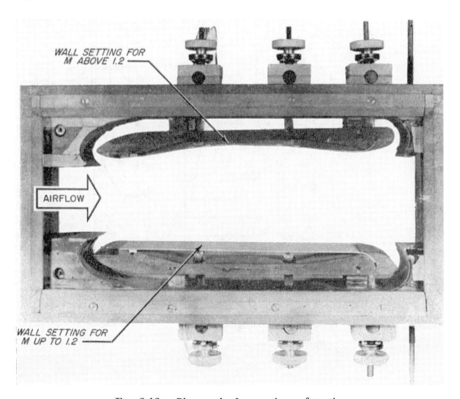

Fig. 2.13a. *Photograph of test section configuration.*

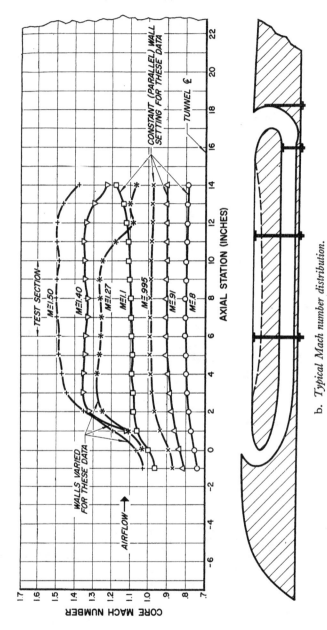

b. *Typical Mach number distribution.*

FIG. 2.13. *UAC transonic two-stream test section (United Aircraft Corporation, 1951).*[11]

Another difficulty of the two-stream wind tunnel is connected with the determination of the velocity corrections in the wind tunnel where the model is installed. Theoretical studies of this problem had previously been carried out for incompressible flow assuming "no-mixing" between the high-speed center core and the low-speed surrounding airflow.[12] However, in view of the numerous parameters involved (as, for instance, the distortion of the dividing line between the inner and outer cores by the model displacement, Mach number, and friction), such wind tunnel corrections become rather inaccurate. Consequently, tests with this type of transonic test section were not carried further than the model tunnel phase in spite of the fact that, in a given test section, the Mach number range from subsonic values to supersonic values as large as Mach number 1.33 could be covered merely by changing the power input to the wind tunnel drive.

b. Closed wind tunnels with air bleed-in for low supersonic operation

During the 1940's the urgent search for wind tunnels which could produce test results in the low supersonic Mach number range led to experiments with closed wind tunnels, both model and full-scale, in which an effective throat was formed at the upstream end of the test section by means of air bleed-in (Fig. 2.14). This scheme was applied in the 10 ft wind tunnel at Wright Field. It produced acceptable Mach number distributions in the range from Mach number 1.06 to 1.12 and, in addition, over the entire subsonic range available in this closed wind tunnel. The supersonic Mach numbers could be established by merely changing the inflow of bleed-air into the test section.[13] With this scheme, it was possible during the late 1940's to obtain supersonic test data on airplane models in a basically subsonic wind tunnel.

However, with the air bleed-in scheme, the power requirements for a wind tunnel became extremely large, particularly in the supersonic Mach number range. This scheme was used only temporarily to provide given wind tunnel test data through the high subsonic and the low supersonic Mach number range, for a given model installation, until such a time as more efficient transonic wind tunnels could be developed.

8. WIND TUNNELS WITH SLOTTED WALLS

a. High subsonic speed range

During the development of the low-speed wind tunnels in the 1930's and later, the combined influence of solid and open wall elements was investigated in detail by numerous researchers.[14,15] It was shown that the wind tunnel velocity correction could be greatly reduced by the proper arrangement of solid and open wall elements.

The demand for wind tunnels with low velocity corrections again became urgent during the 1940's when it became necessary to increase the test velocities, although it was recognized that wind tunnel velocity corrections increased with the third power of the Prandtl factor $\sqrt{(1-M_\infty^2)}$. To keep the magnitude of this wind tunnel correction reasonably small, it was therefore necessary to start with a low correction factor even in the incompressible flow.

At the same time it was again pointed out by several investigators that with a suitable geometry of longitudinal slots it should be possible to completely eliminate the velocity correction in incompressible wind tunnel flow. Moreover, application of the generalized similarity rule for compressible

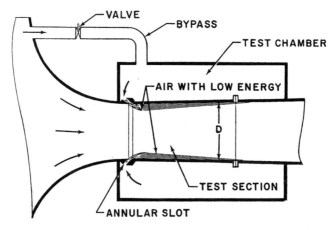

a. *Test section configuration.*

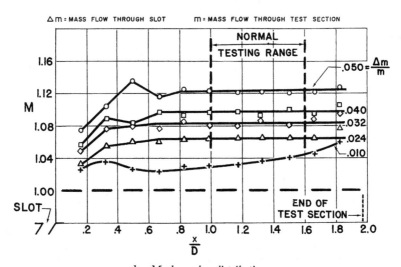

b. *Mach number distributions.*

Fig. 2.14. *Wright Field 10 ft wind tunnel with annular air-bleed-in (1949).*[13]

flow showed that this result would also remain valid at high subsonic velocities as long as the disturbance velocities remained sufficiently small. The calculations on this effect carried out in Germany,[15] Italy and Japan produced theoretical correction-free slot arrangements. These activities, however, did not result in the construction of slotted wind tunnels because of the circumstances connected with and following World War II.

b. Transonic speed range

The first really successful transonic wind tunnel was investigated in the United States in 1947 in tests at the NACA.[16] Based on calculations for the velocity correction in partially open and closed circular wind tunnels, it was determined at what relative slot open area the velocity correction would disappear at subsonic speeds. A small subsonic model tunnel was built with a test section 12 in. in diameter and having eight slots and a slot-open area ratio of 12.5 per cent (Fig. 2.15). The tests not only indicated

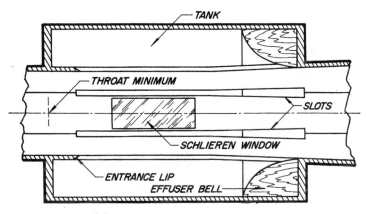

a. *Schematic diagram of transonic slotted tunnel.*

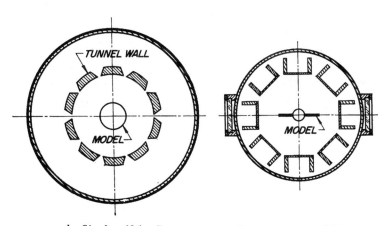

b. *Circular, 12 in. diam.*   c. *Octagonal, 12 in. effective diam.*

FIG. 2.15. *Transonic slotted tunnel configuration of the NACA, U.S.A. (1948).*[16]

that the velocity correction became very small, as intended, but also that the choking was greatly relieved. It was possible to operate the slotted wind tunnel with model blockages of close to 9 per cent at Mach numbers to 0.97. Furthermore, it was discovered that such a wind tunnel could be operated not only at both subsonic and sonic speeds but also at supersonic speeds, merely by raising the power input to the wind tunnel compressor. These experimental results were of extreme importance. They demonstrated that for wind tunnel operation at low supersonic Mach numbers, it is not necessary to have an exchangeable or adjustable nozzle, as in conventional supersonic wind tunnels. In one partially open transonic wind tunnel, useful testing can be accomplished over the entire transonic speed range from high subsonic to low supersonic Mach number.

A considerable number of full-scale slotted wind tunnels have been built, particularly in the United States (for example, Fig. 2.16), according to the principle of the slotted wind tunnel described above. Wind tunnels of this

Fig. 2.16. *NACA 16 ft transonic wind tunnel (1950)*.[19]

Fig. 2.17. *Bell X-1 airplane (first airplane faster than sound) (October 14, 1947)*.

type were also constructed in Great Britain, Switzerland, France, etc. These wind tunnels were developed to a high degree of perfection by means of thorough and detailed experimental work. They were found to provide reliable data in the subsonic speed range up to and including Mach number one. However, it was also shown that they had the serious shortcoming of a very limited potential for cancelling shock-wave reflections. Therefore, the successful use of slotted wind tunnels in the supersonic speed range is restricted to very low supersonic Mach numbers ($M = 1.05$). At higher supersonic Mach numbers these wind tunnels have the same limitations as conventional closed wind tunnels. A typical airplane designed with the knowledge obtained in the initial slotted wind tunnels is the first experimental airplane faster than sound, the rocket-powered Bell X-1 (Fig. 2.17).

9. WIND TUNNELS WITH PERFORATED WALLS

a. Basic design and performance

Investigations around 1950 showed that for effective cancellation of shock waves at wind tunnel walls it is necessary to provide a "small-grain" porous wall which can match the "flow inclination-pressure rise" characteristics of shock waves. These early theoretical investigations were performed mainly by the staff of the Cornell Aeronautical Laboratory in the U.S.A., and were supplemented by some small-scale experiments.[17] The experiments proved that satisfactory shock cancellation can be accomplished when shock intensity and the porosity of the walls are properly matched.

It was found, however, that porous walls would not be very practical from the operational point of view. Not only is it nearly impossible to avoid clogging of the pores of such a wall during operation, but also the effective porosity must be continuously adjusted with each change in Mach number or in intensity of the shock waves. Such a requirement is very hard to satisfy in actual wind tunnel operation.

As a further development, the perforated wall concept was introduced.[17] This wall consists of a large number of small discrete openings in the wind tunnel wall. It was shown both by theoretical calculations and by experiments that properly perforated walls are capable of matching the "no-reflection" requirements over a considerable range of Mach number and shock intensity when the geometry of the perforation is properly selected (Fig. 2.18).

Numerous experiments indicated that such perforated walls are effective in cancelling shock-wave reflections. However, the investigations also indicated that such walls are not basically well adapted to matching the desired characteristics of subsonic flow. This disadvantage of perforated walls at subsonic speeds is acceptable only when the size of the wind tunnel model is properly restricted, that is, when it is kept small enough.

b. Further development of perforated wall wind tunnels

In the course of detailed investigations of perforated wind tunnels, walls with slanted holes were studied in the mid-1950's.[18] It was recognized that normal perforated walls with perpendicular holes do not provide the proper boundary conditions for three-dimensional wave patterns and for two-dimensional or three-dimensional expansion waves. This is the result of the fact that stagnation air from the plenum chamber can easily be sucked into the test section. Consequently, the slanted holes were arranged in such a manner that they offered a larger resistance to inflow into the test section than to outflow out of the test section and in this manner provided a much improved matching for wave cancellation of both the compression and

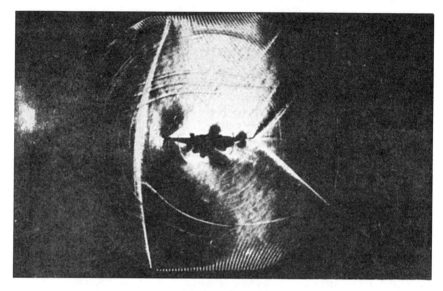

a. *Mach number 1.15.*

b. *Mach number 1.30.*

FIG. 2.18. *Shock reflection on perforated wind tunnel walls (1950).*[17]

expansion-type shock waves (Fig. 2.19). Examples of transonic wind tunnels with perforated walls of the slanted-hole type are the modern AEDC 16 ft transonic and supersonic wind tunnels (Fig. 2.20) and the supersonic

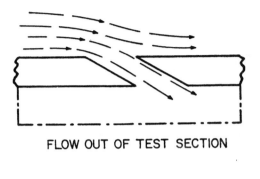

FLOW OUT OF TEST SECTION

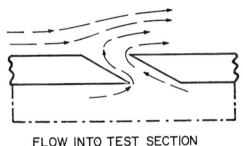

FLOW INTO TEST SECTION

FIG. 2.19. *Streamline pattern for inflow and outflow through a wall with inclined holes.*[18]

8 × 6 ft wind tunnel at the Lewis Laboratory of the NASA. These tunnels are notable for the fact that in their design emphasis was placed on good operational characteristics in the supersonic speed range.

A typical supersonic airplane developed with the aid of extensive transonic wind tunnel testing in both the high-subsonic and low-supersonic speed ranges is the North American F-100A shown in Fig. 2.21.

Many detailed experiments have shown that perforated walls are effective only when the individual holes in the perforation are comparable in size to or larger than the thickness of the boundary layer along the tunnel walls. Therefore, in order to produce the desired ratios between hole size and boundary layer thickness without the necessity for undesirably large holes in the wall, the boundary layer is frequently thinned by plenum chamber suction. Such suction has been found to be a significant method of making the perforated wall effective for wave cancellation. It is also extremely beneficial from the viewpoint of the overall power consumption of a wind tunnel. As a consequence, wind tunnels benefit greatly when a large portion of the total drive power is used for the plenum chamber suction drive system. For example, the 16 ft transonic wind tunnel of the AEDC provides 216,000 hp for the main drive system and approximately 180,000 hp

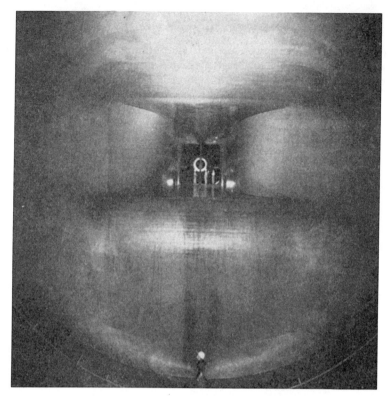

a. *View downstream into the test section.*

b. *Full-scale Army Sergeant in test section.*

FIG. 2.20. *16 ft transonic wind tunnel of Arnold Engineering Development Center, Tullahoma, Tennessee (1956).*

for the plenum evacuation system. It was shown that in the transonic speed range an addition of power to the plenum chamber suction system is frequently two or three times as effective as the addition of the same power to the main drive motors.

Fig. 2.21. *North-American F-100A airplane (1954)*.

10. CONCLUDING REMARKS

In the period after 1950, a large number of transonic wind tunnels have been constructed in the Western countries. These wind tunnels have either slotted or perforated type test sections. As a general rule, the slotted type is preferred whenever the main emphasis is placed on the subsonic speed range to slightly above Mach number one; the perforated type test section is preferred when the emphasis is placed on the supersonic range. Both types of transonic wind tunnels have been successful in providing the information which designers of airplanes and missiles need prior to construction and flight testing. The accuracy of the data obtained in these tunnels has matched the urgency of such data at the time when transonic wind tunnel testing spearheaded airplane development through the transonic speed range.

As is the case for most developments in technical fields, the present partially open wind tunnels still offer numerous opportunities for further refinement of design. For example, improvements in the matching of the wind tunnel boundary conditions with the flow pattern of test models appear possible using suitable combinations of perforated and slotted walls.[18] Also, the investigation of suitable hole size and hole geometry in combination with boundary layer control still offers numerous problems for research.

It should be mentioned that even at the present time it is not possible to calculate reliably the flow pattern, the pressure distribution, and the forces for a model in free transonic flight, except in extremely simplified cases, as, for instance, in the case of a two-dimensional wing. A large field still remains which has not been thoroughly explored by theoretical investigations. Consequently, most of the information required in the transonic speed range has been obtained and remains to be obtained by means of experiments. Since transonic wind tunnels have been developed into a reliable tool, the support required in the development of future airplanes and missiles can be adequately provided by experiments in the transonic wind tunnels which are in operation at the present time.

REFERENCES

[1] GOETHERT, B. and KOLB, A. "Drop Tests to Determine the Drag of Mode SC-50 at High Velocities." DVL, Berlin, Germany. *Techn. Rep. ZWB* 11, 252, 1944.

[2] HART, R. G. "Effects of Stabilizing Fins and a Rear-Support Sting on the Base Pressures of a Body of Revolution in Free Flight at Mach Numbers from 0.7 to 1.3." NACA L52 E06, September 1952.

[3] HABEL, L. W., HENDERSON, J. H. and MILLER, M. F. "The Langley Annular Transonic Tunnel." NACA TR 1106, 1952.

[4] JOHNSON, H. J. "Measurements of Aerodynamic Characteristics of a 35° Sweptback NACA–65–009 Airfoil Model with $\frac{1}{4}$ Chord Plane Flap by the NACA Wing-Flow Method." NACA–RM–L7 F13, 1947.

[5] POLHAMUS, E. C. and KING, T. J., Jr. "Aerodynamic Characteristics of Tapered Wings Having Aspect Ratios of 4, 6, and 8, Quarter-Chord Lines Swept-back 45° and NACA 63_1 A012 Airfoil Sections (Transonic-Bump Method)." NACA–RM L51 C26, June 1951.

[6] WEAVER, JOHN H. "A Method of Wind Tunnel Testing through the Transonic Range"; *J. Aero. Sci.* 15, 28–34, 1948.

[7] NELSON, WARREN H. and McDEVITT, J. B. "The Transonic Characteristics of 17 Rectangular, Symmetrical Wing Models of Varying Aspect Ratio and Thickness (Transonic-Bump Method)." Ames Aeronautical Laboratory, NACA RM A51 A12, May 1951.

[8] GOETHERT, B. H. "Development of the New Test Section with Movable Sidewalls of the Wright Field 10 ft Wind Tunnel (Phase A, Operation with Slots Closed)." WADC Technical Report 52–296, November 1952.

[9] BERNDT, S. B. "On the Influence of Wall Boundary Layers in Closed Transonic Test Sections." Aeronautical Research Institute of Sweden, FFA Report T1, 1957.

[10] PINDZOLA, M. "Establishment of Transonic Flow with Screens in a Two-Dimensional Section." UAC Report No. M–95295–6, March 1950.

[11] TAYLOR, H. D. "UAC Transonic Test Method." United Aircraft Corp. Report R–95295–22, December 1950.

[12] VANDREY, F. "The Influence of the Walls in a Closed Wind Tunnel of Circular Cross-Section with a Discontinuity Surface." AVA Report 45–A–19 (Goettingen, Germany).

[13] GOETHERT, B. H. "Flow Establishment and Wall Interference in Transonic Wind Tunnels." AEDC TR–54–44, June 1954.

[14] KONDO, K. "Boundary Interference of Partially Closed Wind Tunnel." Aeronautical Research Institute, Tokyo XI, Report No. 137, pp. 165–190, 1936.

[15] WIESELBURGER, C. "On the Influence of the Wind Tunnel Boundary on the Drag, Particularly in the Region of the Compressible Flow (German)." DVL, Berlin, Technical Report 1172, December 1939.

[16] WRIGHT, R. H. and WARD, V. G. "NACA Transonic Wind Tunnel Sections." NACA RM L8J06, October 1948.

[17] GOODMAN, T. R. "The Porous Wall Wind Tunnel Part III, Reflection and Absorption of Shock Waves at Supersonic Speeds." Cornell Aeronautical Laboratory, Report No. AD–706–A–1, November 1950.

[18] GOETHERT, B. H. "Physical Aspects of Three-Dimensional Wave Reflections in

Transonic Wind Tunnels at Mach Number 1.20 (Perforated, Slotted, and Combined Slotted-Perforated Walls)." AEDC TR–55–45, March 1956.

[19] WARD, V. G., WHITCOMB, C. F. and PEARSON, M. D. "Airflow and Power Characteristics of the Langley 16 ft Transonic Tunnel with Slotted Test Section." NACA RM L52 E01, July 1952.

BIBLIOGRAPHY

DICK, R. S. "Calibration of the Propulsion Wind Tunnel 16 ft Transonic Circuit with a Full Test Cart Having Inclined Hole Perforated Walls." AEDC TN–58–97, January 1959.

FERRI, A. "Untersuchungen und Versuche im Uberschallwindkanal zu Guidonia." *Jb. dtsch. Luftfahrtf. Ergaenzungsband*, 112, 1938.

MAEDER, P. F. "Note on Deblocked Transonic Test Sections." Brown University, Division of Engineering TR–WT–4, May 1950.

RIEGELS, F. "Correction Factors for Wind Tunnels of Elliptical Cross-Section with Partly Open and Partly Closed Working Sections." *Luftfahrtforsch.* **16**, 26–30, 1939.

STACK, J., LINDSEY, W. F. and LITTELL, R. E. "The Compressibility Burble and the Effect of Compressibility on Pressures and Forces Acting on an Airfoil." NACA Report 646, 1938.

WILDER, JOHN G. "An Experimental Investigation of the Perforated Wall Transonic Wind Tunnel, Phase I." Cornell Aeronautical Lab. AD–706–A–5, August 1951.

CHAPTER 3

FLOW AROUND AIRPLANES IN FREE FLIGHT

Before the specific phenomena of the flow in wind tunnels are discussed, some basic characteristics of the flow around airplane components in free flight will be briefly considered. In this introductory discussion, particular emphasis is placed upon the effect of the displacement of the airplane components because displacement is the source of some major difficulties which occur in wind tunnel testing in the transonic speed range.

1. SUBSONIC SPEED RANGE

In the free flight of an airplane, the flow lateral to the plane's structure must achieve a larger stream density, ρv, in order to compensate for the displacement of the airplane. Consequently, the streamlines bulge out in the vicinity of the airplane as shown in Fig. 3.1a. Near the surface of a wing, for example, the streamlines are curved according to the wing contour, and the velocities near the maximum thickness of the wing are larger and the static pressures smaller than the corresponding values in the undisturbed flow. Because of the curvature of the streamlines, centrifugal forces are produced which increase the static pressure normal to the streamlines according to the well-known relationship:

$$\mathrm{d}p = -\frac{\rho v^2}{R}\mathrm{d}y$$

The elementary centrifugal forces and the resulting pressure gradients compensate for the difference between the pressures at the wing surface and the undisturbed pressure. It should be noted that in subsonic flow the bulging of streamlines extends to infinitely large distances from the wing and that the elementary centrifugal forces in the entire space between the wing and the infinitely distant streamlines contribute to the build-up of the static pressure from the low values at the wing surface to the higher values in the undisturbed flow.

Streamline curvature and velocity increase near the model are correlated in such a manner that the displacement volume of the airplane is compensated for by the larger stream density, ρv, of the flow in the disturbance field around the model. For example, in the plane of the maximum thickness of a two-dimensional symmetrical wing, the continuity condition is satisfied by:

$$\tfrac{1}{2}t(\rho v)_\infty = \int_{t/2}^{\infty}[(\rho v)_y - (\rho v)_\infty]\,\mathrm{d}y = \Delta(\rho v)_{\mathrm{mean}}\Delta y_D$$

This equality is represented in Fig. 3.1b for the incompressible flow case.

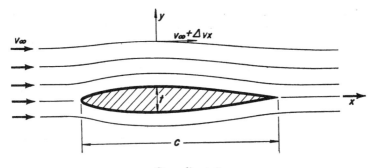

a. *Streamline pattern.*

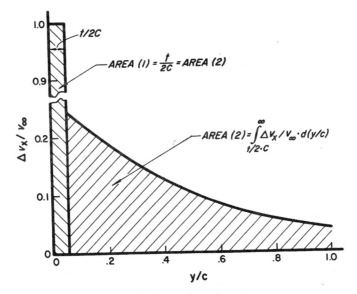

b. *Velocity distribution laterally to the wing.*

Fig. 3.1. *Incompressible flow around an airplane wing in free flight.*

2. MACH NUMBER INFLUENCE IN SUBSONIC SPEED RANGE

When the flight speed is increased in the subsonic speed range, the streamlines of the incompressible flow no longer provide equilibrium between the elementary centrifugal forces and the static pressure gradients in the flow field. Also, the condition of continuity is no longer satisfied. Consequently, two major changes occur. The streamlines at a given distance from the wing will assume larger curvature and adjacent streamlines will become more and more equidistant with the approach of sonic velocity, according to the basic relationship for the stream density of a compressible flow, $\rho v = f(M)$, see Fig. 3.2. On the other hand, the flow between any two streamlines near the model contributes less to the integral in the continuity equation (see Chapter 3, Section 1), since the stream density cannot grow in proportion to the velocity, but approaches a maximum value near the sonic speed.

In order to overcome this two-fold difficulty, that is, larger streamline curvature leading to larger pressure gradients, dp/dy, on one hand, and smaller blockage-relieving margin, $\Delta(\rho v)$, of the flow on the

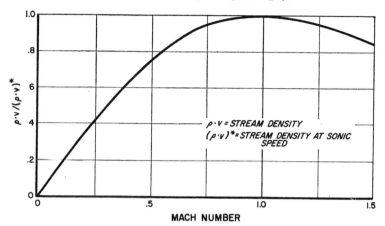

FIG. 3.2. *Relationship between stream density and Mach number in isentropic flow* ($\gamma = 1.4$).

other hand, the velocities, v, near the wing surface are increased and the flow disturbance field is extended farther laterally into the flow with increasing subsonic Mach number (Fig. 3.3 and Ref. 1).

Both relieving effects, however, have definite limitations. When the sonic speed is exceeded locally, the stream density, ρv, no longer increases but instead decreases with increasing velocity as shown in Fig. 3.2. Hence, local supersonic fields around a wing increase the effective wing blockage. Furthermore, when local supersonic speeds are attained, the streamline

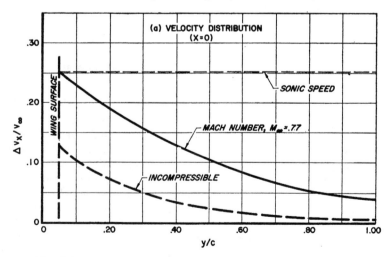

FIG. 3.3a. *Velocity distribution lateral to an airplane wing in subsonic free flight* $M_\infty = 0.77$.

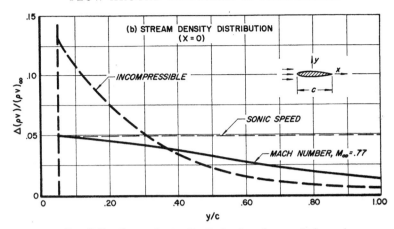

Fig. 3.3b. *Stream density distribution lateral to an airplane wing in subsonic free flight, $M_\infty = 0.77$.*

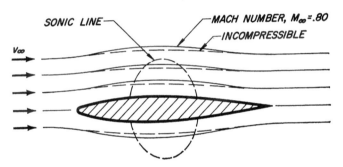

a. *Symmetrical streamline pattern without shock waves, $M_\infty = 0.80$.*

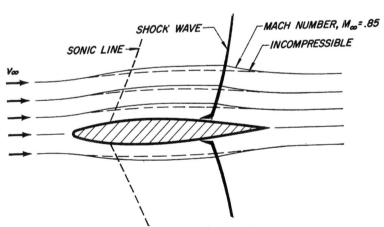

b. *Unsymmetrical streamline pattern with shock waves, $M_\infty = 0.85$.*

Fig. 3.4. *Compressible flow field around an airplane wing in free flight at Mach numbers of $M_\infty = 0.8$ and 0.85.*

curvatures are increased (Fig. 3.4a). Therefore, a more rapid lateral decay of the streamline curvature and a reduction of the mean width, Δy_D, of the disturbance field described in the continuity equation of Chapter 3, Section 1, result (Fig. 3.5). Hence, both terms in the continuity integral are influenced to the detriment of the requirements for compensation of the wing displacement. It is apparent that beginning at a well-defined critical speed close to sonic speed, a symmetric flow pattern (without shock waves) is no longer possible in isentropic flow.[1]

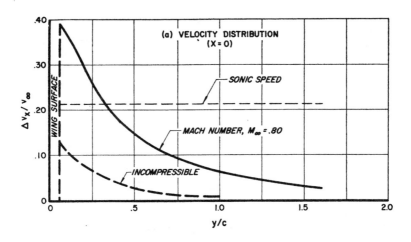

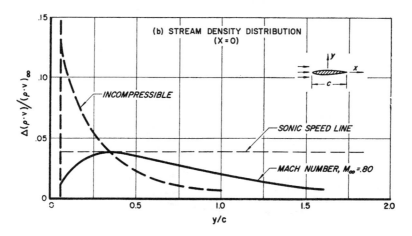

Fig. 3.5. *Velocity and stream density distribution lateral to an airplane wing in subsonic free flight, $M_\infty = 0.80$.*

Only by establishing an unsymmetrical flow field with shock waves (Fig. 3.4b), that is by an increase of the local flow velocity beyond the maximum wing thickness, is the flow again capable of satisfying the continuity requirements in the plane of maximum wing displacement. Such a velocity increase results in a decrease of the local streamline curvature and an increase of the lateral width, Δy_D, of the disturbed flow around the wing (Fig. 3.6).

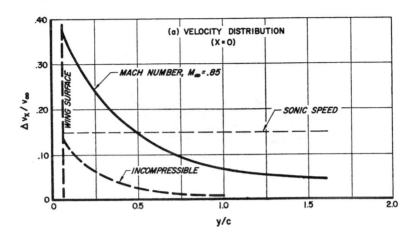

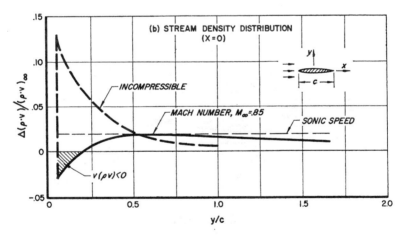

FIG. 3.6. *Velocity and stream density distribution lateral to an airplane wing in subsonic free flight, $M_\infty = 0.85$.*

3. SONIC SPEED

When the undisturbed velocity is equal to the sonic speed, the flow density, ρv, in the disturbed flow region can only be smaller than that of the free flow. Consequently, the continuity condition, formulated in Section 1 of Chapter 3, cannot be satisfied for steady-state flow simultaneously with the condition that the streamlines at infinitely large lateral distances from a two-dimensional wing are not disturbed.[2] In contrast to the usual flow patterns at subsonic velocities, the streamline deflection at sonic speed does not gradually decrease but rather continually increases with increasing distance from a two-dimensional wing (Fig. 3.7a). This results in infinitely large bulging of the streamlines at infinitely large lateral distance.

It should be noted that pure two-dimensional flow is merely an academic type of flow. In free flight the flow is always three-dimensional because at a

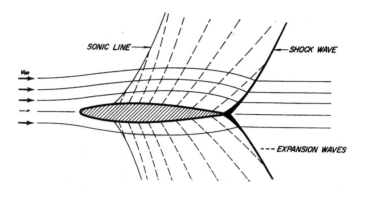

a. *Sonic velocity.*

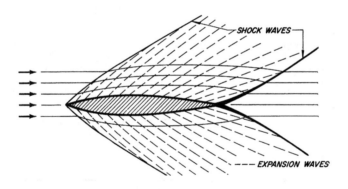

b. *Supersonic velocity.*

FIG. 3.7. *Streamline pattern around an airplane wing in free flight at sonic and supersonic velocities.*

sufficiently large distance, a flying object always acts as a point disturbance. In the case of a three-dimensional body the streamline bulging decreases asymptotically to zero with increasing distance from the body.[2] However, for a three-dimensional body, the total frontal area of streamline bulging,

$$\Delta A_B = (2\pi r)\Delta r_{max}$$

increases with the distance from the body in line with the basic relationship between stream density and velocity (Fig. 3.2) and tends to infinity at infinitely large lateral distances from the body.

4. SUPERSONIC SPEED RANGE

As soon as the free-stream velocity has exceeded Mach number one, it is again possible to establish a steady-state flow solution with undisturbed parallel streamlines at large lateral distances from a two-dimensional wing (Fig. 3.7b). Because of the characteristic shape of the stream density curve, the mean velocity in the disturbance region around the maximum model thickness is smaller than the velocity of the undisturbed flow (Fig. 3.8).

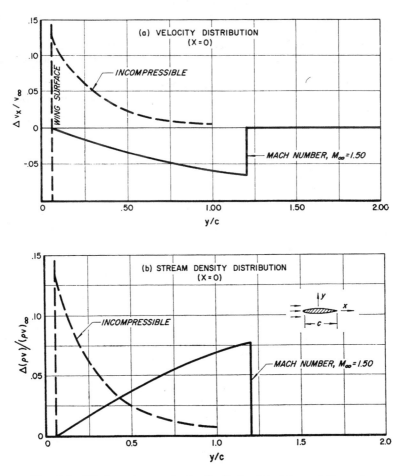

FIG. 3.8. *Velocity and stream density distribution lateral to an airplane wing in supersonic flight, $M_\infty = 1.5$.*

This is the opposite of the conditions in subsonic flow where the mean velocity in the disturbed region around a model is larger than the velocity of the undisturbed flow.

REFERENCES

[1] GOETHERT, B. H. and KAWALKI, K. H. "Compressible Flow of Several Plane Airfoils near the Sonic Speed." DVL Report UM–1471, January, 1945 (German).

[2] BARISH, D. T. and GUDERLEY, G. K. "Asymptotic Forms of Shock Waves in Flows Over Symmetrical Bodies at Mach One." AF Technical Report No. 6660, Wright Air Development Center, January, 1952.

BIBLIOGRAPHY

GOETHERT, B. H. and KAWALKI, K. H. "The Calculation of Compressible Flows with Local Regions of Supersonic Velocity." NACA TM 1114, March, 1947. (Translation of DVL, FB 1794, August, 1943.)

GUDERLEY, G. K. "Singularities at the Sonic Velocity." AF WADC F–TR–1171–ND, June, 1948.

GUDERLEY, G. K. "Axial-Symmetric Flow Patterns at a Free-Stream Mach Number Close to One." AF Technical Report No. 6285, Wright Air Development Center, October, 1950.

GUDERLEY, G. K. "Two-Dimensional Flow Patterns with a Free-Stream Mach Number Close to One." AF Technical Report No. 6343, Wright Air Development Center, 1951.

VINCENTI, W. G. and WAGONER, C. B. "Transonic Flow Past a Wedge Profile with Detached Bow Wave." NACA Report 1095, 1952.

CHAPTER 4

WALL INTERFERENCE CORRECTIONS IN CONVENTIONAL WIND TUNNELS

1. OPEN-JET WIND TUNNELS

a. Subsonic speed range

In an open-jet wind tunnel, the streamlines bulge out around the model in a similar fashion as in free flight (see Fig. 4.1). The free jet is surrounded by a plenum chamber in which the static pressure is essentially constant. Consequently, the static pressure along the free-jet boundary must also be constant and, more specifically, equal to the undisturbed pressure existing upstream of the model in the undisturbed free jet. The mathematical formulation of this statement follows:

Pressure along free-jet boundary:

$$p = \text{constant}$$

and in the case of an infinitely long jet:

$$p - p_\infty = \Delta p = 0$$

or, in first approximation:

$$(v_x - v_\infty)/v_\infty = \Delta v_x/v_\infty = 0$$

After introduction of the velocity potential, $\Phi = \phi_\infty + \phi$, we obtain as the condition for the free-jet boundary:

$$\partial \phi/\partial x = 0$$

along the jet boundary. Since this condition holds true along the bulged-out boundary streamline, as well as in the first approximation for a line $y = \text{constant} = H/2$, the above equation may be integrated to give:

$$\phi = \text{constant}$$

It should be noted that according to the above consideration, the undisturbed free-stream pressure is already established at the finite distance of the jet-boundary streamline. In free flight, however, the undisturbed pressure is not attained before an infinitely large lateral distance from the model. Consequently, in an open wind tunnel the pressure build-up in the flow between model and jet boundary must occur considerably faster; that is, the streamlines will be more highly curved and the resulting centrifugal forces larger than in free flight (Fig. 4.1). Because of the greater bulging-out of the streamlines, the airflow between adjacent streamlines is contained in a larger area, and consequently, the mean velocities are somewhat smaller than in free flight. This effect is usually taken into consideration in the form of a velocity correction which states that the equivalent free-stream velocity of a free jet is somewhat smaller than the undisturbed free-jet velocity.

It should be kept in mind also that in addition to the velocity correction, the streamline pattern around a model in a free-jet wind tunnel is distorted. The only means of reducing such flow distortions in a free jet is to reduce the model size. Only when the model dimensions are sufficiently small in

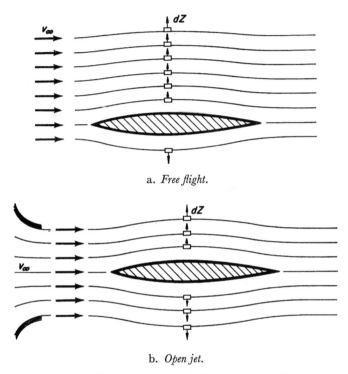

a. *Free flight.*

b. *Open jet.*

FIG. 4.1. *Streamline pattern around airplane wing in free flight and in open-jet wind tunnel.*

comparison to the free-jet dimensions is the simulation of free-flight conditions in a finite-size jet satisfactorily accurate. It has been the objective of numerous mathematical investigations to determine the correction terms of first and higher order for the purpose of predicting from the free-jet test results the behavior of models in free flight.[1]

b. Mach number influence

When the Mach number is increased in a free-jet wind tunnel, the basic deviations between free jet and free flight remain qualitatively the same. However, the bulging-out of the streamlines is increased with increasing Mach number and, consequently, the effective velocity is further reduced or the free-jet velocity correction further increased.[2]

Also, no peculiar difficulties arise in establishing the flow at Mach number one. Unlike free flight or closed wind tunnel conditions, the free-jet boundary streamlines can freely bulge out and thus readily compensate for the inability of the flow to establish flow conditions with larger stream density, ρv, than those occurring in undisturbed flow of Mach number one.

Supersonic flow fields are characterized by the occurrence of discrete disturbance waves of either the compression or expansion type. When the approach velocity in an open-jet wind tunnel is increased to supersonic values, the region of boundary interference is restricted to the region downstream of waves reflected from the free-jet boundary. The disturbance waves impinge upon the free-jet boundary and are reflected to satisfy the requirement of constant undisturbed pressure along the jet boundary. In particular, a compression wave is reflected as an expansion wave of approximately equal but opposite pressure change, and an expansion wave is reflected as a compression wave (Fig. 4.2).

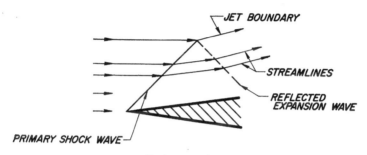

a. *Shock wave reflection.*

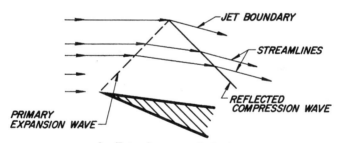

b. *Expansion wave reflection.*

FIG. 4.2. *Wave reflection at open-jet boundary.*

The usual method of testing in supersonic open-jet wind tunnels is to restrict the model length so that the reflected waves do not meet the model but pass behind it. In this case, the free-jet wind tunnel provides test data which are not influenced by boundary effects and are identical to those in free flight. The maximum model length which will yield interference-free flow is determined by the path of that reflected wave which is at the farthest upstream location. This critical reflected wave is normally the bow wave originating at the model nose. The path of the reflected bow wave depends naturally upon the local Mach numbers and the direction of the flow through which it passes, and follows, therefore, approximately the Mach wave direction in the undisturbed flow. Results obtained from numerous wind tunnel tests with typical models are presented in Fig. 4.3 as a guide for determining the maximum interference-free model length.[3]

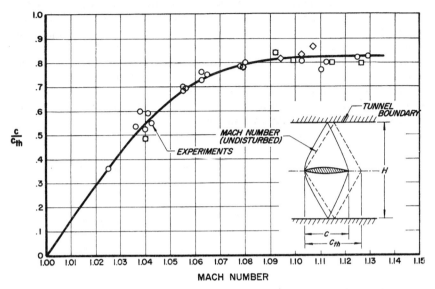

FIG. 4.3. *Maximum model length in supersonic wind tunnels (without shock reflection interference)*.[3]

2. CLOSED WIND TUNNELS

a. Subsonic speed range

In a wind tunnel with solid straight walls, bulging-out of the boundary streamlines is prevented by direct guiding of the walls (Fig. 4.4).

The velocity components normal to the walls, v_n, must vanish at the walls. That is, in mathematical formulation:

$$v_n = 0 \text{ or } \partial\phi/\partial n \simeq \partial\phi/\partial y = 0$$

This condition must be satisfied along the walls, that is, in two-dimensional flow, along lines $y = $ constant.

Closed Wind Tunnels with Straight Walls.—Proceeding from the model outward to the walls, the streamlines are gradually straightened out by the influence of the walls to the extent that near the model surface, they follow the model contour and, at the walls, become completely straight. Consequently, the streamline curvature in the field between model and walls as well as the centrifugal forces and the resulting pressure gradients in this region are reduced in comparison to free-flight conditions. According to the equation for the pressure gradient in curved flow,

$$dp/dn \simeq dp/dy = (\rho v^2)/R,$$

the static pressure at the walls will not reach the undisturbed free-stream value, as in free flight, but remains essentially smaller in the wall area lateral to the model.

Because bulging-out of the streamlines is not possible laterally around the model, the airflow between adjacent streamlines must squeeze through a smaller area than in free flight; that is, in the subsonic speed range the mean flow velocity is increased in the space between model and walls. Consequently, the model in a closed tunnel is exposed to a mean flow with larger velocities than those prevailing far upstream of the model. This influence can be taken into account by a *mean velocity correction*, which for closed test sections is approximately of equal magnitude, but of opposite size, to the corresponding correction in a free jet.

It should be noted that also in the case of a closed wind tunnel, the velocity correction does not perfectly correlate the wind tunnel flow around a model with free-flight conditions. As described before, the streamlines around the model are distorted by the presence of the walls. Therefore,

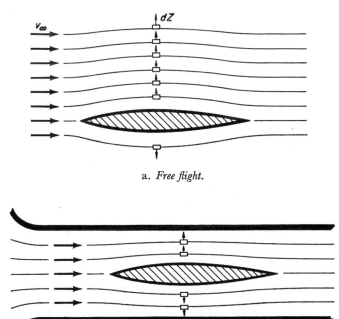

a. *Free flight.*

b. *Closed wind tunnel.*

FIG. 4.4. *Streamline pattern around airplane wing in free flight and in closed wind tunnel.*

also in the case of a wind tunnel with plain straight walls, there is no means of reducing this distortion other than through reduction of the relative model size. It is the purpose of second and higher order correction calculations to modify the wind tunnel test data to correspond more closely to free-flight conditions.[1]

Closed Wind Tunnels with Contoured Walls.—In contrast to the free-jet wind tunnel, the closed wind tunnel offers an additional possibility for simulating free-flight conditions more closely; namely, by elimination of the condition of *straight* rigid walls. The wind tunnel walls could be bulged-out to follow the required streamline shape. If the wall contours were made identical to free-flight streamlines, the correlation would be perfect. It is obvious that such a state cannot be reached without knowing in advance the exact flow around the model; this is generally not before the results of accurate wind tunnel tests are known. Despite the inability to establish perfect wind tunnel contours, it appears feasible to shape the wind tunnel walls according to a few basic types of flow around typical models. For example, the overall displacement of a model could be compensated for by one type of wall shape whereas the detailed local contours of the model would be neglected. In the same manner, the basic characteristics of a lifting surface could be simulated whereas details of the lift distribution would be neglected. It is apparent that successful application of this principle would require special wall shapes for each type of model and even for each model setting (angle of attack) and consequently would become very cumbersome.

Experiments at high subsonic speed in a test section with solid walls which could be bent according to the streamline contours were conducted in a British model wind tunnel. A small 6-in. experimental wind tunnel was used, and some exploratory tests were conducted with a two-dimensional airfoil model. A group of test results were obtained for different wall shapes without having definite proof which condition yielded the most reliable results. Also a large German high-speed wind tunnel (7 ft), constructed at Ottobrun near Munich and completed in 1945, had flexible upper and lower test section walls. No results, however, have been obtained from this tunnel with a model installed and the test section walls contoured according to estimated free-flight streamlines.

It appears that the principle of contouring closed test section walls has been successfully applied only in a few cases in which extremely large models or model supports had to be used, as for example, in the propeller test configurations of the 7 ft wind tunnel of the Cornell Aeronautical Laboratory, Inc. (Ref. 4 and Fig. 4.5).

The extremely large sensitivity of a flow area change in closed wind tunnels, particularly at high subsonic Mach numbers at which relief from flow distortion and choking is most urgently needed, is responsible for the decision of wind tunnel designers to abandon the contoured-wall wind tunnel and to pursue other principles for reducing the flow distortion in high-speed wind tunnels.

b. Mach number influence

Subsonic Speed Range.—With increasing Mach number in the subsonic speed range, the distortion of the streamlines in the constricted area between a model and the walls of a closed wind tunnel remains qualitatively the same. Because of the compressible flow relationship for the stream density, ρv (see Fig. 3.2), the velocity increase caused by the wall effect must increase with Mach number in order to attain the necessary stream density increase,

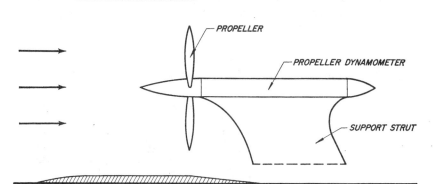

a. *Sketch of tunnel configuration with wall liners.*

b. *Photograph, view in flow direction.*

FIG. 4.5 *Propeller dynamometer installation of Cornell Aeronautical Laboratory, Incorporated.*[4]

$\Delta(\rho v)$. For a small model, the mean velocity increase caused by the model blockage is, in linearized, one-dimensional flow

$$\Delta v / v_\infty = \frac{1}{1-M^2} \cdot \frac{A_M}{A_T}$$

This relationship is shown in Fig. 4.6. It should be noted that in the

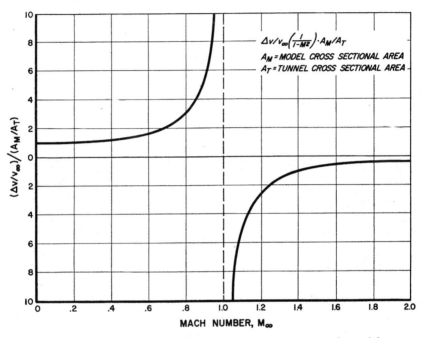

FIG. 4.6. *Mean velocity increase in narrowest sections of closed wind tunnel because of model installation (linearized, one-dimensional flow).*

subsonic speed range the mean velocity is increased to make up for the model blockage, whereas in the supersonic range the mean velocity is decreased. Furthermore, the magnitude of the velocity change is growing to infinitely large values when Mach number one is approached from either the subsonic or supersonic side.

It is only when the model size is drastically reduced that measurements in closed wind tunnels at Mach numbers close to one can be conducted with satisfactory simulation of free-flight conditions. Tests simulating free-flight conditions at Mach number one are not possible in a closed wind tunnel. By nature, the closed wind tunnel does not have either the capability of gradually extending its lateral disturbance region with increasing Mach number as in free flight, or of unlimited streamline bulging-out and the resulting flow area increase as in open wind tunnels. In fact, in closed wind tunnels the well-known choking phenomenon occurs which limits the maximum Mach number to values which correspond to maximum stream density conditions (that is, to Mach number one conditions) in the constricted flow channel between model and walls. The choking Mach numbers for different model sizes with the assumption of one-dimensional isentropic flow in the narrowest section are shown in Fig. 4.7. At a Mach number of

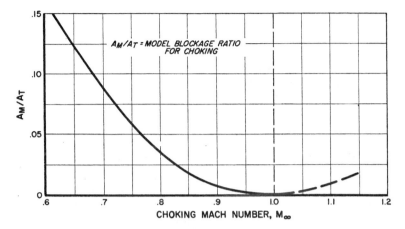

FIG. 4.7. *Choking Mach number in closed wind tunnels as function of Mach number (one-dimensional).*

0.80, a model blockage of $A_M/A_T = 3.7$ per cent causes choking; at Mach number 0.90, the choking blockage is reduced to 0.88 per cent, and at Mach number 0.95, to 0.21 per cent. Realizing that a blockage of 0.21 per cent in a three-dimensional test corresponds to a model with a maximum diameter of no more than 5.5 in. in a 10 ft diameter wind tunnel, it is apparent that transonic testing in a closed wind tunnel is very impractical.

It should also be pointed out that the limit for practical wind tunnel testing has already been reached at Mach numbers somewhat lower than the choking Mach numbers indicated above, since at choking, the flow in the constricted channel between model and test section walls tends to reach the Mach number one condition with its maximum stream density (ρv) the choked flow is normally greatly distorted and bears little similarity to freeflight conditions. As a practical rule, it is frequently recommended that the allowable model blockage be reduced to two-thirds, or better to one-half, of the blockage allowable for one-dimensional flow.[2] The approach of critical choking conditions can be detected by observing the wind tunnel wall static pressure in the vicinity of the narrowest section (see Fig. 4.8). The steep drop of the wall static pressure, as exhibited by a typical airfoil model installed in a closed wind tunnel, signals the approach of critical conditions and the limit for reliable testing.

Supersonic Speed Range.—At supersonic Mach numbers, wave reflections at the tunnel walls introduce disturbances which restrict the useful Mach number testing range in closed wind tunnels in a manner similar to that for openjet tunnels. The wave reflection, however, is different from open-jet reflection since the condition of unchanged flow direction along the wind tunnel walls

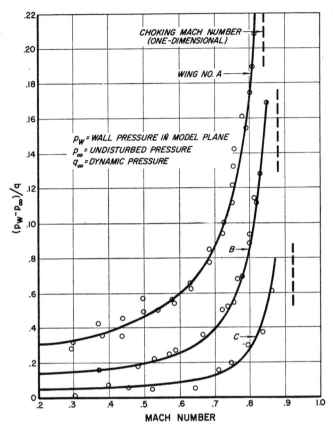

FIG. 4.8. *Wall pressure in plane of maximum model thickness for various wings in closed wind tunnel.*[2]

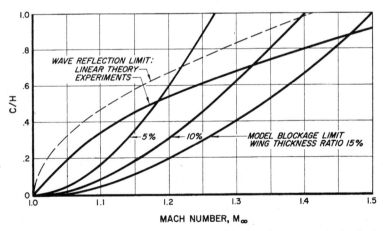

FIG. 4.9. *Maximum chord of two-dimensional model wings in a supersonic closed wind tunnel from considerations of choking and wave reflections (see Fig. 4.3). c = wing chord, H = tunnel height.*

requires that expansion waves be reflected as expansion waves, and compression waves as compression waves (see Fig. 8.2). In order to prevent reflected waves from invalidating the test results, the model length must be kept small enough to cause the reflected waves to pass behind the model. Therefore, the same relationship for maximum model length exists for tests in closed wind tunnels as for tests conducted in open-jet wind tunnels (Fig. 4.3). By comparing the permissible model blockage from the viewpoint of choking elimination with the permissible model size from the viewpoint of wave reflection interference, it can be concluded that the more stringent requirement at low supersonic Mach numbers is the no-choking requirement whereas at high supersonic Mach numbers the no-wave reflection interference is predominant (Fig. 4.9).

c. Analysis of choked flow in closed wind tunnels

As stated previously, in a closed wind tunnel with choked flow, the mean velocity in the constricted cross section between model and test section walls is approximately equal to the speed of sound. Although it is apparent that the flow with choked tunnel conditions does not truly simulate free-flight conditions, it was reasoned that some similarity of the choked tunnel flow and free-flight flow at Mach number one might exist.[5] In both free-flight and choked tunnel conditions, the local Mach numbers will be lower than one in the region somewhat upstream of the model nose, and in the area somewhat upstream of the maximum model thickness the Mach numbers will approach one. Behind the maximum model thickness the flow will be supersonic.

Flow calculations for a two-dimensional double-wedge in free flight and in a closed wind tunnel, with the wall influence introduced into the calculations as a first order disturbance effect, are presented in Figs. 4.10 and 4.11 (Ref. 5). It should be noted that the sonic line, that is, the line separating subsonic and supersonic flow regions, is very similar in both flow cases. Deviations occur mainly near the tunnel wall where the sonic line must be bent into the vertical direction in order to satisfy the non-curved flow conditions near the test section wall, that is:

$$\frac{d v_x}{d y} \simeq -\frac{dp}{dy} = -\frac{\rho v^2}{R}$$

As long as the model length and thickness are kept small enough, the pressure distributions on the double-wedge surface coincide with good approximation for both cases investigated.

For example, even with a model wing of as large as 1.3 per cent blockage, which permits wind tunnel testing only up to approach Mach numbers somewhat smaller than the choking Mach number of 0.88, free-flight conditions at Mach number one are simulated at choked wind tunnel conditions with pressure deviations of no more than $\Delta p/q = 0.027$, which amounts to 4 per cent of the maximum pressure disturbance of the double-wedge airfoil.[5] However, the maximum wing chord in this case must not be larger than 13 per cent of the test section height; that is, the thickness ratio of the double-wedge airfoil must not be smaller than 10 per cent. In view of the generally much more slender model shapes tested in transonic flow, the model size in practical testing must be kept much smaller than indicated in the above example and is governed greatly by wave reflection conditions.

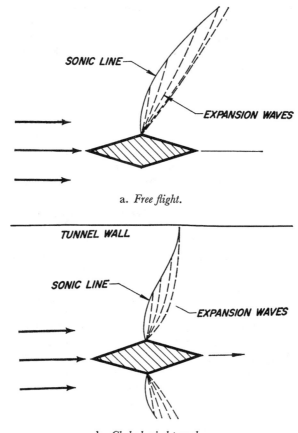

a. *Free flight.*

b. *Choked wind tunnel.*

Fig. 4.10. *Flow fields around double-wedge airfoil in free flight at Mach number one and in a choked closed wind tunnel.*[7,5]

An experimental study of choked wind tunnel flow has been conducted by D. Barish who verified experimentally the validity of the above considerations.[6]

The choked tunnel flow theory discussed in the previous paragraphs makes it possible to obtain approximate experimental information on the free-flight conditions near Mach number one. This Mach number one test point, however, is a singular test point only because no test data can be correlated with free flight in the Mach number range slightly below or above choking Mach number.

3. COMPARISON OF WALL INTERFERENCE CORRECTIONS FOR VARIOUS TYPES OF TEST SECTIONS (SUBSONIC FLOW)

It was shown in the preceding sections that the influence of the flow boundary conditions in open and closed wind tunnels can be taken into consideration by adding correction terms to the local velocities around the model. One of

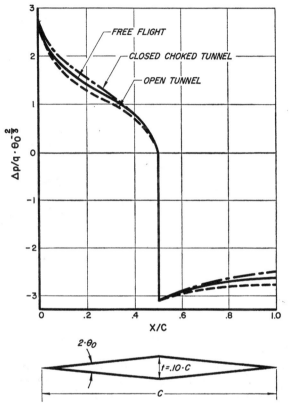

FIG. 4.11. *Pressure distribution of double-wedge airfoil in free flight at Mach number one, in choked closed wind tunnel, and in sonic free-jet.*[5]

the most significant corrections in transonic testing refers to the effective flow velocity which corresponds to the flight velocity in unlimited air space. Various theoretical methods have been successfully employed to determine the correct wind tunnel velocity and its dependence on Mach number.[2]

For open wind tunnels, for example, the velocity corrections due to the solid model displacement was found to be in a first order approximation:

$$\Delta v_x / v_\infty = \frac{1}{(1 - M_\infty^2)^{3/2}} \cdot K \cdot \frac{\text{Vol}}{A_T^{3/2}}$$

where M_∞ is the *corrected* Mach number and K is a constant depending upon tunnel as well as model geometry. The velocity corrections according to the above equation have been determined for many tunnel and model configurations with the assumptions that the compressibility effect can be linearized and that the model dimensions are small in comparison to the tunnel dimensions and, therefore, second and higher order terms of the ratio of wing chord/tunnel height, C/H, may be neglected. For a circular cross-sectional wind tunnel, the velocity corrections according to the above equation are shown in Fig. 4.12.

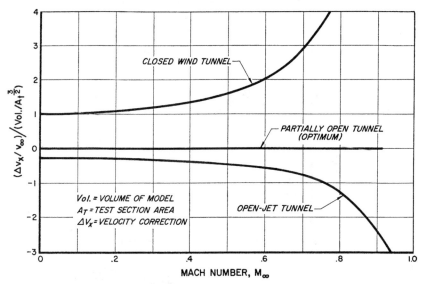

Fig. 4.12. *Velocity corrections in open, closed and partially open wind tunnels as function of Mach number.*[2]

For a closed wind tunnel, theoretical investigations for the model displacement influence resulted in first order wind tunnel corrections which follow the same equation as for open tunnels. However, the sign is opposite to the open-tunnel corrections and the value of the constant K is different. Again, for a circular cross section, the closed wind tunnel correction was calculated and is represented in Fig. 4.12. It should be noted that the closed tunnel correction is four times as large as the open tunnel correction and that both corrections tend asymptotically to infinity when the Mach number approaches one. The magnitude of the velocity correction indicates that the first order correction calculations are no longer reliable near Mach number one, and hence, higher order correction calculations with the resulting cumbersome application become necessary even for relatively small models. It must also be realized that the compressibility effect has been considered in the above correction equation only in first order; calculations for second or higher order compressibility correction as are needed near Mach number one are extremely complicated. Even for the much simpler free-flight flow, such refined calculations have been made only in few special cases, and are found to be impractical for general wind tunnel testing.

In order to avoid this complication, wind tunnel testing near Mach number one has been frequently accomplished with very small models which can be expected to stay within linearization limits of wind tunnel corrections. However, the small-model test method is very uneconomical and, in addition, not a very reliable method because the Reynolds numbers of the tests are generally too small.

To overcome these difficulties, the most successful efforts for reliable transonic wind tunnel testing have been directed toward the development of correction-free tunnels, that is, wind tunnels in which, by means of proper arrangement of open as well as closed wall elements, the velocity correction disappears on the basis of first order calculations.

For a wind tunnel with zero-velocity correction due to model displacement in *incompressible* flow, it can also be expected that at high subsonic Mach numbers the velocity correction remains zero with good approximation as long as the large flow disturbances around the model do not extend to the vicinity of the tunnel walls. In such a case (see Fig. 4.13), the

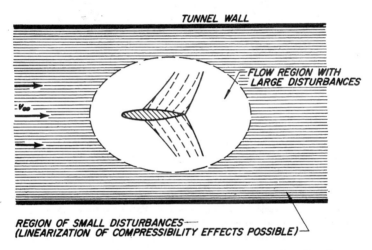

FIG. 4.13. *Model in wind tunnel with flow regions of small and large disturbances.*[2]

flow in the proximity of the walls, that is, the flow which principally determines the wind tunnel boundary correction, is still in the velocity range of linearized compressible flow and will produce, with the same wall geometry, zero-velocity correction independent of the size of the model displacement. A typical curve for a correction-free wind tunnel is included in Fig. 4.12. The wind tunnel correction remains zero independent of Mach number and model size according to considerations presented previously and in Ref. 2.

It must also be realized, however, that in partially open wind tunnels the zero-velocity correction results begin to become questionable when Mach number one is approached very closely. Close to Mach number one, the velocity disturbance near the tunnel walls reaches values which can no longer be considered small enough for linearization of the compressibility effect. This critical value has not been reliably determined by means of theoretical studies because of the extreme complexity of the compressible flow around a model in a mixed wall-type wind tunnel. However, numerous comparative wind tunnel tests have been carried out in various types of partially open wind tunnels to determine the maximum model size which will produce acceptably reliable results (see Chapter 10).

REFERENCES

[1] RIEGELS, F. W. "Wind Tunnel Corrections for Incompressible Flow." MAP Voelkenrode—VG 258—T, AVA Monographs, June, 1947.
[2] GOETHERT, B. H. "Wind Tunnel Corrections at High Subsonic Speeds Particularly for an Enclosed Circular Tunnel." (Translation of DVL, FB–1216, May, 1940) NACA TM–1300, February, 1952.
[3] RITCHIE, V. S. and PEARSON, A. O. "Calibration of Slotted Test Section of the Langley 8-ft Transonic Tunnel and Preliminary Experimental Investigations of Boundary Layer-Reflected Disturbancs." NACA RML–51K14 July, 1952.
[4] KLEBER J. "Preliminary Calibration of a Mock-Up of the Propeller Dynamometer Installation in the Cornell Aeronautical Laboratory 12-ft Variable Density Wind Tunnel." CAL Report AB–625–W–2, May, 1950.
[5] MARSCHNER, B. W. "The Flow Over a Body in a Choked Wind Tunnel and in a Sonic Free Jet." *J. Aero. Sci.* **23**, 368–376, April, 1956.
[6] BARISH, D. T. "Interim Report on a Study of Mach One Wind Tunnels." WADC TR–52–88, April, 1952.
[7] GUDERLEY, G. K. "The Wall Pressure Distribution in a Choked Tunnel." WADC TR–53–509, December, 1953.

BIBLIOGRAPHY

ACUM, W. E. A. "Wall Corrections for Wings Oscillating in Wind Tunnels of Closed Rectangular Section." Part 1, A.R.C. 19,593, October, 1957. Part 2, A.R.C. 19,756, January, 1958.
ACUM, W. E. A. "Approximate Wall Corrections for an Oscillating Swept Wing in a Wind Tunnel of Closed Rectangular Section." A.R.C. 16,512, January, 1954.
EVANS, J. Y. G. "Corrections to Velocities for Wall Constraint in a 10×7 Rectangular Subsonic Wind Tunnel." R.M. No. 2662, April, 1949.
GLAUERT, H. "Wind Tunnel Interference on Wings, Bodies, and Airscreens." British A.R.C., R. & M.–1566, 1933.
GOETHERT, B. H. "Experimental Facts on High-Speed Aerodynamics and Brief Comparison with Theory, Part I." ATI–45169, 1948.
GOETHERT, B. H. "Development of the New Test Section with Movable Side Walls of the Wright Field 10-ft Wind Tunnel (Phase A, Operation with Slots Down)." WADC TR–52–296, November, 1952.
O'HARA, J. C. "A Design Study of Wall Liners to Give a Mach Number of 1.20 for the WADC–CAL Propeller Dynamometer Installation." CAL Report AB–625–W–6, March, 1952.
PRANDTL, L. "Wind Tunnel Corrections." AVA Report Vier Abhandlungen, Zur Hydrodynamik and Aerodynamik, Goettingen, 1927.
RITCHIE, V. S., WRIGHT, R. H. and TULIN, M. P. "An 8-ft Axisymmetrical Fixed Nozzle for Subsonic Mach Numbers up to 0.99 and for a Supersonic Mach Number of 1.2." NACA—RM–L50A03a, February, 1950.
SPREITER, J. R., SMITH, D. W., HYETT, B. J. "A Study of the Simulation of Flow with Free-stream Mach Number One in a Choked Wind Tunnel." NASA TR R–73, 1960.
THOM, A. "Tunnel Wall Effect from Mass Flow Considerations." British A.R.C Report 11,004, November, 1947.

CHAPTER 5

SUBSONIC FLOW IN WIND TUNNELS WITH LONGITUDINAL SLOTS

1. PHYSICAL ASPECTS AND WALL INTERFERENCE IN SLOTTED WIND TUNNELS (WITHOUT LIFT)

a. Wind tunnels with few wide slots

Since open and closed wind tunnels produce deformations of the streamlines around models with opposite signs and thus require velocity corrections with opposing signs, wind tunnels with mixed open and closed wall elements can be expected to eliminate flow distortions and velocity corrections if the open and closed wall elements are properly distributed. Accordingly, in theoretical investigations, Wieselberger determined various wall configurations which would produce a zero-velocity correction for nonlifting bodies of revolution (Fig. 5.1). He demonstrated that for zero-velocity correction the ratio of open to closed wall elements must differ, depending on the arrangement of the open areas and the cross-sectional shape of the wind tunnel.[1] In the specific case of a rectangular cross section, the two open sidewalls must be shorter than the solid upper and lower walls to produce a zero-velocity correction (side ratio $b/a = 1.17$) at the plane of the model.

As noted in Chapter 4, the streamlines around a model in the neighborhood of open wall elements bulge more than in free flight. This difference results from the necessity of building up the static pressure in the test section from the low values around the model to an undisturbed pressure along the free-jet boundary. On the other hand, the streamlines along the closed wall portions of a slotted wall straighten out to follow the plane of the wall. Because of the large relative size of the individual open or closed wall elements in tunnels with few slots, the correct streamline curvature exists only in a small region around the tunnel centerline. Outside this region, streamline deformation typical of either the open or the closed wall will be predominant.

Wind tunnels with wide longitudinal slots of the Wieselberger type have serious disadvantages for practical transonic testing. First, the power requirements are considerably larger than those for closed wind tunnels since the mixing losses along free-jet boundaries are several times larger than the friction losses along closed walls. This fact is documented by a comparison of the power requirements for typical open and closed wind tunnels (Fig. 5.2). Open wind tunnels require at least twice as much power to achieve the same velocity as closed wind tunnels of similar size. Secondly, in open wind tunnels flow pulsations can become very disturbing because of the periodic shedding of vortices from the mixing zone at the free-jet boundaries. The pressure waves resulting from these boundary vortices become steeper

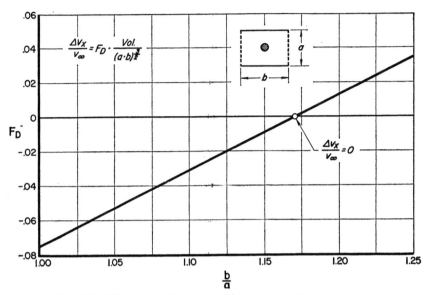

Fig. 5.1. *Velocity corrections in rectangular tunnel with open sides for small spherical model.*[1]

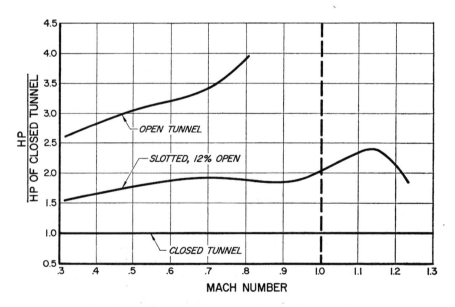

Fig. 5.2. *Comparison of drive power for open, closed and slotted wind tunnel (12 in. diameter).*[2]

as the local velocity reaches sonic speed. They propagate themselves upstream with the relative velocity, $v_{\text{rel}} = a - v_\infty$, that is, they tend to become stationary when the local flow velocity approaches the sonic speed. A good demonstration of the type of disturbance to be expected from such vortex shedding is shown in the Schlieren picture of a two-dimensional airfoil model with a blunt trailing edge (see Fig. 5.3). It is clear that the pressure

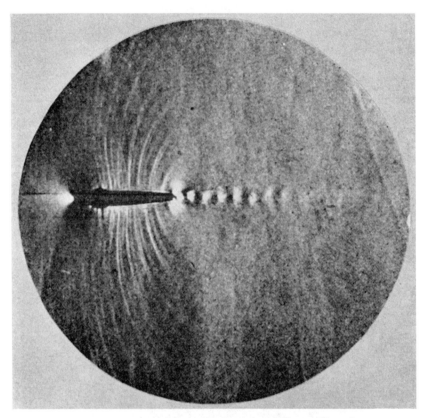

FIG. 5.3. *Schlieren photograph of airfoil with blunt trailing edge, illustrating periodic vortex formation at $M = 0.70$.*[12]

waves produced by the periodic shedding of trailing-edge vortices become stationary when they travel into the high-subsonic velocity field around the model mid-chord point. When local supersonic fields with terminating shock waves occur around models in high-subsonic flight, the model-produced pressure waves will produce the well-known type of fluctuating shock waves. However, the shock-wave fluctuations produced by the model itself must not be confused with the shock-wave fluctuations produced by disturbance waves originating along the free-jet boundary or at the wind tunnel diffuser. Since both types of fluctuations frequently occur simultaneously, it is difficult to separate them, and the danger of gathering erroneous wind tunnel data exists.

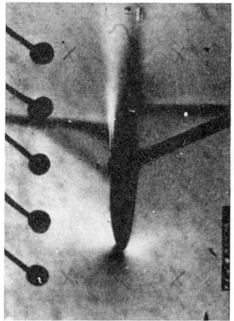

Without sonic throat *(fluctuating shocks)*   With sonic throat *(stable shock formation)*

FIG. 5.4. *Influence of sonic throat behind closed test section on shock-wave stabilization at Mach number 0.78 (photographs by Th. Zobel, DFL, Braunschweig).*

In closed wind tunnels, the upstream movement of pressure waves originating in the tunnel diffuser can be successfully prevented by a controllable second throat having sonic flow (see Fig. 5.4). With open-jet wind tunnels or mixed-type wind tunnels with large open wall elements, no such positive method of preventing the upward movement of pressure waves exists.

b. Wind tunnels with many narrow slots

The disadvantages of large open-area wind tunnels can be effectively reduced when the open wall elements are broken up into numerous narrow slots. In this case, the power requirements are also reduced since the large vortices are replaced by numerous small vortices, and the total open area, that is, the sum of the areas of all individual slots, is essentially reduced for the case of correction-free flow.

The early work done on multi-slot test sections by the NACA became widely known in 1948.[2] In the case of longitudinally slotted wind tunnels with ten slots, it was found that an open-to-total area ratio of approximately 12 per cent was necessary to eliminate the velocity correction required by three-dimensional model displacement (Fig. 5.5). This ratio compares with

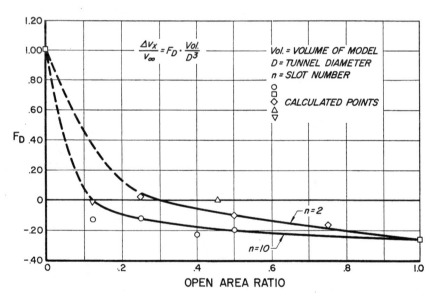

FIG. 5.5. Calculated velocity corrections in slotted wind tunnel based on various approximation methods as function of open-area ratio.[2]

the ratio of 36 per cent required by the Wieselberger test section with two open side walls, a demonstration of the significant reduction in the open area required for correction-free wind tunnels when the number of slots is increased.

The early NACA work on slotted wind tunnels[2] was based on potential flow calculations in which the condition of $v_y = 0$ along the closed wall parts and of $\Delta v_x = 0$ along the open wall parts was satisfied by a disturbance flow field superposed in the conventional manner. These flow disturbances are strong in the proximity of the slots, but they tend to eliminate each other at large distances from the slotted walls, since the radial extent of the non-uniform flow region in the proximity of the slotted wall is proportional to the distance between slot centerlines.

If the number of slots is increased, the width of the region of non-uniform flow near the wall is reduced at the same rate as the decrease in distance between slot centerlines. In the case of a fine-grain test section, the region of non-uniform flow shrinks to a narrow zone. Inside this zone, the influence of each individual slot is noticeable; outside there is only the uniform overall effect of a partially open wall (Fig. 5.6). Furthermore, if the open-to-closed

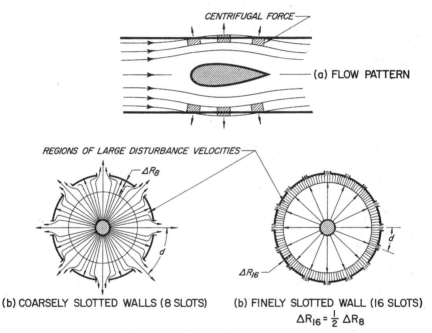

FIG. 5.6. *Influence of number of slots (equal open-area ratio, s/d) on width of the regions with large disturbance velocities.*[13]

area ratio is chosen correctly, the flow inside the narrow near-wall strip coincides with the flow in free flight, that is, streamline shapes and local disturbance velocities are identical with those occurring in free flight. When this is the case, the static pressure along the inside boundary of the narrow non-uniform flow region is, in general, different from the pressure of the undisturbed free flow, and the static pressure on the plenum chamber side of the slotted test section remains equal to the undisturbed pressure. Consequently, the static pressure must change rapidly within the narrow strip close to the slotted wall. In particular, the pressure build-up in this zone of non-uniform flow must be equal to the entire pressure build-up which would occur in free flight beyond the test section between the tunnel wall and infinity. Since the pressure build-up lateral to the streamlines is produced by centrifugal forces, the centrifugal forces in the narrow strip near the wall must be magnified by the arrangement of the slots so that they exceed significantly the centrifugal forces ordinarily occurring in free flight at this distance from the model.

c. Pressure build-up in flow through slotted walls

The mechanism by which the flow through a slotted wall magnifies the centrifugal forces can be readily understood from a consideration of Fig. 5.6. Figure 5.6a shows a typical streamline pattern in the axial direction; Fig. 5.6b shows a cross section of the region of non-uniform flow near the slotted wall. Because of continuity requirements, the radial velocities at the slot bases are considerably larger than those along the inner boundary of the non-uniform region. In the case of small velocity gradients in the main flow direction (x-direction), the radial velocities are inversely proportional to the open-area ratio of the wall, that is:

$$v_{r\,\text{slot}} = \frac{d}{s} \cdot v_{r\,\text{tunnel}}$$

Since the radius of curvature can be expressed as:

$$R_c = \frac{v}{\partial v_r / \partial x} \simeq \frac{v_\infty}{\partial v_r / \partial x}$$

the total local velocity can be replaced, as a first approximation, by the undisturbed velocity v_∞. Then the local pressure gradient at the slot bases is:

$$(\partial p/\partial r)_{\text{slot base}} = -\rho \frac{v^2_{\text{slot}}}{R_c} = -\frac{\rho v_\infty^2}{v_\infty} \left(\frac{\partial v_r}{\partial x}\right)_{\text{tunnel}} \cdot \frac{d}{s}$$

$$= \frac{d}{s}\left(\frac{\partial p}{\partial r}\right)_{\text{tunnel}}$$

The magnification of the pressure gradients at the slot base, therefore, is directly proportional to the open-area ratio of the slotted wall, since

$$d/s = A_{\text{total}}/A_{\text{slot}}$$

At other locations within the disturbed flow annulus, the local centrifugal forces are different from those at the slot base. Their distribution within the disturbance region depends only upon the slot open-area ratio when points of equivalent locations, that is, constant n_s/d and t_s/d, are considered, and the main flow parameters, such as velocity, density, and flow curvature, are constant. The local pressure gradients at any position within the region near the wall can therefore be written:

$$(\partial p/\partial r)_{n_s,t_s} = -\frac{\rho v_\infty^2}{v_\infty}\left(\frac{\partial v_r}{\partial x}\right)_{\text{tunnel}} \cdot f\left(\frac{d}{s}, \frac{n_s}{d}, \frac{t_s}{d}\right)$$

where

$$\left(\frac{\partial(v_r/v_\infty)}{\partial x}\right)_{\text{tunnel}}$$

is the streamline curvature at the inner boundary of the annular disturbance zone.

The total pressure build-up in the non-uniform flow region of the slotted wall is determined by the summation of all individual centrifugal forces over the radial width of the non-uniform flow area. Since at a constant open-area ratio the radial width is proportional to the distance between slots, a total pressure build-up through the non-uniform flow region can be expected as follows:

$$\Delta p = (\partial p/\partial r)_{\text{mean}} \Delta n = \text{const} \, (\partial p/\partial r)_{\text{slot base}} \, d$$

or finally:

$$\Delta p/q = \frac{\Delta p}{(\rho/2)v_\infty^2} = -\frac{\partial(v_r/v_\infty)}{\partial(x/d)} f(s/d)$$

$$= \frac{d}{R_c} f(s/d)$$

Detailed mathematical calculations[3,4] resulted in the determination of the function $f(d/s)$ when slowly changing flow curvature in the axial direction and small values of d/R, as is generally the case, are assumed. With these results inserted and realizing that $v_r/v_\infty \simeq \theta_w$, the above equation can be written finally as:

$$\Delta p/q = \frac{2}{\pi} \log \sin\left(\frac{\pi}{2}\frac{s}{d}\right) \frac{\partial \theta_w}{\partial(x/d)}$$

$$= -2f\left(\frac{s}{d}\right)\frac{\partial \theta_w}{\partial(x/d)}$$

It has been shown that this equation is valid for subsonic, transonic and supersonic flow as long as the disturbance velocities in the slotted wall region are kept sufficiently small in relation to the free-stream velocity.[3]

The above equation defines the pressure build-up which occurs in the non-uniform flow region near the slotted wall. For multi-slotted test-section walls, the width of this region is small (see Fig. 5.6). Since in the case of wind tunnel correction calculations, the flow at some distance from the wall is of main significance the actual flow around the individual slots may be disregarded, and only the overall effect of the slotted wall need be considered. It is therefore permissible in such calculations to replace the actual wall with a fictitious wall which produces the same effect as the real wall at some distance from the wall. According to the preceding equations, the boundary condition which must then be satisfied by the fictitious wall is as follows: the pressure disturbance produced by the model in the wind tunnel must be compensated by the pressure rise through the fictitious wall so that at the plenum chamber side of the wall the undisturbed free-stream pressure is maintained. That is:

$$\Delta p_{\text{fictitious}} = \Delta p_{\text{slotted wall}}$$

By using the linearized flow relationship,

$$\Delta p/q = -2\frac{\Delta v_x}{v_\infty} = -2\phi_x/v_\infty$$

the above equation can be written:

$$\phi_x + f(s/d)\,d\phi_{xy} = 0$$

with

$$f(s/d) = -\frac{1}{\pi} \cdot \log \sin\left(\frac{\pi}{2}\frac{s}{d}\right)$$

This equation* represents the boundary conditions to be satisfied by the fictitious slotted wall. It should be noted that, in comparison with the boundary condition for open walls, $\phi_x = 0$, and for closed walls $\phi_y = 0$, the boundary condition represented by the equation for slotted wall contains one more term; this fact makes it possible to influence the flow field in the wind tunnel by means of additional parameters, that is, by the slot width and the distance between slots.

d. Influence of slot width

It was concluded in Chapter 5, Section 1c that the pressure build-up occurring in curved flow as it approaches a slotted wall is directly proportional to the distance between slot centers. This fact is readily understood when it is realized that, for a given wall open-area ratio, the radial extent of the high-curvature region near the wall is proportional to the distance between slots. Also, since the distribution of the local velocities and centrifugal forces must be the same relative to slot width and distance, the total build-up of the static pressure in the high-curvature region must be proportional to the radial extent of the high-curvature region or, as stated before, to the distance between slot centers. Thus, to maintain a desired value of the additional pressure build-up in the case of a large number of slots, it is necessary to reduce the slot width and thus to compensate for the smaller width of the high-curvature region by increasing the curvature of the flow in the slot region.

Slotted walls with a large number of narrow slots therefore offer the significant advantage of a correction-free wind tunnel with a small wall open-area ratio and consequently permit a reduction in the power requirements and flow fluctuations associated with free-jet boundaries (see Fig. 5.2). The theoretical limit of this reduction of slot width is reached when the normal flow in the slots approaches sonic velocities and begins to choke. However, in reality, the validity limit of the theory presented is reached much sooner; that is, whenever the normal velocity through the slots is no

---

* Subscripts x and y indicated differentiations vs. x and y, respectively.

longer small in relation to the free-flow velocity (as required for first order approximations).

Theoretical results showing the necessary open-area ratio for various slotted walls obtained from the above "fictitious-equivalent wall" theory are presented in Fig. 5.7 as a function of slot number* for three-dimensional

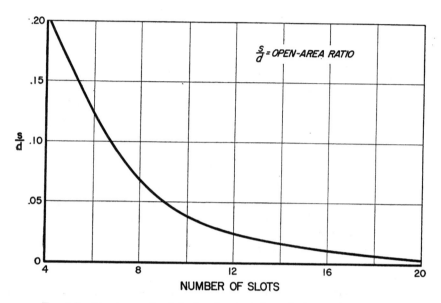

FIG. 5.7. *Open-area ratio of circular slotted wind tunnels for vanishing velocity corrections as function of slot number.*[4]

models in correction-free circular wind tunnels.[4] The rapid decrease in open-area ratio required with increasing slot number is apparent. For example, with eight slots, an open-area ratio of 6.6 per cent is required; with sixteen slots, this ratio is reduced to 0.7 per cent.

For two-dimensional models and tunnels, the required slot width, considered as a function of slot number, exhibits a similar trend[5]. Figure 5.8 shows, however, that in this case considerably smaller open-area ratios are required to produce correction-free wind tunnel flow. At a distance between slot centerlines of 40 per cent of the tunnel height, which is approximately the same relative distance between slots as in the three-dimensional case with eight slots distributed over the circumference of a circular tunnel, an open-area ratio of only 0.57 per cent is required instead of the 6.6 per cent needed for the three-dimensional case.

The influence of deviations from the correct open-area ratio is shown in Fig. 5.9 for circular wind tunnels with eight slots. It is apparent that deviations towards more "closed wall" configurations have considerably more

---

* The mathematical methods of calculating the velocity corrections are generally the same as those used for the downwash corrections for lift. Other methods are indicated in more detail in Section 5 of this chapter.

influence than deviations towards more "open wall" configurations. Therefore, it is advisable to select a somewhat "too-open" slot width in order to avoid overshooting into the more sensitive "too-closed" region.

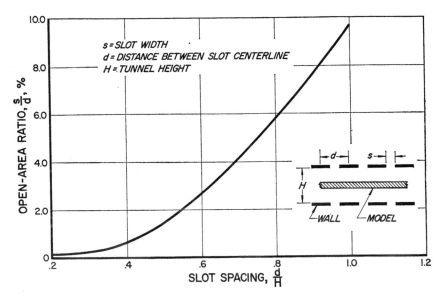

FIG. 5.8. *Open-area ratio for zero-velocity correction for small two-dimensional model in two-dimensional slotted tunnel as function of slot spacing.*[5]

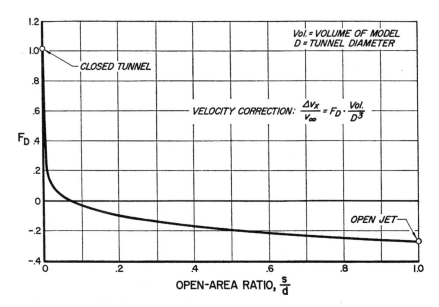

FIG. 5.9. *Velocity correction for circular tunnel with eight longitudinal slots as function of open-area ratio.*[4]

2. CHARACTERISTICS OF SPECIAL SLOT CONFIGURATIONS

a. Slotted walls with protruding slats

The effectiveness of a slotted wall diminishes rapidly when the distance, d, between slot centers is reduced, for example, when the number of slots is increased for a constant wall open-area ratio. It is possible to compensate for this loss of effectiveness by using solid wall portions (slats) which protrude into the test section and by equipping the slots with solid or perforated sidewalls to guide the flow (see Fig. 5.10a). With such a wall, the flow between the sidewalls of the slots is curved, with the radius:

$$R_c = \frac{s}{d} \frac{v_\infty}{\partial v_r/\partial x}$$

where v_r is the mean radial velocity component of the flow in the vicinity of the slotted wall. Consequently the additional pressure build-up of flow within slots of a height, l, is calculated with the assumption of potential flow in the slots:

$$\Delta p/q = -2\frac{l}{s}\frac{\partial v_r/v_\infty}{\partial (x/d)}$$
$$= -2\frac{l}{s}\frac{\partial \theta_w}{\partial (x/d)}$$

the entire pressure build-up of such a wall is:

$$\Delta p/q = 2\frac{\partial \theta_w}{\partial (x/d)}\left\{\frac{1}{\pi}\log\sin\left(\frac{\pi s}{2d}\right) - \frac{l}{s}\right\}$$

It is evident from this equation* that increases in the slot height, l, support the slot effect. Consequently, slotted walls with protruding slats can be made more open than conventional slotted walls without changing the wall characteristics.

Varying the slot height is an additional means of adjusting the slot effect to the required values so that matching of the wall and model characteristics can be more readily achieved without approaching critical conditions resulting from choking or from excessively large radial velocity components to which the linearized theory can no longer be applied. When it is realized that the protruding slat effect can also be accomplished by guide plates which extend into the test-section flow (Fig. 5.10b), it is apparent that a slotted wind tunnel with controllable effectiveness could be simply built by providing for adjustability of the slot sidewalls. In spite of these apparent advantages, no experiments with such special slotted walls are known in the literature.

b. Slotted walls with perforated cover plates

In the preceding paragraphs it was shown that an ideal slotted wall produces a pressure change in the cross flow through the wall, mainly as a result of the flow curvature which is amplified by the slots. The pressure

---

* Note: $\log \sin[(\pi/2) \cdot (s/d)]$ is always a negative term.

(a) SLOTS WITH EXTERNAL SIDE PLATES

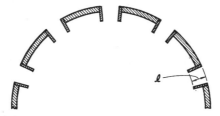
(b) SLOTS WITH INTERNAL SIDE PLATES

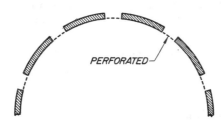
(c) SLOTS WITH PERFORATED COVER PLATES

d. *Photograph of slotted test section with perforated cover plates (model of Wright Field 10 ft wind tunnel).*

FIG. 5.10. *Slotted walls with different slot configurations.*

drop in frictionless cross flow without curvature is determined approximately by the Bernoulli equation:

$$\Delta p/q = \left(\frac{d}{s}\right)^2 \left(\frac{v_y}{v_\infty}\right)^2 = \left(\frac{d}{s}\right)^2 \theta_w^2$$

This second-order term for the pressure drop is, generally, small in comparison with the term representing the curvature effect and is usually neglected in wind tunnel correction calculations.

However, experiments show that in actual flow, even in the case of non-curved streamlines, a pressure drop occurs which, at moderate cross-flow velocities, is approximately linearly proportional to the cross-flow velocity, that is:

$$\Delta p/q = K\frac{v_y}{v_\infty} = K\theta_w$$

This pressure drop can be explained as the result of friction or local separation zones (see Fig. 5.11); it can also be increased artificially by covering

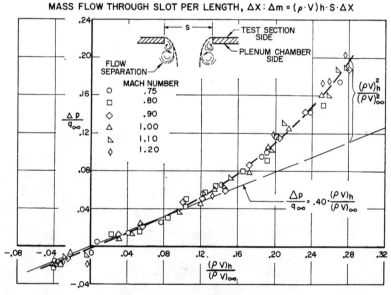

FIG. 5.11. *Cross-flow characteristics of single longitudinal slot (unpublished data of AEDC transonic model tunnel by Gardenier and Chew).*

the slots with perforated plates (Fig. 5.10c). Generally, perforated plates in oblique flow have a pressure drop approximately linearly proportional to the cross-flow velocity. The factor K in the above pressure equation then depends upon the geometry of the cover plates and the open-area ratio of the basic slotted wall.

Thus the pressure change in a curved flow crossing a slotted wall which has protruding slots and perforated cover plates can be written finally as:

$$\Delta p/q = 2\frac{\partial \theta_w}{\partial x/d}\left\{\frac{1}{\pi}\cdot \log \sin\left(\frac{1}{\pi}\frac{s}{d}\right) - \frac{l}{s}\right\} - K\frac{d}{s}\theta_w$$

In this equation, $\Delta p/q$ represents the effective pressure difference between the test section and the plenum chamber (positive when plenum chamber pressure is higher).

The same relationship can also be expressed as a differential equation in terms of the velocity potential, ϕ:

$$\phi_x - \phi_{xy}\, d.\left\{\frac{1}{\pi}\cdot \log \sin\left(\frac{1}{\pi}\frac{s}{d}\right) - \frac{l}{s}\right\} - \frac{1}{2}K\frac{d}{s}\phi_y = 0$$

This equation* must be satisfied along the wall which, for the purpose of these interference calculations, has been assumed to be homogeneous.

A wall such as that described by the above equation has four geometrical parameters which can be selected independently of each other:

1. Open-area ratio of basic slots, s/d.
2. Distance between slot centers in relation to the tunnel diameter, d/D.
3. Height of slot sidewalls in relation to slot width, l/s.
4. Pressure drop constant of cover plates, K.

The first three parameters influence the magnitude of the flow curvature term, and, hence, the calculations for conventional slotted walls can be applied to determine the interference effects of slotted walls. The fourth parameter is basically of a different nature, and the determination of the resulting wind tunnel flow requires other methods (see Chapter 6).

The final equation shown above indicates the wall boundary condition resulting from the combined influence of all four wall parameters. It can be used profitably when the degree of matching between model flow in free flight and wall characteristics is to be determined. For example, for a typical model in a *subsonic* wind tunnel, the addition of perforated cover plates would not serve a useful purpose since, as will be shown in Chapter 6, no improvement of the matching between model requirements and wall characteristics would result. On the other hand, for a model in a *supersonic* wind tunnel an improvement in matching would result from the addition of suitably selected cover plates (see Chapter 9).

3. DISTRIBUTION OF VELOCITY CORRECTIONS ALONG MODEL
a. Slots with constant width

The velocity corrections for coarsely slotted and fine-grain multi-slotted wind tunnels discussed previously refer to the center point of the test models. In the special case of a correction-free tunnel with slots of constant width, the velocity correction disappears only at the mid-point of the model; at other points along the surface the velocities differ from free-stream values. This variation of wall-produced disturbance velocities is shown in Fig. 5.12

---
* Subscripts x and y indicate differentiations vs. x and y, respectively.

for a small model whose displacement is simulated by a doublet.[4] The flow around the model is somewhat distorted in a manner similar to the distortion of the flow around models in open and closed wind tunnels. However,

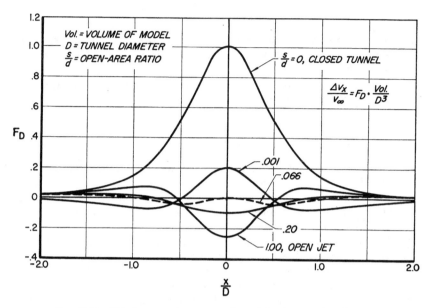

Fig. 5.12. *Velocity corrections at centerline of circular tunnel with eight longitudinal slots.*[4]

in the slotted wind tunnel, the magnitude of the disturbance velocities is considerably less than in the open or closed wind tunnel. This difference in magnitude remains, even after a suitable correction for the free-stream velocity is applied, that is, even when the actual velocity at the model mid-point is considered equal to the corrected free-stream velocity and is used as the reference point.

b. Slots with varying width

With a slotted wind tunnel it is possible to vary the slot in the flow direction and, in this way, to reduce or even eliminate the wall-produced disturbance velocities at the model mid-point and also at other stations. When the slot width changes only gradually in the flow direction, it is permissible to apply the previous equations to determine slot effectiveness at each axial station along the wall and the slot width required for proper matching of model-produced disturbances and wall geometry.

Complete matching of model disturbance and wall geometry is obtained when the slot-produced pressure build-up along the wall is equal to the model pressure disturbance, that is:

$$\left(\frac{D^3}{Vol}\right)\left(\frac{\Delta p}{q}\right)_{slot} = \frac{2}{\pi} \cdot \log \sin\left(\frac{\pi}{2}\frac{s}{d}\right) \frac{\partial \theta_w}{\partial x/d} \frac{d}{D} = -\left(\frac{D^3}{Vol}\right)\left(\frac{\Delta p}{q}\right)_{model}$$

SUBSONIC FLOW IN WIND TUNNELS WITH LONGITUDINAL SLOTS

This equation was evaluated assuming a small model in a circular tunnel with eight longitudinal slots of different constant open-area ratios. The model was assumed small enough to allow simulation by a single doublet. It is apparent from Fig. 5.13 that the model disturbances change in such a manner that only in restricted regions along the wall can the open-area ratio of the slots be properly selected. Also, there is one region in particular in which the model disturbances have a sign opposite to that of the slot-produced pressure build-up; a slot configuration with characteristics

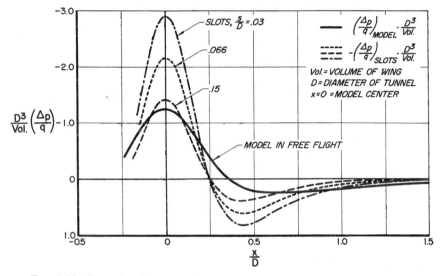

Fig. 5.13. *Comparison between model- and slot-produced disturbance pressures along wall of circular tunnel with eight slots.*

opposite to those of regular slots would therefore be required. Such an "opposite wall effect" is physically impossible since it would be necessary to assume a reversal in the direction of the centrifugal forces.

With a slot open-area ratio of 6.6 per cent, determined previously to be the correction-free open-area ratio at the mid-point of the model (see Chaper 5, Section 1d), the slot pressure build-up matches the model disturbance pressures in free flight only in certain areas. However, since the tunnel interference effect along the tunnel centerline is the integrated effect of the wall along its entire length, this integrated- or mean wall-effect should match the requirements much better than in the local areas represented in Fig. 5.13.

It is also noted that the slot-produced pressure build-up is symmetrical with respect to the x-axis, that is, is symmetrical around the mid-point of the model. This fact represents a significant advantage of the slotted wall over the perforated wall. The perforated wall exhibits an unsymmetrical pressure build-up proportional to the local cross-flow velocity, v_r/v_∞, which has a fundamentally different characteristic than the model-produced disturbance pressure along the wall (see also Fig. 6.3).

73

In view of the results shown in Fig. 5.13, it would be advantageous to narrow the slot width upstream and downstream of the model in order to enlarge the region of correction-free flow instead of using a uniform slot width over the entire length of the tunnel.

4. MACH NUMBER INFLUENCE ON THE EFFECTIVENESS OF WALLS WITH LONGITUDINAL SLOTS (WITHOUT LIFT)

The preceding calculations concerning the effectiveness of slotted walls were made assuming incompressible flow. However, it should be realized that the velocity components perpendicular to a slotted wall are, in actual practice, always of low subsonic velocity, and the edges of the longitudinal slots may be considered "subsonic leading edges" with the extreme sweep-back angle of 90°. Consequently, the basic assumption of the preceding calculations holds true for both subsonic and supersonic flow. Furthermore, the flow pattern in any plane perpendicular to the wall is of the same type if the flow inclination changes only very gradually in the direction of the main flow.

Thus, as in the slender body theory, the flow field in any plane perpendicular to the main flow may be treated as incompressible flow, whether the velocity of the main flow is subsonic or supersonic (see Ref. 6). This statement can be readily understood when the linearized continuity equation is considered:

$$(1 - M_\infty^2)\frac{\partial v_x}{\partial x} + \frac{\partial v_y}{\partial y} + \frac{\partial v_z}{\partial z} = 0$$

If the flow field changes very gradually in the x-direction; that is, in the direction of the main flow, then the expression $(1 - M_\infty^2) \cdot \partial v_x/\partial x$ is considerably smaller than the terms: $\partial v_y/\partial y$ and $\partial v_z/\partial z$. This is particularly true in the vicinity of the slotted wall where large changes of v_y and v_z occur as the flow approaches the individual slots. Also, in the range near Mach number one, the factor $(1 - M_\infty^2)$ is substantially smaller than one and thus contributes to making v_y and v_z the predominant terms in the linearized continuity equation. Consequently, the continuity equation for flow fields which change only very gradually in the x-direction may be written approximately:

$$\frac{\partial v_y}{\partial y} + \frac{\partial v_z}{\partial z} = 0$$

This equation is the familiar continuity equation for incompressible flow in the y-z planes. Since it no longer contains the Mach number of the main flow (which might be subsonic or supersonic), the calculations concerning the slot effect presented in the preceding paragraphs are also valid for both subsonic and supersonic flow, within the approximation described above.

In the case of slotted walls with perforated cover plates, the factor K, which describes the pressure drop through perforated cover plates, may change with Mach number and must be introduced as the particular value for the Mach number under consideration (see Chapter 5, Section 2b).

5. LIFTING WING IN CIRCULAR WIND TUNNEL WITH LONGITUDINAL SLOTS

a. General discussion of the lift influence

In transonic wind tunnel testing the flow distortions due to lift are generally not as serious as those produced by the displacement of the model. This is readily understood when the choking phenomenon is considered. Choking is caused by the inability of the sonic flow to increase its stream density, ρv, and thus to squeeze through the constricted passage existing between the model and the tunnel wall. In contrast, the lift of a model is mainly connected with changes in flow direction and resulting effects on the mean stream density, ρv; there is no direct effect on a sensitive parameter of the sonic flow.

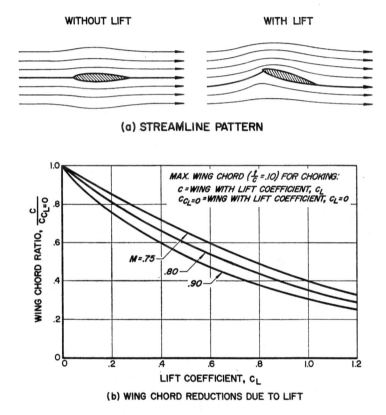

FIG. 5.14. *Influence of lift coefficient on choking Mach number of two-dimensional wing with thickness ratio, $t/c = 0.10$, in closed wind tunnel.*[7]

Only when an unsymmetrical diversion of the flow from one side of a lifting wing to the other occurs does the choking effect become somewhat more severe (see Fig. 5.14 and Ref. 7).

The fact that lift is not as critical a parameter as model displacement is also evident from the fact that in the subsonic speed range the velocity corrections in a wind tunnel due to *model displacement* grow with the third-power of the Prandtl factor,[8] that is:

$$(\Delta v_x/v_\infty)_{\text{compr}} = \left(\frac{1}{(1-M_\infty^2)^{\frac{1}{2}}}\right)^3 (\Delta v_x/v_\infty)_{\text{inc}}$$

$$= \left(\frac{1}{(1-M_\infty^2)^{\frac{1}{2}}}\right)^3 \delta_D \cdot \frac{\text{Vol}}{D^3}$$

where, for slender models, $\delta_D = 1.00$. The flow inclination corrections due to *lift* do not change with Mach number at all when the lift coefficient is kept constant, that is:

$$\Delta\alpha_{\text{compr}} = \Delta\alpha_{\text{inc}} = \frac{\delta}{8} \frac{A_{\text{wing}}}{A_{\text{tunnel}}} C_L$$

where, for small wings in circular closed tunnel, $\delta = 1.00$.

b. *Flow inclination corrections at the center-of-pressure of a lifting wing*

The following discussion is based first on the assumption of incompressible flow. It will be shown later that the results can be readily applied to compressible subsonic flow by means of a simple transformation.

The magnitude of the flow inclination corrections required at the center-of-pressure of a lifting wing can be determined in a simple manner when the Trefftz theorem is applied. This theorem states that the downwash correction required at the center-of-pressure of a lifting wing is exactly one-half the downwash correction required at infinitely large distances behind the wing.[9] This theory assumes that all free vortices which produce the downwash correction start in the plane of the wing. This Trefftz theorem is valid, not only for the main wing free vortices, but also for all image vortices required to satisfy the boundary conditions. The determination of the downwash at infinitely large distances behind the wing in the so-called Trefftz plane is therefore greatly simplified since it is reduced to a two-dimensional problem.

For the purposes of this survey, the basic singularity of a lifting wing is assumed to consist of a horseshoe vortex with its two free vortices starting at the wing tip and extending infinitely far downstream. In the case of a wing span which is small in relation to the tunnel diameter, the two free vortices may be replaced by a single doublet line. The mathematical problem is then to determine at the centerline of the wind tunnel for $x \to \infty$, the downwash velocity produced by the interference potential required to satisfy the mixed wall boundary conditions (Fig. 5.15).

The general solutions for the velocity field of the basic doublet and the interference flow can be written in the form of a series, in complex variables, (see Ref. 10), as follows:

$$W = \frac{A_0 + \sum_{m=1}^{\infty} \{A_{n\pm m} \cos[(n \pm m)\zeta] + B_{n\pm m} \sin[(n \pm m)\zeta]\}}{R^2 \zeta [\sin^2(nv) - \sin^2(n\zeta)]^{\frac{1}{2}}}$$

or with

$$Z = Re^{i\zeta} = Rz$$

$$W = \frac{A_0 + \sum_{m=1}^{\infty} \{\tfrac{1}{2}A_{n\pm m}[z^{n\pm m} + z^{-(n\pm m)}] + \tfrac{1}{2}B_{n\pm m}[z^{n\pm m} - z^{-(n\pm m)}]\}}{\tfrac{1}{2}R^2[1 - 2z^{2n}\cos(2\nu) + z^{4n}]^{\tfrac{1}{2}} z^{-(n-1)}}$$

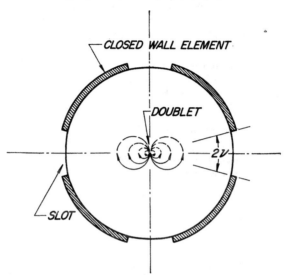

FIG. 5.15. *Doublet replacing wing-tip vortices in Trefftz-plane behind small-span wing in slotted tunnel.*

where:
- W = Complex velocity of doublet with interference field
- $2n$ = Number of slots of equal width equally-spaced
- 2ν = angle indicating width of slots (see Fig. 5.15)
- Z = complex coordinate (distance from tunnel center) = $Rz = Re^{i\zeta}$.
- θ = direction angle of vector Z
- R = tunnel radius
- $m = 1, 2, 3,...$ = constant
- A, B = constants (real numbers).

It can be readily verified that the above equation satisfies the boundary conditions of the slotted wall. At the wall $Z = Re^{i\theta}$; hence, $\zeta_{\text{wall}} = \theta$ is a real value.

For the *open* parts of the wall

$$n\nu > n\zeta$$

and the term

$$[\sin^2(n\nu) - \sin^2(n\zeta)]^{\tfrac{1}{2}}$$

is a real value. Consequently, the velocity vector has the direction of the local coordinate, z, that is, it is perpendicular to the tunnel boundary.

On the other hand, when $(n\nu) < (n\zeta)$, as in the case of the solid wall parts, the root becomes imaginary, and the velocity vector is oriented perpendicular to z. These two velocity directions satisfy the boundary condition.

For the area around the tunnel center ($z \to 0$), the denominator of the above equation for W is directly proportional to $z^{-(n-1)}$. Hence, the velocity W in this area can be represented by a simple power series of z. The constants A and B must be determined in such a way that at the tunnel center only singularities with terms z^{-2} remain (they represent the singularity of the original doublet) and that the symmetry of the velocity field is properly considered.

Calculations based on these requirements are described in Ref. 10; the results for the downwash correction at the wing ($x = 0$) in tunnels of various configurations are presented in the form:

$$\alpha_{corr} = \alpha_{geom} + \Delta\alpha = \alpha_{geom} + \frac{C_L}{8}\frac{A_W}{A_T}\delta$$

Tunnel with Two Slots ($n = 1$) with the Open Slots Located near the Wing Tips.—In the case of a tunnel with open slots located near the wing tips, the value δ is then:

$$\delta = \cos(\pi c)$$

where $c =$ the ratio of open to total wall area.

Thus for $c = 0$ (closed tunnel), $\delta = 1$; for $c = 1$ (open tunnel), the result is $\delta = -1$. In order to make the correction zero ($\delta = 0$), the tunnel wall must be exactly half open and half closed (see Fig. 5.16).

Tunnel with Two Slots ($n = 1$) with the Open Wall Areas Located above and below the Wing.—In the case of a tunnel with two slots with the open wall areas located above and below the wing, the value of δ is:[10]

$$\delta = -2\int_0^{\pi/2}\frac{1-2\sin^2[(1-c)\pi/2]\sin^2\theta}{\{1-\sin^2[(1-c)\pi/2]\sin^2\theta\}^{\frac{1}{2}}}d\theta \Bigg/ \int_0^{\pi/2}\frac{d\theta}{\{1-\sin^2[(1-c)\pi/2]\sin^2\theta\}^{\frac{1}{2}}} + \cos[(1-c)\pi]$$

Again the limits for open and closed tunnel walls are satisfied by $\delta = -1$ and $\delta = +1$, respectively. However, it is found that the downwash correction becomes zero, that is $\delta = 0$, when an open-area ratio of approximately

$$c = 5 \text{ per cent}$$

is established. It is noteworthy that only a very small open slot area at the top and bottom of the tunnel is needed to make the necessary downwash correction disappear (see Fig. 5.16). This is in decided contrast to the arrangement of two slots located at the tunnel sides.

Tunnel with $2n$ Slots of Equal Width Equally Distributed ($n > 1$) disregarding location of slots on top and bottom or on sides.—In the case of a tunnel with $2n$

SUBSONIC FLOW IN WIND TUNNELS WITH LONGITUDINAL SLOTS

slots of equal width equally distributed, disregarding the location of the slots on the top and bottom or on the sides, the value of the correction factor δ is:[10]

$$\delta = - \int_0^{\pi/2n(1-c)} \frac{\cos[(n+1)\theta]\,d\theta}{\{\sin^2[(1-c)\pi/2] - \sin^2 n\theta\}^{\frac{1}{2}}} \bigg/ \int_0^{\pi/2n(1-c)} \frac{\cos[(n+1)\theta]\,d\theta}{\{\sin^2[(1-c)\pi/2] - \sin^2 n\theta\}^{\frac{1}{2}}}$$

For any value of $n > 1$, the conditions for $c = 0$ (closed tunnel), and $c = 1$ (open tunnel) are satisfied by the above relationship (see Fig. 5.17).

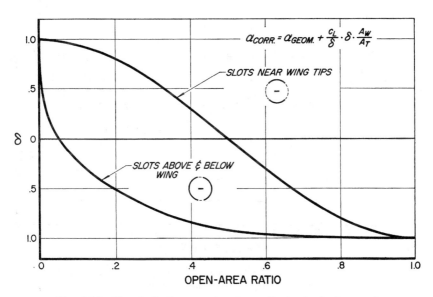

FIG. 5.16. *Flow inclination corrections for small wing in circular tunnel with two slots.*[10]

It can also be seen from this equation that the open-area ratio, c, for correction-free tunnels, that is, for $\delta = 0$, meets the following requirements:

$$c < 1/(n+1)$$

Hence, with increasing slot number, the open-area ratio required for zero interference becomes smaller, and, with infinitely many slots, approaches an infinitely small value (see Fig. 5.17). These results are in agreement with the results discussed in Section 1d of this chapter, that is, in order to maintain its effectiveness, a multi-slotted wall must be provided with a progressively smaller open-area ratio as the slot number is increased. On the other hand, for any finite open-area ratio, a slotted wind tunnel acts as an open wind tunnel when the slot number is increased to large values, $n \to \infty$.

For a circular tunnel with eight slots, an open-area ratio of 1.2 per cent is required to eliminate the flow inclination correction at the wing.

c. Flow inclination in vicinity of lifting wing

The necessary flow inclination correction differs for different points in a slotted wind tunnel; in particular, the values of the downwash corrections given in the preceding paragraph are valid only for the center-of-pressure of a lifting wing simulated by a single horseshoe vortex. However, by suitable superpositions of the horseshoe vortices, it is possible to determine

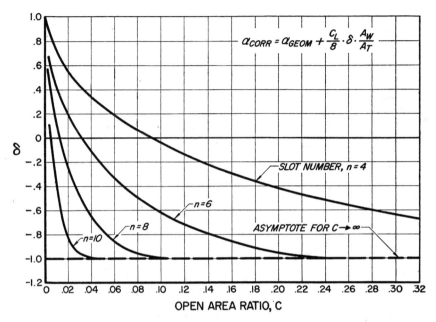

FIG. 5.17. *Flow inclination corrections for small wing in circular tunnel with several longitudinal slots.*[10]

the deviations from the single horseshoe theory. Moreover, the change of the flow inclination correction in the flow direction is important since it determines the flow curvature correction required along the main wing, as well as the downwash correction necessary at the horizontal tail. It is not possible to determine these corrections using the simplified analysis in the Trefftz plane; more complex calculations are necessary. By introducing the simplified boundary condition of a multi-slotted wall, that is, by replacing the slotted wall with an equivalent homogeneous wall (see Section 1c of this chapter) the variation in the necessary downwash correction along the centerline of the wind tunnel can be determined.

For the purpose of these calculations, the velocity potential of the original horseshoe vortex with the small span was developed into the series[4]:

$$\phi_1 = \frac{C_L}{32} \frac{A_W}{A_T} (v_\infty R) \cos \omega \left\{ \frac{R}{r} + \int_0^\infty -H_1^{(1)}\left(i\lambda \frac{r}{R}\right) \sin\left(\lambda \frac{x}{R}\right) d\lambda \right\}$$

SUBSONIC FLOW IN WIND TUNNELS WITH LONGITUDINAL SLOTS

where:

C_L = lift coefficient of wing
R = radius of wind tunnel
r, x, ω = cylindrical coordinates
λ = variable.

The terms $H_1^{(1)}$ indicate Hankel functions of the first order, that is, special types of the Bessel functions, as defined, for instance, in Ref. 11. The Hankel function has the proper characteristics of disappearing at large values of the argument, $r \to \infty$; and at small values, $r \to 0$, they develop a singularity, $H_1^{(1)} \to \infty$.

The variation of the potential in the x-direction is reflected in the above equation by the Fourier development with the non-symmetric terms $\sin(\lambda x/R)$. Thus the variation of the potential in the r-direction is properly represented by the selected Hankel function since the above expression for ϕ_1 satisfies the continuity equation:

$$\phi_{xx} + \phi_{rr} + \frac{\phi_{ww}}{r^2} + \frac{\phi_r}{r} = 0$$

It should be noted that in the above equations the potential ϕ_1 is divided into a symmetrical and a non-symmetrical term. The term, R/r, represents the symmetrical part, that is, the potential of a pair of infinitely long vortex lines with opposite directions of rotation. The second term with the Hankel function represents the non-symmetrical part, that is, the potential of a pair of vortex lines which change their directions of rotation at the origin $x = 0$. Since for small values of r, the Hankel function can be written:

$$\lambda H_1^{(1)}(i\lambda r) \simeq \frac{2}{\pi}\frac{1}{r}$$

and since the Fourier development of a non-symmetrical step function is:

$$\frac{1}{\pi}\int_0^\infty \frac{1}{\lambda}\sin\left(\frac{\lambda x}{R}\right) d\lambda$$

the correctness of the above equation for ϕ_1 is apparent.

The velocity potential ϕ_2 of the wall interference field can also be expressed as a series in a similar manner:

$$\phi_2 = \frac{C_L}{32}\frac{A_W}{A_T}v_\infty R \cos\omega\left\{C_1\frac{r}{R} + \int_0^\infty C_{(\lambda)}\left[-i\mathcal{J}_1\left(i\lambda\frac{r}{R}\right)\right]\sin\left(\lambda\frac{x}{R}\right)d\lambda\right\}$$

The Bessel function, $i\mathcal{J}_1(i\lambda r/R)$, is used in this development since this function is regular and finite at all points within the tunnel, $r \leqslant R$, and since it also satisfies the continuity equation. The constants C_1 and $C_{(\lambda)}$, must be determined in such a manner that the boundary condition of the slotted wall is satisfied by the total potential, $\phi_1 + \phi_2$:

$$\phi_x + \phi_{xy}f(s/d)\,d = 0$$

with:
$$f(s/d) = -\frac{1}{\pi}\log\sin\left(\frac{\pi}{2}\frac{s}{d}\right)$$

By introducing ϕ_1 and ϕ_2 into the equation for the boundary conditions and solving for the constants C, the final result for the wall interference potential is obtained[4]:

$$\phi_2 = \frac{C_L}{32}\frac{A_W}{A_T}v_\infty R \cos\omega\left\{\frac{(2\pi/n)f(s/d)-1}{(2\pi/n)f(s/d)+1}\cdot\frac{r}{R}\right.$$

$$- \int_0^\infty \frac{[1-(2\pi/n)f(s/d)][-H_1^{(1)}(i\lambda)] - (2\pi/n)f(s/d)\lambda i H_0^{(1)}(i\lambda)}{[1-(2\pi/n)f(s/d)][-i\mathcal{J}_1(i\lambda)] + (2\pi/n)f(s/d)\lambda\mathcal{J}_0(i\lambda)}$$

$$\left.\times\left[-i\mathcal{J}_1\left(i\lambda\frac{r}{R}\right)\sin\left(\lambda\frac{x}{R}\right)\right]\mathrm{d}\lambda\right\}$$

With this equation it is possible to determine the interference velocities inside the wind tunnel, as desired.

For example, Fig. 5.18 represents the downwash correction necessary behind a wing in a circular wind tunnel with eight slots. The same correction may be applied in the case of circular tunnels having a different number of slots by varying the open-area ratio in such a manner that the term $f(s/d)/n$ is kept constant, that is:

$$\frac{1}{n}\log\sin\left(\frac{\pi}{2}\frac{s}{d}\right) = \text{constant}$$

The curves of Figs. 5.17 and 5.18 show that an eight-slotted wind tunnel requires an open-area ratio of approximately 1.2 per cent for zero-downwash correction at the wing itself and at large distances behind the wing. However, at small distances behind the wing a downwash correction is necessary in the direction of the closed tunnel. It might therefore be more practical in actual wind tunnel operation to increase the open-area ratio slightly, for instance, to approximately 5 per cent. At the larger open-area ratio, the downwash correction needed at the plane of the wing is still small, only approximately one-fifth that required for the open wind tunnel case. Furthermore, the required downwash correction in the vicinity of the wing remains essentially constant so that a curvature correction for the main wing is not necessary.

It should be again noted that for a given slot number the open-area ratio required for zero-downwash correction is smaller than that required to eliminate the velocity correction.

The curvature corrections in the plane of the main wing, that is $\mathrm{d}\Delta\alpha/\mathrm{d}(x/R)$ are shown in Fig. 5.19 for a circular slotted wind tunnel with eight slots. Again, the results are applicable to wind tunnels which have a different number of slots when the open-area ratio is varied according to:

$$\frac{1}{n}\log\sin\left(\frac{\pi}{2}\frac{s}{d}\right) = \text{constant}$$

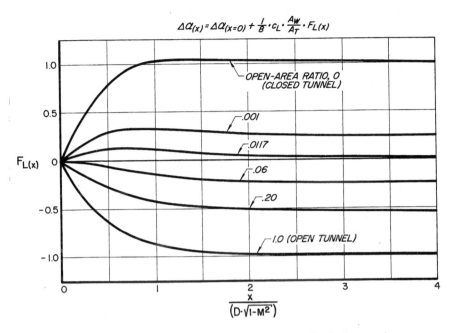

FIG. 5.18. *Flow inclination corrections behind wing in circular tunnel with eight slots.*[4]

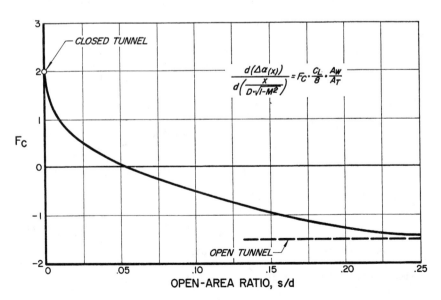

FIG. 5.19. *Flow curvature corrections behind wing in circular tunnel with eight slots.*[4]

d. Mach number influence

Within the validity of the Prandtl–Glauert approximation, the downwash correction discussed in the preceding paragraph can be readily applied to the subsonic compressible flow range. According to Ref. 8 and the general discussions preceding this section, the required downwash correction as a function of the lift coefficient remains unchanged with Mach number when the wind tunnel flow is distorted according to the given similarity rules.

For example, all dimensions perpendicular to the flow direction must be reduced by the Prandtl factor while the dimensions in the flow direction remain unchanged. Thus, the open-area ratio of the tunnel remains the same, since both slots and slats are distorted by the same factor. However, the tunnel radius is reduced by the Prandtl factor while the distances behind the wing remain unchanged. Therefore the graphs (Figs. 5.18 and 5.19) can be generalized by using $x/[R(1-M^2)^{\frac{1}{2}}]$ as a universal parameter. It is apparent that with increasing Mach number, the necessary curvature correction $d\Delta\alpha/d(x/R)$ is increased if all other geometrical parameters are kept constant.

REFERENCES

[1] WIESELBERGER, C. "On the Influence of the Wind Tunnel Boundary on the Drag, Particularly in the Region of the Compressible Flow" (in German). *Luftfahrtforsch.* **19**, 124–128, 1942.

[2] WRIGHT, R. H. and WARD, V. G. "NACA–Transonic Wind Tunnel Test Sections." NACA–RM–L8JO6, October, 1948.

[3] GOETHERT, B. H. "Properties of Test Section Walls with Longitudinal Slots in Curved Flow for Subsonic and Supersonic Velocities (Theoretical Investigations)." AEDC–TN–55–56, August, 1957.

[4] GUDERLEY, G. K. "Wall Corrections for a Wind Tunnel with Longitudinal Slots at Subsonic Velocities." WADC–TR–54–22, January, 1954.

[5] MAEDER, P. F. "Theoretical Investigations of Subsonic Wall Interference in Rectangular Slotted Test Sections." Brown University, TR–WT–11, September, 1953.

[6] JONES, R. T. "Properties of Low-Aspect-Ratio Pointed Wings at Speeds Below and Above the Speed of Sound." NACA–TN–1032, March, 1946.

[7] GOETHERT, B. H. "Choking Mach Number in High Speed Wind Tunnel with Consideration of the Model Lift." AF–TR–5666, March, 1948.

[8] GOETHERT, B. H. "Wind Tunnel Corrections at High Subsonic Speeds Particularly for an Enclosed Circular Tunnel." (Translation of DVL, FB–1216, May, 1940), NACA–TM–1300, February, 1952.

[9] TREFFTZ, E. "Zur Prandtlschen Tragflachen Theorie." *Math. Ann.* **82**, 306–319, 1921.

[10] PISTOLESI, E. "On the Interference of a Wind Tunnel with a Mixed Boundary." *Comment. Pontif. Acad. Sci.* **4** (9), 1940. (Translated by J. V. Foa, Cornell Aeronautical Laboratory, Inc., December, 1949.)

[11] JAHNKE, E. and EMDE, F. *Table of Functions.* Dover Publications, New York, 1945.

[12] STACK, JOHN. "Compressible Flows in Aeronautics." *J. Aero. Sci.* **12**, No. 2, 127–148, April, 1945.

[13] GOETHERT, B. H. "Physical Aspects of Three-Dimensional Wave Reflections in Transonic Wind Tunnels at Mach Number 1.2." AEDC–TR–55–45, March, 1956.

BIBLIOGRAPHY

BAZJANAC, D. "Investigations by Means of Electrical Analogy on the Influence of the Flow Boundaries in Wind Tunnels" (in German). Technical University, Zurich, Switzerland, 1943. (Dissertation Printed by AG. Gebr. Leemann & Co.)

BRESCIA, R. "Study of Interference in Wind Tunnels with Slotted Sides in Compressible Flow." *Atti Accad. Torino*, **88**, 1953–1954. (Brutcher Translation No. 3533.)

CHEN, C. F. and MEARS, J. W. "Experimental and Theoretical Study of Mean Boundary Conditions at Perforated and Longitudinally Slotted Wind Tunnel Walls." AEDC–TR–57–20, December, 1957.

DAVIS, D. D. and MOORE, D. "Analytical Study of Blockage and Lift Interference Corrections for Slotted Tunnels Obtained by the Substitution of an Equivalent Homogeneous Boundary for the Discrete Slots." NACA–RM–L53EO7b, June, 1953.

GOETHERT, B. H. "Plane and Three-Dimensional Flow at High Subsonic Speeds." NACA–TM–1105, October, 1946.

GOETHERT, B. H. "Flow Establishment and Wall Interference in Transonic Wind Tunnels." AEDC TR–54–44, June, 1954.

GUDERLEY, GOTTFRIED. "Simplifications of the Boundary Conditions at a Wind Tunnel Wall with Longitudinal Slots (at Subsonic Speeds)." WADC–TR–53–150, April, 1953.

HARRIS, WILLIAM G. and LESKO, J. S. "Development of a New Test Section with Movable Sidewalls, Wright Field 10 ft Wind Tunnel. Part 2. Operation with Slots Open." WADC–TR–52–296, May, 1954.

MAEDER, P. F. and WOOD, A. D. "Transonic Wind Tunnel Test Sections." *Z. Angew. Math. Phys.* **7**, 1956.

THEODORSEN, T. "Theory of Wind Tunnel Interference." NACA–TR–410, 1931.

RITCHIE, V. S. and PEARSON, A. O. "Calibration of the Slotted Test Section of the Langley 8 ft Transonic Tunnel and Preliminary Experimental Investigations of Boundary Layer-Reflected Disturbances." NACA–RM–L51–K14, July, 1952.

ROY, M. "The Large Transonic Wind Tunnel at Modane-Avrieux (La Grande Soufflerie de Modane-Avrieux est Devenue Transsonique)." *Rech. Aéro.* No. 49, January–February, 1956.

SCHNEIDER, S. "Transonic Experiments at Modane (Essais Transsoniques à Modane)." *Rech. Aéro.* No. 72, September–October, 1959.

TIRUMALESA, D. "Wall Interference in Transonic Wind Tunnels." (Part of thesis presented for the degree of Doctor of Science at the University of Paris.) July, 1956.

WEINIG, F. *Die Stroemung um die Schaufeln von Turbomaschinen, Beitrag zur Theorie, Axial Durchstroemter Turbomaschinen.* Johann Ambrosius Barth, Leipzig, 1935 (or J. W. Edwards, Ann Arbor, Michigan, 1948).

CHAPTER 6

SUBSONIC FLOW IN WIND TUNNELS WITH PERFORATED WALLS

1. BASIC CHARACTERISTICS OF PERFORATED WALLS

a. Pressure change in cross flow through perforated walls

Perforated wind tunnel walls, in contrast to the wind tunnel walls with longitudinal slots discussed in Chapter 5, are characterized by a large number of openings which may be either transverse slots or holes with circular or other cross-sectional shapes. The most essential aerodynamic difference between these two types of walls is the arrangement of lift-producing leading edges. The perforated wall has numerous openings, each of which forms the leading edge of a lift-producing wall element. In the extremely simplified case of transverse slotted walls (Fig. 6.1), each solid wall element acts as an individual wing so that the entire wall is, in effect, a lattice with a stagger angle of 90°. A longitudinally slotted wall may also be considered to consist of numerous elementary wings of infinitely small aspect ratio; however, within first order calculations, such wings are known to produce no lift at large distances behind their leading edges. The transverse slotted wall, like a lattice, produces a lifting force which is in first order, directly proportional to the angle between the wall surface and the approaching flow.

For wind tunnel wall correction calculations, the flow around each individual slot is of little interest; the integrated effect of the multi-slotted wall at some distance from the wall determines the amount and type of wall interference. Hence, it is sufficient for such calculations to replace the real wall, with many discrete slots, with a fictitious *homogeneous wall*. This fictitious wall is to produce the same total lift as the real wall. Because the fictitious wall is assumed to be homogeneous, the lift is evenly distributed over the wall, and, consequently, a uniform pressure drop or rise in the wall cross flow is produced. Since a perforated wall acts like a lattice and since the lifting force of a lattice is proportional to the flow angle, the pressure drop produced by the fictitious perforated wall will also follow this relationship. Hence, in the first approximation:

$$\Delta p/q = K\theta$$

where:
- Δp denotes the pressure difference between plenum chamber and test section
- q is the dynamic pressure of the flow
- θ is angle between the wall surface and the approaching flow.

The constant K depends upon the geometry of the wall and the Mach number and is discussed in more detail in Chapter 11.

It is noteworthy that in the case of outflow, the flow is attached to the wall surface only at the test-section side of the perforated wall; at the plenum chamber side it is usually completely separated. The main result of this difference is a change in the magnitude of the constant K of the preceding equation. Experiments have shown that, even in the case of

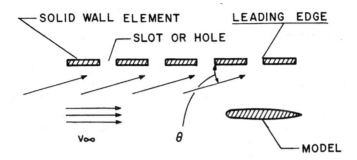

Fig. 6.1. *Wall with transversal slots or perforation in oblique flow.*

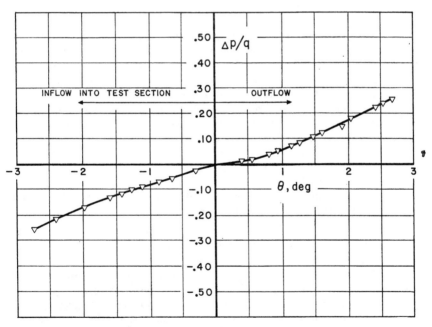

Fig. 6.2. *Pressure drop characteristics of perforated wall in cross flow for incompressible flow (wall thickness 0.022 in.; hole diameter 0.024 in.; open-area ratio 20 per cent).*[6]

small inflow into the section, the wall retains its essentially linear characteristics (see Fig. 6.2).

Though the previous discussion is concerned with walls with transverse slots, it has been shown by experiments that any wind tunnel wall having transverse or oblique slots or many individual holes of an arbitrary shape follows the general relationship:

$$\Delta p/q = K\theta + K_2\theta^2 + K_3\theta^3 + ... \simeq K\theta$$

The linear term in the above equation is generally the predominant one in the range of flow angle near a wind tunnel wall and is the only term used in wind tunnel correction theory in existing publications.

b. Comparison between perforated walls in pure isentropic flow and in highly viscous flow

The above linear relationship between pressure drop and small cross-flow angles is a good description of the characteristics of perforated walls when the flow approaches the conditions of potential flow, that is, when the friction effects are negligibly small. In the case of large friction effects, for example, when the boundary layer is thick in comparison with the hole size, the similarity between the flow through a lattice in potential flow and a perforated wall is no longer maintained. This fact is confirmed by wall characteristic tests which showed large non-linear effects of a sometimes irregular nature when the wall spacings were made small in relationship to the boundary layer thickness.

On the other hand, when the walls are formed of porous material with many small channels, the friction forces are the predominant forces, and, again, the pressure drop is a linear function of the cross-flow velocity. It is interesting that both extremes—perforated walls with extremely small friction effects and porous walls for which friction is the dominating factor—lead to purely linear relationships between the pressure drop and the flow angle. However, a major difference exists in the characteristics of the two walls. At constant flow angle, θ, the pressure drop of perforated walls increases as the dynamic pressure of the approaching flow increases, that is:

$$\Delta p = K\theta q_\infty = K\theta \frac{\rho_\infty}{2} v_\infty^2$$

In the case of porous walls, the pressure drop increases only as the velocity component, v_y, perpendicular to the wall, that is[1]:

$$\Delta p \simeq v_y = K_P \frac{\rho_\infty}{2} \theta v_\infty = \frac{K_P}{v_\infty} \theta q_\infty$$

In order to maintain the same relationship between model-produced disturbance pressures and wall-produced pressure changes, the porous wall must be continuously adjusted as the test velocity changes. The perforated wall provides the desired relationship automatically. A similar disadvantage in wall characteristics also exists in the case of longitudinally slotted walls if the slots are very narrow and very deep.

2. FLOW DISTORTIONS IN PERFORATED WALL WIND TUNNELS

When a model is installed in a perforated wall wind tunnel, the perforated wall characteristics do not match the type of disturbance which occurs with a model in free flight. Figures 6.3a and b present the local pressure disturbances and the local flow inclinations of a wing in free flight along a line

y = constant; also the characteristic line of a perforated wall is shown for comparison in Fig. 6.3b. The discrepancy between the two curves indicates that at subsonic speeds no perfect simulation of the free-stream conditions is possible in perforated wall wind tunnels. The corresponding curves for a slotted wall (see Chapter 5, Fig. 5.13) show a much closer match between model and wall characteristics. Therefore, it can be concluded that for *subsonic* testing the longitudinally slotted wind tunnel is generally superior to the perforated wall wind tunnel.

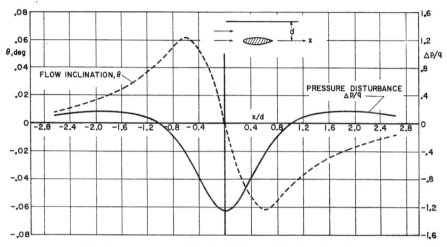

a. *Disturbance pressure and flow inclination as function of coordinate x.*

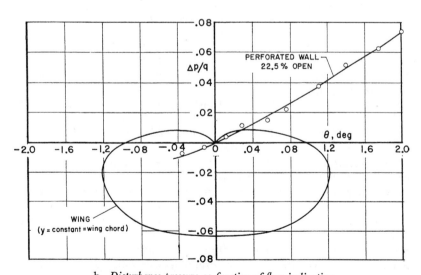

b. *Disturbance pressure as function of flow inclination.*

FIG. 6.3. *Flow disturbances of wing in free flight (incompressible) along line v = constant = wing chord $(t/c = 0.28)$.*

The character of the flow distortion can be most clearly recognized by considering a model in a wind tunnel which initially has closed walls, and then drilling holes in selected areas of these walls. A model in a closed wind tunnel produces a flow-area constriction in the region of its maximum thickness so that at high subsonic Mach numbers "choking" of the flow occurs. In free flight, choking is avoided by means of the bulging of the streamlines. To simulate free-flow conditions and to circumvent the choking phenomenon in the wind tunnel, provisions must be made for the flow to expand through suitable openings into the surrounding plenum chamber. When such openings are provided in the neighborhood of maximum model thickness (for example, if the initially closed walls were replaced by perforated walls in this area) and the surrounding plenum chamber pressure is maintained at the undisturbed pressure level, *there would be no flow out of the test section* into the plenum chamber. On the contrary, there would be flow from the plenum chamber *into the test section*. Consequently, the flow constriction, and thus the choking difficulty, become even more severe (Fig. 6.4). This

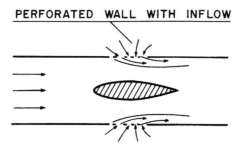

Fig. 6.4. *Model in closed wind tunnel with perforated wall section in model region at subsonic speeds.*

inflow into the test section is the result of the fact that the pressure around the maximum thickness of a model is smaller than the free-stream pressure; thus air is sucked into the test section instead of flowing to the outside, as required for free-flight simulation.

Since inflow and outflow are controlled by the pressure difference between the test section and the plenum chamber, it is obvious that outflow from perforated test sections can be accomplished only by placing the wall openings in regions in which the static pressure is larger than the plenum chamber pressure. Such regions exist at large distances upstream and downstream of the model. For example, when the wall upstream of the low pressure region of the model is replaced by perforated material, some flow enters the plenum chamber ahead of the model and thus by-passes the critical area in which flow constriction normally occurs and choking takes place at high Mach numbers (Fig. 6.5).

When the entire wall of a wind tunnel is perforated, it is obvious that both effects described above occur simultaneously. Ahead of the model, some air is diverted into the plenum chamber. In the low pressure regions, that is, upstream and downstream of the maximum model thickness, some air is sucked into the test section. In the high pressure region downstream of the model, air again flows into the plenum chamber (Fig. 6.6). The overall-flow picture, therefore, is highly unsymmetrical in contrast with free-flight conditions. Despite this lack of symmetry, it has been shown that, with a proper open-area ratio of the perforated walls, the flow velocity at the mid-point of the model can be made equal to that of free flight so that no correction for the mean effective free-stream velocity is required. It can also be shown that choking of the flow between the model and the perforated wall can be eliminated. However, in such a case it is even more essential

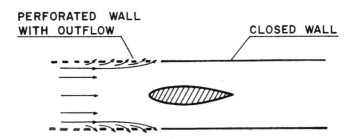

Fig. 6.5. *Model in closed wind tunnel with perforated wall sections upstream of model at subsonic speeds.*

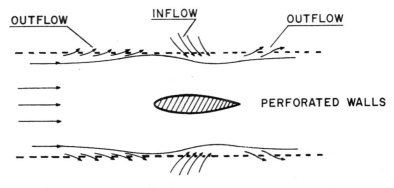

Fig. 6.6. *Schematic outflow- and inflow-pattern for model in perforated wind tunnel at subsonic speeds.*

than with other types of transonic wind tunnels that the partially open walls be extended far enough upstream of the model to produce the required outflow into the plenum chamber ahead of the model.

The above considerations show, as stated before, that for subsonic testing the perforated wall is not the most suitable type of partially open wall. The reason so much attention is given it in transonic testing is the fact that in the supersonic speed range perforated wall tunnels are superior to other types of wind tunnels (see Chapters 8 and 9). Furthermore, in practice, the flow distortions in the range of subsonic operation of a perforated wall wind tunnel are small enough so that it is permissible either to neglect them completely in the case of small models or to apply suitable corrections in the case of larger models (see Chapter 7).

3. VELOCITY CORRECTIONS IN THE VICINITY OF ONE PERFORATED WALL (TWO-DIMENSIONAL, INCOMPRESSIBLE FLOW)

a. General relationships

In the case of two-dimensional incompressible flow, for example, around a cylinder or a lifting wing with an infinitely large span, a simple relationship can be derived by which to determine the interference velocity in the vicinity of one perforated wall.[2] The term ϕ may denote the velocity potential, ψ the steam function, and Φ the complex stream function $\Phi = \phi + i\psi$. Furthermore, the subscript ∞ may indicate the condition of the undisturbed parallel flow, the subscript 1 the primary singularity (doublet in free flight, etc.), and the subscript 2 the additional singularity required to satisfy the boundary condition of a perforated wall. The subscript $2c$ may indicate the correction singularity required for a closed wall.

Derivation of Universal Relationships.—For one perforated wall the boundary conditions required along the wall are, then:

$$\Delta p_{\text{model}} + \Delta p_{\text{wall}} = 0$$

or, in terms of the velocity potential, with only first order terms considered:

$$(\phi_{1x} + \phi_{2x}) + \tfrac{1}{2} K(\phi_{1y} + \phi_{2y}) = 0$$

where K is the perforated wall constant, according to the definition:

$$\Delta p_{\text{wall}} = K \theta q_\infty$$

It should be noted that the above relationship is valid when the wall is located at a positive distance y with respect to the primary singularity.

The above equation for the boundary condition of a perforated wall can be transformed to:

$$\frac{\phi_{1y} + \phi_{2y}}{\phi_{1x} + \phi_{2x}} = -\frac{2}{K} = \arctan(-\delta/2) = \arctan(\pi - \delta/2)$$

Hence, the streamlines of the total disturbance flow field, $\phi_1 + \phi_2$, must intersect the perforated wall at a constant inclination angle $-\delta/2$ or $(\pi - \delta/2)$, respectively. The angle $\delta/2$ is a function of the wall porosity only:

$$\tan \delta/2 = 2/K$$

A schematic of the streamline is shown in Fig. 6.7.

The special case, $K \to \infty$, indicates the closed wall, and the above equation is simplified to:

$$\phi_{1y} + \theta_{2y} = 0$$

It is known that the boundary conditions of a closed wall can be fulfilled by image singularities referenced to the closed wall, since for an image, the singularity is (see Fig. 6.8):

$$v_{y_{2c}} = -v_{y_1}$$

and

$$v_{x_{2c}} = v_{x_1}$$

SUBSONIC FLOW IN WIND TUNNELS WITH PERFORATED WALLS

It was shown by Kassner[2] that the following simple relationship exists between the complex stream function of a perforated wall, Φ_2, and that of a closed wall, Φ_{2c} (see Appendix VI):

$$\Phi_2 = e^{i\delta} \Phi_{2c}$$

Thus, the conjunct velocity vectors are connected in the following manner:

$$\frac{d\Phi_2}{d\mathcal{Z}} = e^{i\delta} \frac{d\Phi_{2c}}{d\mathcal{Z}}$$

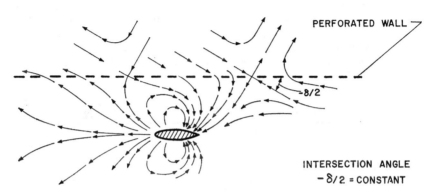

Fig. 6.7. *Streamline pattern of disturbance flow, $\phi_1 + \phi_2$, around model in vicinity of a perforated wall.*[2]

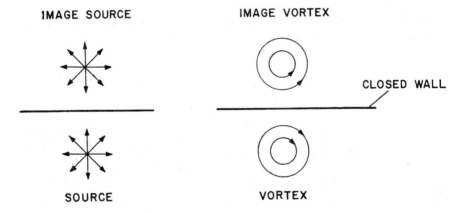

Fig. 6.8. *Source and vortex with image singularities in vicinity of one closed wall.*

The remarkably simple fact exists that the conjunct velocity vector, $d\Phi_2/d\mathcal{Z} = v_{x_2} - iv_{y_2}$, is found by merely rotating the conjunct velocity vector of the closed wall image singularity by the angle δ in the *counterclockwise* direction. For the velocity vectors themselves this manipulation corresponds to a rotation by the angle δ in the *clockwise* direction.

From simple geometrical considerations it is readily apparent that the rotation of the velocity vector of the image singularity flow field, Φ_{2c}, by the angle δ in the clockwise direction results in a rotation of the total disturbance velocity vector of the flow field, $\phi_1 + \phi_{2c}$, by $\delta/2$ in the clockwise direction. Since, for the closed wall case the wall streamline is parallel to the wall, this rotation makes all disturbance flow streamlines intersect with the perforated wall at the constant angle, $-\delta/2$, or $(\pi - \delta/2)$, as required.

In the case of the closed wall, which is a special case of the perforated wall with $K \to \infty$, the angle δ is found to be zero from the previous relationship, $\tan \delta/2 = K/2$, as required. For an open wall, that is, a perforated wall with $K = 0$, the velocity vector rotation angle is $\delta = 180°$, confirmation of the fact that the image singularity of an open-jet boundary changes its polarity in comparison with the closed wall boundary. For all perforated walls between the two extremes of the open and closed wall, the angle δ has values between $0°$ and $180°$ (see table below):

Rotation Angle, δ, for Various Wall Porosities

δ, deg	0	30	60	90	120	150	180
K	∞ closed wall	7.46	3.46	2.00	1.16	0.536	0 open wall

The velocity diagrams for some typical cases of a doublet in the vicinity of various perforated walls are presented in Fig. 6.9.

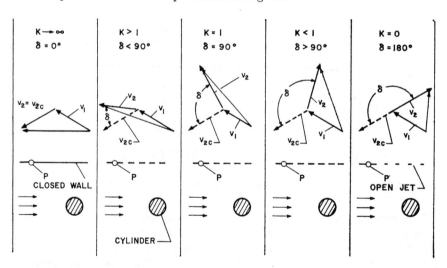

FIG. 6.9. *Vector diagrams of disturbance velocities at wall-point, P, for cylinder (doublet) near various perforated walls.*[2]

SUBSONIC FLOW IN WIND TUNNELS WITH PERFORATED WALLS

b. Specific singularities near one perforated wall

Doublet.—When the primary singularity is a doublet which represents a cylinder in parallel flow, the complex stream function of the image singularity for a closed wall is:

$$\Phi_{2c} = \frac{v_\infty r_0^2}{\mathcal{Z}} = \frac{v_\infty r_0^2}{r} e^{-i\theta}$$

with:

r_0 = radius of cylinder
$\mathcal{Z} = r\, e^{i\delta}$ = complete variable ($\mathcal{Z} = 0$ at center of image cylinder).

The complex stream function for the singularity required to satisfy the boundary condition along a perforated wall is then:

$$\Phi_2 = e^{i\delta}\Phi_{2c} = \frac{v_\infty r_0^2}{r} e^{-i(\theta-\delta)}$$

In this case the resulting singularity is again a doublet; the orientation however, is turned *counter-clockwise* by the angle δ (see Fig. 6.10).

Source.—When the primary singularity is a source, the image singularity for a closed wall is again a source and its complex stream function is:

$$\Phi_{2c} = \frac{Q}{2\pi} \log \mathcal{Z} = \frac{Q}{2\pi} \log(r\, e^{i\theta})$$

with Q = source intensity.

The complex stream function of the interference flow is then:

$$\Phi_2 = e^{i\delta}\Phi_{2c} = \frac{Q}{2\pi} e^{i\delta} \log \mathcal{Z}$$

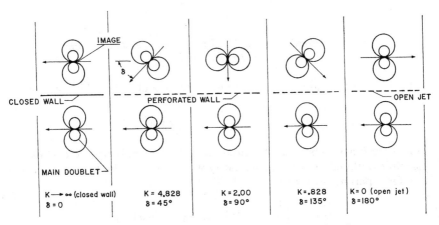

FIG. 6.10. *Doublet with image near various perforated walls.*[2]

Obviously the interference singularity is *not* a source. By means of further mathematical manipulation

$$\Phi_2 = \left(\frac{Q}{2\pi}\cos\delta\right)\log Z + \left(\frac{Q}{2\pi}\sin\delta\right) i\log Z$$

This equation indicates that the interference flow singularity consists of a source with the intensity ($Q\cos\delta$) and of a vortex with the circulation ($-Q\sin\delta$) (see Fig. 6.11).

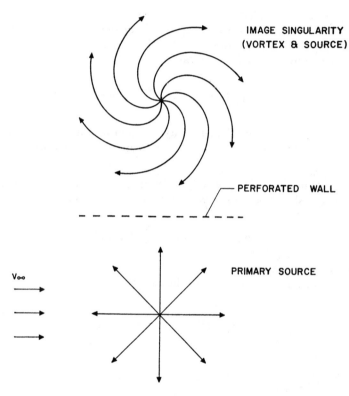

Fig. 6.11. *Streamline pattern for source and image singularity in vicinity of perforated wall.*

Vortex.—In this case, the image singularity for a closed wall is given by:

$$\Phi_{2c} = -\frac{\Gamma}{2\pi} i \log Z = -\frac{\Gamma}{2\pi} i \log(r\, e^{i\delta})$$

representing a vortex with the circulation, Γ. By applying the calculation principle outlined above the singularity which satisfies the boundary condition for a perforated wall is:

$$\Phi_2 = -\left(\frac{\Gamma}{2\pi}\cos\delta\right) i \log Z + \left(\frac{\Gamma}{2\pi}\sin\delta\right) \log Z$$

Hence, a vortex in the vicinity of a perforated wall requires a vortex with the circulation, $\Gamma \cos \delta$, and a source with the intensity, $\Gamma \sin \delta$, to satisfy the boundary conditions.

c. Mach number influence

The preceding calculations can be made applicable to compressible subsonic flow by changing the flow field according to the general similarity rule based on linearized compressible flow.[3] According to this rule, all dimensions of models and streamlines in the flow direction remain unchanged, and all dimensions perpendicular to the flow direction are reduced by the factor $(1-M_\infty^2)^{\frac{1}{2}}$. In addition, all disturbance velocity components in the flow direction, as well as the static pressure disturbances, are reduced by the square of the Prandtl factor, that is by $(1-M_\infty^2)$.

With the above similarity rule, the calculations and graphs developed for both two-dimensional and three-dimensional *incompressible* flow, can also be used for *compressible subsonic* flow when the following parameters are introduced:

Model (three-dimensional)

Cross-sectional area	$A_M(1-M_\infty^2)^{\frac{1}{2}}$
Volume of the model	$Vol(1-M_\infty^2)$

Wind Tunnel Dimensions

Coordinate in x direction	x
Coordinates perpendicular to the flow direction	$y(1-M_\infty^2)^{\frac{1}{2}}$
Tunnel height	$H(1-M_\infty^2)^{\frac{1}{2}}$
Tunnel diameter	$D(1-M_\infty^2)^{\frac{1}{2}}$

Flow Disturbances

Disturbance velocity component in flow direction	$\Delta v_x(1-M_\infty^2)$
Disturbance velocity component perpendicular to flow direction	$v_y(1-M_\infty^2)^{\frac{1}{2}}$
Local flow angle	$\theta(1-M_\infty^2)^{\frac{1}{2}}$
Local pressure disturbance	$(\Delta p/q)(1-M_\infty^2)$
Wall cross-flow parameter	$K(1-M_\infty^2)^{\frac{1}{2}}$

With increasing Mach number, the cross-flow parameter K decreases with the Prandtl factor $(1-M_\infty^2)^{\frac{1}{2}}$. In the limiting case when M approaches one, the wall parameter approaches zero, that is, approaches the value for an open wall. Hence, a perforated wall which exhibits a linear relationship between pressure drop and flow inclination angle acts like an open jet when Mach number one is approached. However, this situation does not occur in reality since near Mach number one the validity of the linear theory is restricted to very small flow disturbances. Furthermore, the wall parameter K does not approach infinitely large values as predicted by the strict application of the linear theory.

In the case of a quadratic relationship between pressure drop and cross-flow velocity, that is:

$$\Delta p/q = K'\theta^2$$

the cross-flow parameter K remains constant with increasing Mach number within the validity range of the similarity rule. Consequently a perforated wall which shows a quadratic relationship near Mach number one will experience no change in its wall characteristic parameter through the entire range in which the quadratic relationship holds true.

It may be concluded, therefore, that in the main subsonic Mach number range, particularly when the cross-flow characteristics are linear, a perforated wall tends to act more and more like an open wall when the Mach number is increased. Near Mach number one, where the cross-flow relationship has a tendency towards a quadratic relationship, a perforated wall produces a boundary effect which is distinctly different from that of an open jet.

In view of the complexity of transonic flow theory, the validity of the above conclusions from theory must be carefully investigated by experiments (see Chapter 7). As a general rule, for practical testing in perforated wind tunnels, it was found sufficient to select a wall parameter which would give velocity corrections close to zero in the high subsonic Mach number range. It was then assumed that any deviations from the correction-free conditions were small enough to be neglected. Naturally this is true only for a limited range of model size (see Chapters 7 and 10).

d. Two-dimensional wing without lift

The influence of the wall open-area ratio was examined more closely with regard to the interferences produced by the model displacement. Within linear theory, particularly for small models, the displacement effect of a two-dimensional wing may be examined by introducing a single doublet into the calculations. Hence the relationships of Section 3b of this chapter may be applied directly when the doublet strength is adjusted according to the size and shape of the model. Since a doublet in parallel flow represents a cylinder (radius $= r_0$), the radius of the equivalent cylinder must be selected to satisfy the following relationship:

$$r_0^2 = \frac{A_M}{2\pi}\lambda_v$$

The coefficient, λ_v, is a form factor that has a value of two in the case of a cylinder and may be sufficiently well approximated by $\lambda_v = 1 + t/c$ in the case of slender wings (see Ref. 4).

The distribution of the vertical velocity component v_y and of the local pressure disturbance along a perforated wall was calculated, and the results are shown in Figs. 6.12a and b. In an open jet, $K = 0$, the curve for the velocity component v_y is anti-symmetrical with respect to the x-direction (see Fig. 6.12a). The shape of the curve is very similar to the shape of the curve for free flight. However, in free flight the maximum vertical velocity disturbances are only approximately one-half as large as in an open jet.

When a perforated wall with a gradually decreasing open-area ratio corresponding to $K > 0$ is considered, the distribution curves for v_y gradually deviate more and more from the free-jet curve (see Fig. 6.12a). The location of the maximum velocity disturbances shifts forward, but the magnitude of

SUBSONIC FLOW IN WIND TUNNELS WITH PERFORATED WALLS

the disturbances is gradually decreased. It is significant that at $x = 0$ the vertical velocity component at the wall no longer disappears as in free flight; instead negative vertical velocities occur which reach maximum values for wall parameters of $K \simeq 2$. This fact is an indication of inflow through the wall into the test section at the model mid-station, as discussed in detail in Section 2 of this chapter. Naturally, for a closed wall, that is, for $K \to \infty$, the disturbance velocities v_y disappear along the wall.

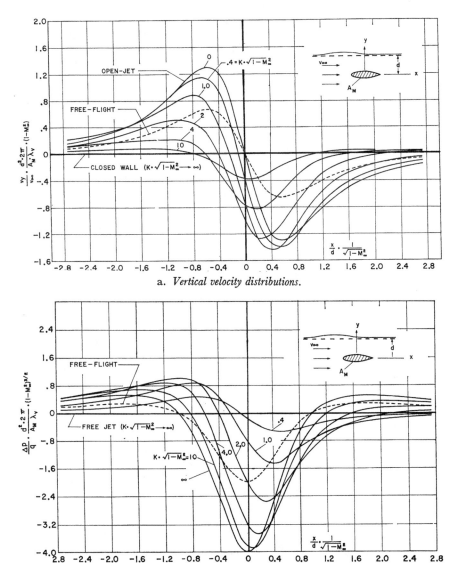

a. *Vertical velocity distributions.*

b. *Disturbance pressure distribution.*

Fig. 6.12. *Flow disturbances along wall for slender model near one perforated wall.*[2]

The pressure disturbances along the perforated wall are presented in Fig. 6.12b. For a free jet ($K = 0$), the pressure disturbances along the jet boundary disappear as required by the characteristics of a free jet. For a closed wall, ($K \to \infty$), negative disturbance pressures occur which are twice as large as the disturbance velocities in free flight. By opening the wall, that is, by reducing gradually the wall parameter K, the maximum pressure disturbances along the wall are also reduced. However, the disturbance pressure distribution is no longer symmetrical in the x-direction as in the case of the closed wind tunnel and the free-flight case. It is only possible to create an approximate matching of free-flight conditions in that region of the wall which is considered to be the most significant. However, it is impossible to match even approximately the disturbances over the entire length of the perforated wall.

4. FLOW DISTURBANCES IN A PERFORATED TWO-DIMENSIONAL WIND TUNNEL (TWO PERFORATED WALLS)

a. Method of calculation

For a model wing, the solution represented by a doublet placed between two perforated walls can be readily found by applying the method of suitable consecutive image arrangement based on the solutions for a single wall. For instance, in Fig. 6.13a the principle of consecutive image formation is represented for a doublet between two closed walls, and, in Fig. 6.13b, by a doublet in an open jet. When the same principle of image formation

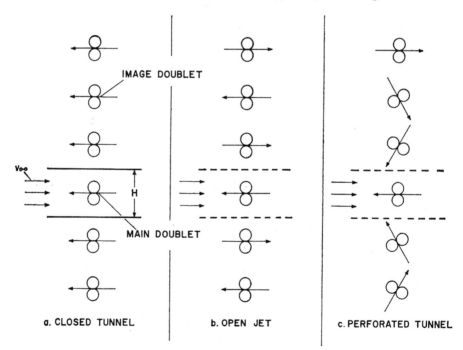

Fig. 6.13. Doublet between two perforated walls of various open-area ratio (two-dimensional).

is applied to a doublet between two *perforated* walls, an arrangement of doublets is obtained such as is indicated in Fig. 6.13c. Since all image singularities are again doublets and have their alignment in space rotated by given angles, a mathematical solution can be readily found.

The complex stream function for all image doublets required to satisfy the boundary conditions for a doublet between two perforated walls is then (see Ref. 2):

$$\Phi_2 = v_\infty r_0^2 \left\{ \sum_{-\infty}^{\infty} \frac{e^{in\delta}}{Z - inH} - \frac{1}{Z} \right\}$$

$$= v_\infty r_0^2 \left\{ \frac{\pi}{H} \cotan H\left(\frac{\pi}{H} Z\right) - \frac{1}{Z} \right\}$$

By evaluating this equation, that is, by separating the real velocity potential from the above complex stream function, the disturbance velocities in the entire field can be calculated.

The above equations have been derived for incompressible flow. By application of the similarity rule (see Section 3c of this chapter), the results can be made applicable to compressible flow in the subsonic speed range within the validity of linearized theory.

b. Interference due to displacement of model between two perforated walls

The preceding equations for a doublet placed between two perforated walls were evaluated, and the results are presented in Fig. 6.14 for typical interference parameters.

The distribution of the vertical velocity along one of the perforated walls shows a pattern similar to that found previously for a doublet in the vicinity of one perforated wall (Fig. 6.14a). For gradually decreasing open-area ratios of the wall, that is, for K proceeding from 0 to 0.4, 1.0, etc., the curves for the vertical velocity distribution assume more and more unsymmetrical shapes. In particular the vertical velocity component at the wall in the region of the maximum model thickness ($x = 0$) is no longer zero. On the contrary, a gradual increase of inflow velocity occurs, which reaches a maximum for a flow parameter of approximately

$$K(1 - M_\infty^2)^{\frac{1}{2}} = 3$$

With the open-area ratio smaller than $K(1 - M_\infty^2)^{\frac{1}{2}} = 3$, the inflow velocity tends to decrease again until finally, for a closed wall, no cross-flow velocities occur along the wall, as required for physical reasons.

The particular distribution of the vertical disturbance velocities along the wall results in a distortion of the streamlines in the vicinity of the wall, as indicated in Fig. 6.14b. The open-jet streamline is symmetrical in the x-direction and exhibits a shape similar to the corresponding streamline in free flight. As the wall open-area ratio decreases, the streamlines become more and more unsymmetrical, and their maximum deflection point shifts further upstream from the mid-point of the model. If the streamlines are normalized, for instance, at a station

$$\frac{x}{H} \frac{1}{(1 - M_\infty^2)^{\frac{1}{2}}} = -1.5$$

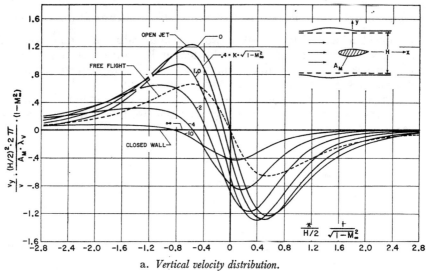

a. *Vertical velocity distribution.*

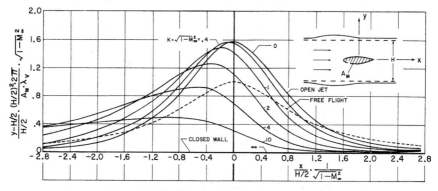

b. *Coordinates of wall streamline.*

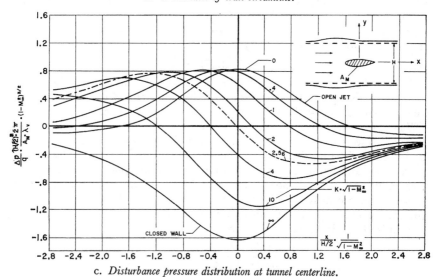

c. *Disturbance pressure distribution at tunnel centerline.*

FIG. 6.14. *Disturbance parameters due to wall interference for slender model between two perforated walls.*[2]

SUBSONIC FLOW IN WIND TUNNELS WITH PERFORATED WALLS

representing a test section with its upstream end located at this point, such a test section would experience more inflow than outflow along its perforated walls between stations

$$\frac{x}{H}\frac{1}{(1-M_\infty^2)^{\frac{1}{2}}} = -1.5 \text{ and } +1.5$$

Consequently, such a streamline pattern can be maintained only when some backflow is provided from the test section into the plenum chamber (see Chapter 12). Such outflow from the test section usually occurs in a very restricted region at the downstream end of the test section and does not greatly disturb the velocity distribution in the test section proper (see Fig. 12.14). As a result, the distributions presented in Figs. 6.14a and b for an infinitely long test section may also be applied to test sections of finite length when the test-section region near the downstream end is excluded.

The unsymmetrical interference pattern is also recognizable in the unsymmetrical distribution of the disturbance pressures along the centerline of the wind tunnel (see Fig. 6.14c). The distribution curves for both the open jet and the closed wall wind tunnel are symmetrical in the x-direction; however, the maximum disturbance pressures have a different magnitude and sign. With a gradually decreasing wall open-area ratio, a gradual transition from the open-jet curve to the closed-wall curves occurs with an accompanying development of lack of symmetry in the flow. It is noticeable that for a wall parameter of $K(1-M_\infty^2)^{\frac{1}{2}} = 2.56$, the pressure disturbance at the model center disappears. Therefore, for sufficiently small models, the velocity correction in a perforated wall wind tunnel also disappears for walls with this characteristic. However, even at this specific wall parameter $K(1-M_\infty^2)^{\frac{1}{2}} = 2.56$, a pressure gradient develops in the flow direction which, particularly in the case of long slender models, might require consideration of second-order correction terms for pressure and drag measurements. This pressure gradient cannot be detected during empty-tunnel calibrations since it is produced by the model installation itself.

5. VELOCITY CORRECTIONS IN CIRCULAR PERFORATED WALL WIND TUNNELS

a. Method of calculation

In the case of a cylindrical perforated wall wind tunnel, it is not possible to satisfy the wall boundary conditions by simply adding images, as in the case of two-dimensional or square wind tunnels. It is necessary to utilize the basic continuity equation for potential flow (note: $\phi = \phi_1 + \phi_2$):

$$\frac{\partial^2 \phi}{\partial x^2} + \frac{\partial^2 \phi}{\partial r^2} + \frac{1}{r}\frac{\partial \phi}{\partial r} = 0$$

The boundary condition along the perforated wall of the circular tunnel is then

$$\frac{\partial \phi}{\partial r} + \frac{2}{K}\frac{\partial \phi}{\partial x} = 0$$

Doublet.—The displacement of a small model in a wind tunnel can be simulated by a three-dimensional doublet with the velocity potential:

$$\phi_1 = \frac{m}{4\pi} \frac{x}{(x^2+r^2)^{3/2}}$$

When this expression is introduced into the continuity equation, it requires the addition of an interference potential ϕ_2 which, according to Goodman,[5] can be written in general form as:

$$\phi_2 = \sum I_0\left(\frac{n\pi(r/R)}{l/R}\right)\left\{B_n \sin\frac{n\pi(x/R)}{l/R} + C_n \cos\frac{n\pi(x/R)}{l/R}\right\}$$

This equation can be modified by means of mathematical manipulations, and the following result is finally obtained for the disturbance velocity in the x-direction at the axis of a circular wind tunnel:

$$\Delta v_x = \phi_{2x} = \frac{m}{2\pi^2} \frac{1}{R^3} \int_0^\infty \frac{(I_1(q).K_1(q) - (K/K^2).I_0(q).K_0(q))q^2}{I_1^2(q) + (K/K^2)I_0^2(q)} dq$$

and finally:

$$\frac{\Delta v_x}{v_\infty} = F_D \frac{Vol}{D^3}$$

Small Lifting Wing.—The simplest method of introducing a lifting wing into the calculation is by means of a horseshoe vortex. Assuming a horseshoe vortex of infinitely small span and following the method of calculation used for the doublet in the preceding paragraph, the velocity potential of the wall interference can be determined. Then the flow inclination correction at the wing is[5]:

$$\Delta\alpha = \frac{\delta F}{4\pi R^2}\left\{1 - \frac{2/\pi}{K} \int_0^\infty \frac{dq}{I_1^2(q) + (K/K^2)I_1^2(q)}\right\} = F_L \times \tfrac{1}{8}C_L \frac{A_M}{A_T}$$

Mach Number Influence.—The equations in the preceding paragraph are written for incompressible flow. By application of the similarity rule, which is valid within the limitations of linear theory, the calculations and its solutions can be adapted to compressible subsonic flow by means of the general relationships indicated in Section 3c of this chapter.

b. Wall interference due to model displacement and model lift

The preceding calculations were evaluated with respect to the velocity correction in the x-direction and the angle-of-attack correction at the lifting wing. The velocity correction in an open wind tunnel was found to have a magnitude of only one quarter of that in a closed wind tunnel and was of the opposite sign (see Fig. 6.15).

It should be noted that at a value of:

$$K(1 - M_\infty^2)^{\frac{1}{2}} = 2.42$$

the velocity correction disappears.

SUBSONIC FLOW IN WIND TUNNELS WITH PERFORATED WALLS

The flow inclination correction at the position of a small lifting wing has the same absolute magnitude in an open jet as in a closed wind tunnel; the corrections, however, are of opposite sign (see Fig. 6.16). The flow inclination correction disappears when a wall cross-flow coefficient is provided with the magnitude:

$$K(1-M_\infty^2)^{\frac{1}{2}} = 2.27$$

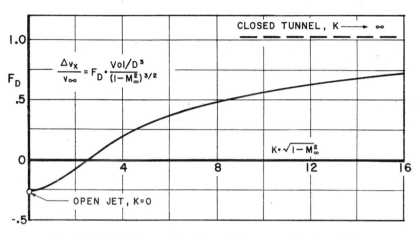

Fig. 6.15. *Velocity corrections due to displacement of slender model in circular perforated wind tunnel.*[7]

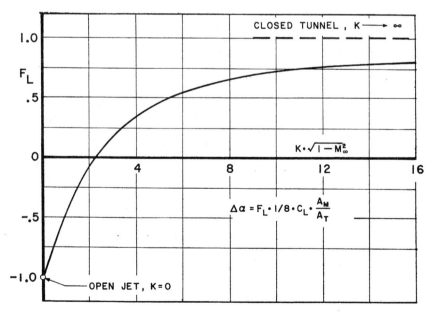

Fig. 6.16. *Flow inclination correction due to lift of a small wing in circular perforated wind tunnel.*[7]

The wall parameters for zero correction in perforated wall wind tunnels are only slightly different for the various types of correction under consideration. For example, for a small two-dimensional model between two perforated walls, $K(1-M_\infty^2)^{\frac{1}{2}} = 2.56$; for a small model in a circular perforated wind tunnel, $K(1-M_\infty^2)^{\frac{1}{2}} = 2.42$; and for a small horseshoe vortex in a circular perforated wind tunnel, $K(1-M_\infty^2)^{\frac{1}{2}} = 2.27$. These small differences make it possible for one constant geometry to approximately eliminate both the velocity correction and the flow inclination correction at the mid-point of the model.

This possibility represents a significant advantage of the perforated wall wind tunnel in comparison with longitudinally slotted wind tunnels. In the latter case, different wall parameter values, resulting in different open-area ratios, are required to make either the displacement velocity correction or the flow inclination correction disappear. In particular, for a circular wind tunnel with eight slots, an open-area ratio of 7.0 per cent is required to eliminate the velocity correction; an open-area ratio of 1.2 per cent is required to eliminate the flow inclination correction (see Chapter 5, Sections 1d and 5c).

It should be stressed once more that the values indicated above for zero correction of velocity and flow inclination hold true only for the area around the center of the model. As was the case for longitudinally slotted wind tunnels, the corrections vary along the surface of the model, particularly in the flow direction. Higher order correction terms can be calculated based on the general solutions discussed. However, the general practice is to select the open-area ratio of the perforated wall in such a manner that the corrections disappear at the center of the model, making it likely that at other points of the model the corrections will be immaterial. However, these assumptions need to be examined more closely experimentally.

6. COMPARISON OF THE FLOW AT LARGE DISTANCE BEHIND A LIFTING WING IN SLOTTED AND PERFORATED WIND TUNNELS

In the preceding paragraph, the downwash corrections for a lifting wing in a perforated wind tunnel were determined at the position of the wing ($x = 0$). In order to determine the downwash at other points, for instance very far behind the wing, general solutions of the type presented[5] must be numerically evaluated. Even for points infinitely far behind the wing it is not possible to determine the downwash using the "Trefftz-plane assumption" that the downwash at infinitely large distances ($x \to \infty$) is twice as large as that at the place of the wing. This fact indicates a basic difference in the behavior of slotted and perforated test sections and can be readily understood by considering the boundary conditions of the two types of walls.

Slotted Wall.—As shown in Chapter 5, the boundary condition of a slotted wall is:

$$\phi_x + d \cdot f(s/d) \cdot \phi_{xy} = 0$$

At large distances behind the wing, the pressure and the velocity disturbances ϕ_x disappear because this area ($x \to \infty$) is influenced only by the disturbance of the infinitely long wing-tip vortices. From the above equation, it can also be seen that the flow curvature ϕ_{xy} must disappear at $x \to \infty$, as is plausible. However, no information can be gained on the vertical velocity component v_y from this simple consideration.

The required information on v_y can be obtained from application of the "Trefftz theory". This theory states, as noted above, that the downwash at large distances behind a wing is twice as large as that at the wing itself. With the downwash angle known at the wing and from the equations presented in Section 5b of Chapter 5, the downwash at large distances behind a wing can be readily determined.

Perforated Wall.—The boundary condition of a perforated test section is according to the equation in Section 3 of Chapter 6:

$$\phi_x + \tfrac{1}{2} K \phi_y = 0$$

where K is the pressure drop coefficient of the wall. At large distances behind a lifting wing, the term ϕ_x must again disappear because of symmetry considerations. Consequently, according to the above equation, the normal velocity component ϕ_y must also disappear in this area. Thus the perforated wind tunnel at $x \to \infty$ behaves like a closed wind tunnel in which the streamlines are straightened out completely. Hence, the Trefftz condition that the downwash at large distances behind a wing is twice as large as the downwash at the plane of the wing does not hold true. The physical reason for this behavior is that the perforated wall acts like a lattice of elementary wings which generate free vortices not only in the plane of the wing ($x = 0$) but throughout. The Trefftz condition holds true for each elementary vortex, but because of the lack of symmetry of the vortex pattern in the x-direction, it does not hold true for the vortex system as a whole. To date, no numerical calculations have been published concerning the downwash as a function of the distance behind a lifting wing in a perforated wind tunnel.

This discussion indicates a significant basic difference between the behavior of slotted and perforated wind tunnels at large distances behind a lifting wing.

APPENDIX

(Ref. Section 3a)

Mathematical proof for $\Phi_2 = e^{i\delta} \Phi_{2c}$

Assuming the above relationship to be correct, the conjunct velocity vectors of the flow fields Φ_2 and Φ_{2c} are found by differentiation in the complex plane:

$$\frac{d\Phi_2}{d\mathcal{Z}} = e^{i\delta} \frac{d\Phi_{2c}}{d\mathcal{Z}}$$

and by introducing the velocity components v_x and v_y

$$v_{x_2} - i v_{y_2} = e^{i\delta}(v_{x_{2c}} - i v_{y_{2c}})$$

Since the closed-wall singularity is represented by the image of the primary singularity, the above equation for points along the perforated wall can also be written:

$$(v_{x_2} - i v_{y_2})w = e^{i\delta}(v_{x_1} + i v_{y_1})w$$

and with

$$\delta/2 = \arctan 2/K$$

after some transformations:

$$(v_{x_2} - i v_{y_2})w = \frac{\tfrac{1}{2}K + i}{\tfrac{1}{2}K - i}(v_{x_1} + i v_{y_1})w$$

The imaginary terms of this equation result in:

$$(v_{x_1} + v_{x_2})w + \tfrac{1}{2}K(v_{y_1} + v_{y_2})w = 0$$

This relationship is identical with the boundary conditions previously noted and indicates the compatibility of the relationship $\Phi_2 = e^{i\delta}\Phi_{2c}$ with the boundary requirements.

REFERENCES

[1] GOODMAN, T. R. "The Porous Wall Wind Tunnel, Part II: Interference Effect on a Cylindrical Body in a Two-Dimensional Tunnel at Subsonic Speeds." Cornell Aeronautical Laboratory, Inc., AD–594–A–3, November, 1950.

[2] KASSNER, R. R. "Subsonic Flow over a Body between Porous Walls." WADC–TR–52–9, February, 1952.

[3] GOETHERT, B. H. "Plane and Three-Dimensional Flow at High Subsonic Speeds." NACA–TM–1105, October, 1946.

[4] GOETHERT, B. H. "Wind Tunnel Corrections at High Subsonic Speeds Particularly for an Enclosed Circular Tunnel." (Translation of DVL, FB–1216, May, 1940.) NACA–TM–1300, February, 1952.

[5] GOODMAN, T. R. "The Porous Wall Wind Tunnel, Part V: Subsonic Interference on a Lifting Wing between Two Infinite Walls." Cornell Aeronautical Laboratory, Inc., AD–706–A–3, August, 1951.

[6] MAEDER, P. F. and STAPELTON, J. F. "Investigation of the Flow through a Perforated Wall." Brown University TR–WT–10, May, 1953.

[7] GOODMAN, T. R. "The Porous Wall Wind Tunnel, Part IV: Subsonic Interference Problems in a Circular Tunnel." Cornell Aeronautical Laboratory, Inc., AD–706–A–2, August 1951.

BIBLIOGRAPHY

BRESCIA, R. "Wall Interference in a Perforated Wind Tunnel (Studio dell'Interferenze delle Gallerie Aerodinamiche con Pareti a Fessure)." NACA–TM–1429. (Translation from *Atti. Accad. Torino* 87, 1952–1953.)

CARROLL, J. B. and RICE, J. B. "Analytical and Experimental Studies of Wall Interference in Perforated Wind Tunnels." WADC TR–56–240, June, 1956.

MAEDER, P. F. "Investigation of the Boundary Condition at a Perforated Wall." Brown University TR–WT–9, May, 1953.

MAEDER, P. F. "Some Aspects of the Behavior of Perforated Transonic Wind Tunnel Walls." Brown University TR–WT–15, September, 1954.

MAEDER, P. F. and HALL, J. F. "Investigation of Flow over Partially Open Wind Tunnel Walls, Final Report." AEDC–TR–55–67, December, 1955.

RITCHIE, V. S. and PEARSON, A. O. "Calibration of the Slotted Test Sections of the Langley 8 ft Transonic Tunnel and Preliminary Experimental Investigation of Boundary Layer-Reflected Disturbances." NACA RM–L51K14, July, 1952.

CHAPTER 7

EXPERIMENTS IN SLOTTED AND PERFORATED WIND TUNNELS AT SUBSONIC AND SONIC MACH NUMBERS

1. INTRODUCTION

IN Chapters 5 and 6, the characteristics of slotted and perforated wind tunnels, and in particular their superior features compared with conventional closed or open wind tunnels, were discussed. It is the purpose of this chapter to give a brief survey of experimental investigations in order to show to what extent the theoretical expectations have been borne out by experimental evidence.

The survey is mainly concerned with two-dimensional airfoils and models of rotational symmetry, with occasional reference made to combinations of fuselage and wing. A more thorough discussion of experimental data obtained in transonic wind tunnels will be presented in Chapter 10 where the emphasis is placed on the testing of complete airplane models over the transonic and supersonic speed ranges. In contrast, the following discussion is restricted to the subsonic and sonic speed ranges to which the theoretical investigations of Chapters 5 and 6 apply. In order to establish points of reference, some model test results obtained in closed and open jet wind tunnels are also cited.

2. TWO-DIMENSIONAL DOUBLE-WEDGE AIRFOIL MODELS IN CONVENTIONAL AND SLOTTED WIND TUNNELS

Several experimental investigations have been carried out using symmetrical double-wedge airfoils. This particular contour was selected by several investigators because theoretical pressure distribution data for this shape are available from the transonic theory presented in Ref. 1.

a. Test results from conventional wind tunnels

A series of symmetrical double-wedge airfoils with blockage ratios of 2.8, 3.6 and 4.4 per cent were investigated in the $7\frac{1}{2} \times 9$ in. closed wind tunnel of Brown University.[2] The configuration of the wind tunnel and the three models, with the locations of the pressure orifices, are shown in Figs. 7.1a and b. The double-wedge airfoils were built as semi-models and were placed close to the bottom wall so that larger models, and hence larger Reynolds numbers, could be used in the wind tunnel. The boundary layer along the lower wall was eliminated by a boundary bleed gap. The upper wall was solid for the experiments under discussion. However, it could be replaced by other test walls, for example, by slotted walls, as discussed later.

The models simulated double-wedge airfoils with a thickness ratio of

12 per cent. Each airfoil was equipped with ten pressure orifices which were positioned at similar locations so that a direct comparison between the pressure readings was possible.

The experimental pressure readings from four forward and four rearward pressure orifices are presented in Fig. 7.2a as a function of the *uncorrected* Mach number. The results indicate that, as expected, model size has a continuously increasing influence when the Mach number is increased. This

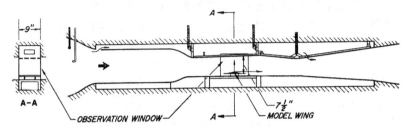

a. *Wind tunnel with model*

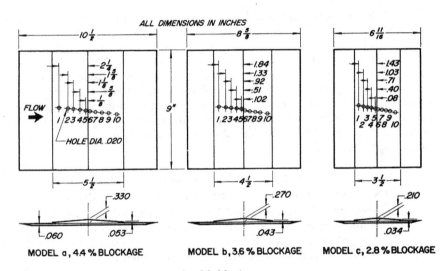

b. *Model wings.*

FIG. 7.1. *Wind tunnel configuration and double-wedge model wings.*[2]

was particularly true for the rearward pressure orifices which extended into the supersonic Mach number region at the higher test Mach numbers. Also, as predicted by theory, the large model approached choking of the closed test section at a Mach number somewhat less than 0.8.

The same experimental data shown in Fig. 7.2a, corrected according to the correction method given in Ref. 2 of Chapter 4, are presented in Fig. 7.2b. It is apparent that the correlation between the data for the three different model sizes was greatly improved after the corrections were applied, particularly in the case of the pressure orifices in the forward region of the

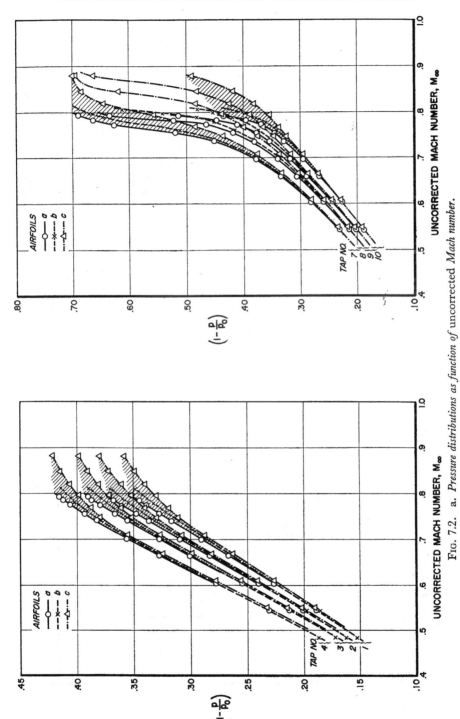

Fig. 7.2. a. *Pressure distributions as function of uncorrected Mach number.*

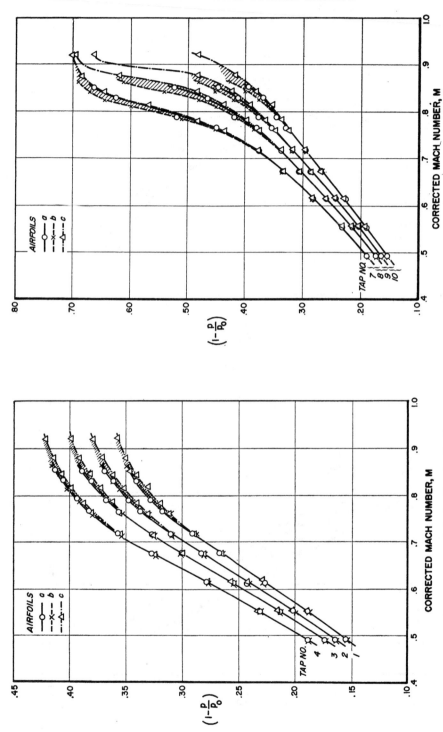

b. *Pressure distributions as function of corrected Mach number.*

FIG. 7.2. *Pressure distribution for double-wedge model wings as function of Mach number in closed test section (Figs. 7.1a and b).*[2]

wings. The rearward orifices are located in a much more sensitive region having flow fields in which the shock-wave locations shift rapidly.

These experimental data demonstrate clearly the limitations of a closed wind tunnel. The maximum Mach number is limited by the choking Mach number, and, even more so, by the flow distortion which occurs at Mach numbers somewhat below choking.

b. Experiments in slotted wind tunnels

The same experimental setup described above was used for the study of slotted wall wind tunnel interference after the upper wall of the closed tunnel had been replaced by experimental slotted walls (see Fig. 7.1a). In this case, an arrangement with three individual slots was selected.[2] According to theory (see Fig. 5.8), a slot width of approximately 0.006 in. is required to eliminate blockage and the need for velocity correction for a two-dimensional model. However, this theoretical result holds true only for non-viscous flow. In reality, the effective open area of such a narrow slot will be greatly reduced by the boundary layer build-up along the slot sidewalls. In order to allow for this viscosity effect, the investigators selected considerably wider slots of 0.05 in. width.

The pressure measurements for this investigation are presented in Fig. 7.3 as a function of Mach number in the same manner as the closed wind tunnel pressure measurements. No velocity correction was applied. It is apparent that in most of the Mach number range the agreement among the measurements for the three airfoil models with blockages varying between 2.8 and 4.4 per cent is very good. In the forward region of the airfoil, and also in the rearward region, all three airfoils produced similar results up to a Mach number of 0.98, the highest attainable in the experimental setup. However, the largest model with 4.4 per cent blockage was somewhat too large, as indicated by its somewhat lower pressures.

The influence of slot width was then explored by repeating the same experiments in a test section which had three slots $\frac{1}{4}$ in. wide; that is, the slot width was arbitrarily increased by a factor of five. The results, presented in Fig. 7.4 for two typical orifices of the smallest airfoil (2.8 per cent blockage) indicate good agreement. Evidently the slot width can be increased without penalty considerably beyond the theoretical value, at least in the case of pressure distribution measurements of symmetrical airfoil models without lift. This result is encouraging since it indicates that the width of relatively narrow slots is not a particularly sensitive parameter. Slotted walls will also yield reliable data for models without lift, even when significant slot width deviations occur in the direction of the open wind tunnel.

c. Comparison of slotted wind tunnel data with theory for Mach number one

The results of the slotted wall wind tunnel tests were replotted in a more conventional manner, that is, pressure distribution versus the orifice station for various selected Mach numbers (see Fig. 7.5). Also indicated on this graph is the theoretical pressure distribution for Mach number one according to Ref. 2. It is apparent that a smooth transition takes place from typical subsonic Mach numbers to supercritical Mach numbers with shock formation

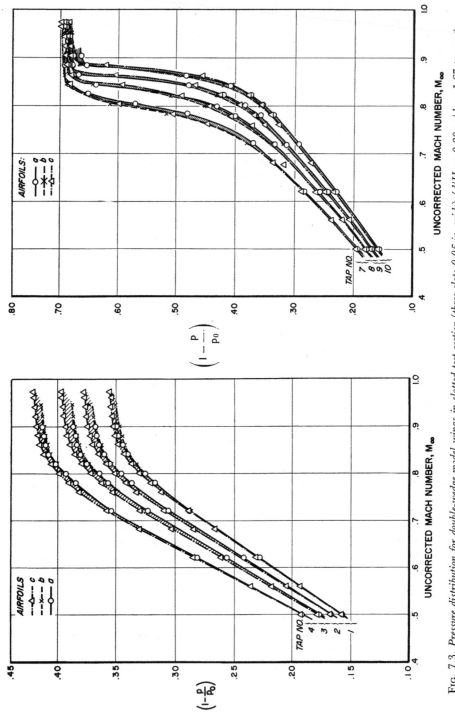

Fig. 7.3. Pressure distribution for double-wedge model wings in slotted test section (three slots 0.05 in. wide) ($d/H = 0.20$; $s/d = 1.67$ per cent) (Figs. 7.1a and b).[2]

and gradual downstream shifting of the terminating shock until the experimental distribution finally has a close resemblance to the theoretical distribution. The comparison is particularly good in the region downstream of the maximum thickness of the airfoil. In the upstream region, the experimental pressures at high Mach numbers are somewhat smaller than theory predicts. It cannot be definitely stated that this deviation is caused by an interference effect in the slotted wind tunnel. It should also be remembered that transonic theory utilizes a series development of flow parameters around the sonic point, that is around $1 - p/p_0 = 0.472$.

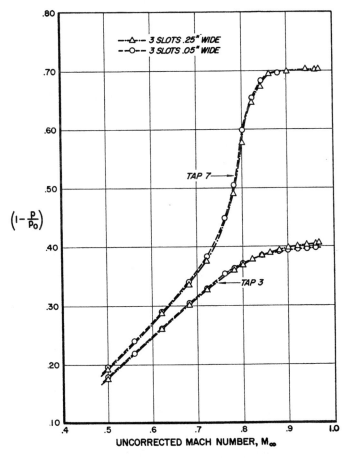

FIG. 7.4. *Pressure distribution for double-wedge model wing with 2.8 per cent blockage in slotted test section with different slot configurations (Figs. 7.1a and b).*[2]

In summary, it can be concluded from these early experiments that a test section with properly selected slot geometry is suitable for tests in the critical Mach number range to Mach number one since choking is practically eliminated and the typical trends of the flow field and the rapid changes in the shock-wave location are properly simulated.

3. TWO-DIMENSIONAL AIRFOIL MODEL IN DIFFERENT SLOTTED WIND TUNNELS

Another comprehensive investigation of slotted and perforated wall wind tunnel characteristics was carried out in the high-speed model wind tunnel of the National Physics Laboratory in England.[3] This tunnel was $7\frac{1}{2} \times 3$ in. in cross section and was arranged so that the upper and the lower walls could be modified. The results were checked in the NPL closed 20×8 in. tunnel in order to obtain a comparison with nearly interference-free data. It cannot be assumed that in the 20×8 in. tunnel the small model wing experienced no interference effects at all; however, the results in this tunnel were sufficiently close to free flow and may be used as a reasonably good guide.

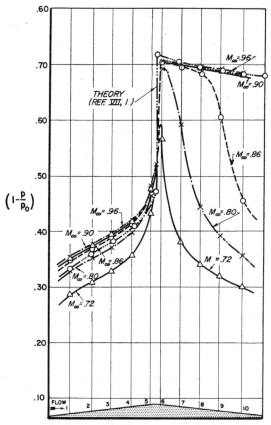

FIG. 7.5. *Pressure distributions for double-wedge model wing with 2.8 per cent blockage in slotted test section (three slots 0.05 in. wide) and comparison with theory (Figs. 7.1a and b).*[2,1]

a. *Pressure distribution at zero lift*

In the NPL investigation, an RAE-104 model wing with a thickness ratio of 10 per cent was used. The model was sized to give 2.67 per cent blockage in various test sections. A slotted wall with a slot centerline spacing of $d/H = 0.20$ and an open-area ratio of 4 per cent produced the best results. Some typical pressure distributions at various Mach numbers are presented in Fig. 7.6. The distributions in the larger 20×8 in. closed wind tunnel are also indicated. At the lower Mach numbers, to approximately Mach number 0.85, the comparison between the data from the two tunnels is very good. At the higher Mach numbers, however, the data deviate, the slotted model tunnel being somewhat too open.

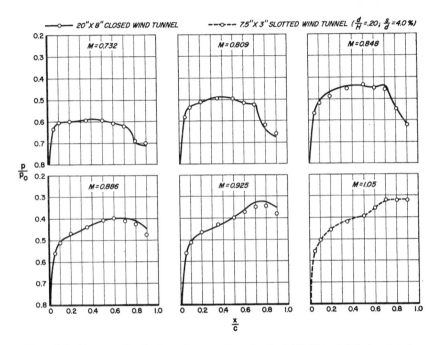

FIG. 7.6. *Pressure distributions for two-dimensional RAE-104 airfoil in slotted wind tunnel (2.67 per cent blockage) and in large closed wind tunnel.*[3]

Again, the open-area ratio of 4 per cent is much larger than that predicted by theory for non-viscous flow. The latter ratio would be on the order of 0.2 per cent, but such narrow slots would be largely blocked by the boundary layer along the walls, as in the case discussed in Section 2b of this chapter.

b. Blockage correction at zero lift

To obtain an indication of the velocity correction which might be required in a slotted wind tunnel because of incorrect selection of the open-area ratio of the slots, the pressure distributions for the airfoil in the slotted NPL tunnel were compared in detail with those for the same airfoil in the assumedly correction-free 20×8 in. tunnel. For this comparison, the test Mach number and the pressure distribution in the 20×8 in. tunnel were considered to be correct. The Mach number correction for the slotted tunnel was then determined, assuming that for each individual orifice the same pressure ratio would result from the tests in the small slotted wall and big closed wind tunnels. Corrected Mach numbers were determined in this manner for a number of orifices along the airfoil chord; the results are presented in Fig. 7.7. The data for all orifices scatter around a single mean curve which coincides with the uncorrected Mach number to approximately Mach number 0.87 but starts to deviate toward an open tunnel curve with increasing Mach number.

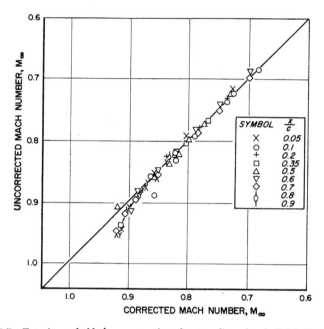

FIG. 7.7. *Experimental blockage correction for two-dimensional RAE-104 airfoil ($\alpha = 0°$) with 2.67 per cent blockage in slotted wind tunnel, based on comparison with tests in 20 in. × 8 in. closed wind tunnel.*[3]

It is of interest to note (Fig. 7.8) how the corrected Mach numbers for slotted walls of different open-area ratios compare with those for a closed or an open jet tunnel. The data for the slotted wall even cross the curve of the open jet tunnel at Mach numbers in the vicinity of Mach number one. In the case of the closed wind tunnel the Mach number correction produces values higher than the uncorrected test Mach number; in the case of the slotted test sections and, naturally, in that of the open jet tunnel also, the correction produces values lower than the test Mach number. The slotted wall with an open-area ratio of 4 per cent and much narrower slot spacing ($d/H = 0.033$) acts much like an open jet tunnel.

It would be interesting to apply this correction procedure to tests in large slotted wall wind tunnels and to compare the results with interference-free data.

c. Lift curve slope

The same RAE-104 model airfoil was also investigated in various slotted wall model tunnels of different configurations (2.67 per cent blockage at zero lift) and in the larger closed wind tunnel at an angle of attack of 2°. Again, the slotted walls with a $d/H = 0.20$ and 4 per cent open-area ratio yielded results which came closest to the data obtained in the 20×8 in. closed wind tunnel (see Fig. 7.9). For this comparison the Mach number

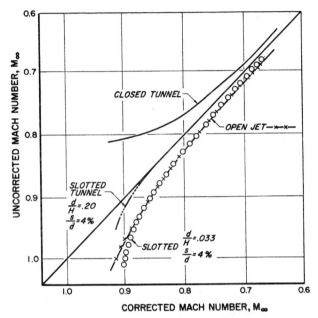

Fig. 7.8. *Experimental blockage correction for two-dimensional RAE-104 airfoil ($\alpha = 0°$) with 2.67 per cent blockage in different wind tunnels, based on comparison with tests in 20 in. $\times$ 8 in. closed wind tunnel.*[3]

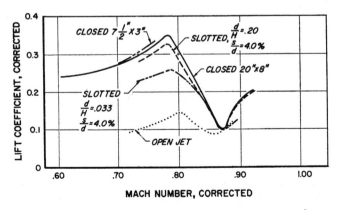

Fig. 7.9. *Lift coefficient of two-dimensional RAE-104 airfoil ($\alpha = 2°$) with 2.67 per cent blockage in different wind tunnels (data corrected for blockage according to Fig. 7.8).*[3]

was corrected according to the blockage correction described previously and represented in Fig. 7.7. No angle-of-attack correction was applied. It was noticeable that when the slot number was increased and the same total open-area ratio was maintained the walls rapidly approached open jet conditions. The deviation is apparent in spite of the fact that the Mach number has been corrected for blockage. It is obvious, therefore, that an additional angle-of-attack correction is required; without it the data obtained from this multi-slotted wall are much like data obtained from an open jet tunnel, not only with respect to blockage, but also with respect to the flow inclination correction.

For comparison, the lift curve slopes of the model airfoil are also presented for the closed and open jet wind tunnels with a $7\frac{1}{2} \times 3$ in. cross section. These curves have been corrected for blockage but not for lift; the result is an angle-of-attack change. Large deviations remain uncorrected in the open jet results.

4. BODY OF REVOLUTION IN SLOTTED WIND TUNNELS AND IN FREE FALL

To obtain further insight into the interference effect in slotted wind tunnels, a slender body of revolution was tested in both the Langley 16 ft transonic tunnel and in free fall.[4] The Langley tunnel was equipped with eight longitudinal slots for a total open-area ratio of 12 per cent, and a model of rotational symmetry was built with a length of 10 ft and a fineness ratio of 12 (Fig. 7.10). The blockage ratio amounted to 0.27 per cent of the

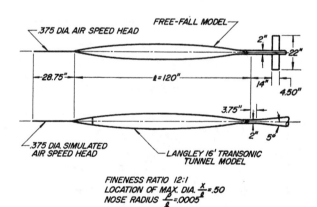

Fig. 7.10. *Models investigated in the Langley 16 ft transonic tunnel and in free-fall tests.*[4]

16 ft wind tunnel. The free fall model was equipped with horizontal and vertical tail surfaces to produce the required stability. The Reynolds number for both the wind tunnel tests and the drop tests was approximately 38 million.

Some typical pressure distribution data from this series of tests are shown in Fig. 7.11 for Mach numbers between 0.85 and 1.003. It is apparent that the general trend of the pressure distribution is identical for both the wind tunnel and the free fall tests. There are, however, two marked deviations between the two sets of data. Particularly at the lower Mach numbers the static pressures measured in the 16 ft tunnel are somewhat lower than those of the free fall tests. This difference has a magnitude of $\Delta p/q = 0.03$, corres-

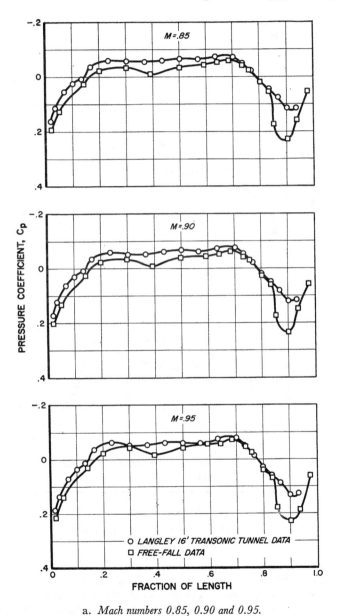

a. *Mach numbers 0.85, 0.90 and 0.95.*

FIG. 7.11. *Pressure distributions for body of revolution (Fig. 7.10) from Langley 16 ft transonic tunnel (0.27 per cent blockage) and from free-fall tests.*[4]

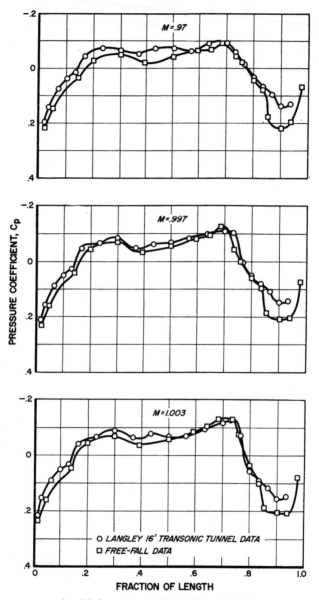

b. *Mach numbers 0.97, 0.997 and 1.003.*
Fig. 7.11. *Concluded.*

ponding to a change in the free-stream Mach number of approximately 1.5 per cent. It cannot be definitely stated whether this difference is the result of wind tunnel interference effects or of an inaccurate assessment of the accuracy of the flight tests.

The second difference occurs near the downstream end of the model, between the stations at 0.85–0.95 of the model length. The cause of this difference is also not known exactly. However, the shift toward higher pressures in this area in the free fall tests might be the result of differences in the fairings between the sting support and the model end. Since the latter explanation is very plausible, the discrepancy in the rearward regions may be dismissed as unessential as far as wall interference is concerned.

In summary, it can be stated that the overall trends of the pressure distribution changes in the high subsonic and sonic speed ranges were equally well represented by the data obtained in the Langley 16 ft transonic wind tunnel and that obtained in the free fall tests.

5. EXPERIMENTS WITH BODIES OF REVOLUTION IN PERFORATED TEST SECTIONS

a. Model with fineness ratio of 12

A 6×6.25 in. model test section equipped with perforated walls having an open-area ratio of 22 per cent was used to investigate a fuselage model with a fineness ratio 12 and with the maximum thickness located at approximately 40 per cent of body length behind the nose.[5] Practically interference-free data for the same model were obtained by experiments in an 8 ft closed tunnel in which the model blockage amounted to only 0.012 per cent. In contrast, the blockage of the fuselage model in the 6×6.25 in. perforated model tunnel was 2.09 per cent.

Some of the results are presented in Fig. 7.12 for the selected Mach numbers of 0.89 and 0.95. In these experiments, the sidewall setting of the perforated wall wind tunnel was changed from 0 to 30′ converged and finally to 60′ converged. The deviations among the three wall settings are relatively small.

The agreement between the data from the perforated wall tunnel and the interference-free data is excellent at Mach number 0.89. At Mach number 0.95, however, a significant discrepancy exists. The pressure distribution in the large closed tunnel changed rapidly between Mach numbers 0.89 and 0.95. This change is not readily explainable, and it is possible that the data in the large closed wind tunnel, rather than the data in the perforated wall wind tunnel, may be erroneous (see for comparison the test results for a similar model shown in Fig. 7.11). Such reasoning appears to be supported by the fact that the tests in the closed wind tunnel were conducted close to the choking Mach number, that is, in a region in which testing in a closed wind tunnel is extremely sensitive.

No indication of lack of symmetry of the pressure distribution due to symmetrical interference effects can be noted.

b. Cone-cylinder model

Extensive experiments in a large 16 ft transonic wind tunnel (perforated) and in a 1 ft transonic wind tunnel (perforated) were conducted at the Arnold Engineering Development Center with cone-cylinder models having a cone angle of 20°. In these experiments, as well as in theoretical studies, it was found that transonic tests of cone-cylinder models are extremely sensitive in the Mach number range slightly below Mach number one and in the low supersonic range. This sensitivity is caused by the abrupt turning of the flow around the sharp shoulder at the transition from the cone to cylinder which leads to concentrated expansion waves that are very difficult for either slotted or perforated walls to absorb. It was found in these test series that a wind tunnel with fixed perforated wall geometry is not capable of producing interference-free data for cone-cylinder models in the critical transonic range.[6]

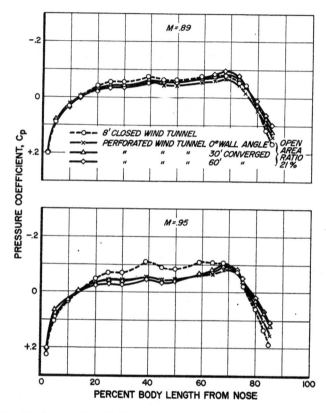

Fig. 7.12. *Fuselage model with fineness ratio of 12 in perforated model test section (2.09 per cent blockage) and in large closed test section.*[5]

In these experiments the correlation between the results obtained in the perforated wall wind tunnel and interference-free data could be made reasonably close only when the open-area ratio of the walls in the vicinity of Mach number one was drastically reduced to values as low as 1.5 per cent. In addition, it was necessary to use walls with inclined holes slanted 60° in the flow direction to make the perforated walls more capable of absorbing concentrated expansion waves.

Some results obtained with the 2 per cent blockage cone-cylinder model are presented in Fig. 7.13 for the most critical Mach numbers of 0.95, 1.00 and 1.05 (Ref. 7). By using small open-area ratios, proper wall settings, and inclined holes, the correlation could be made reasonably close. It should be mentioned that at Mach numbers below 0.95, cone-cylinder models exhibit smoother pressure distribution curves and are less difficult to test in transonic wind tunnels. Also, in the case of the more common model shapes which do not have sharp shoulders but are smoothly curved like airplane fuselages, the test conditions are considerably less severe than those shown for the cone-cylinder model. However, the Mach number range immediately around Mach number one remains a region of particular difficulty for wind tunnel testing.

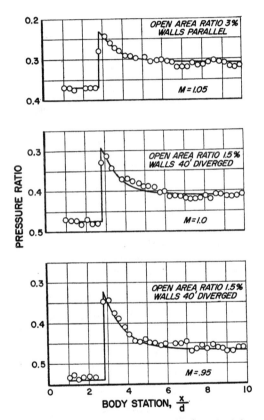

Fig. 7.13. 20° cone-cylinder model with 2 per cent blockage in 1 ft × 1 ft perforated test section having 60° inclined holes and optimum open-area ratio.[7]

6. EXPERIMENTS WITH WING–FUSELAGE MODELS IN PERFORATED TEST SECTIONS

a. Pressure distribution measurements in model test section

In the absence of any systematic test series performed with two-dimensional airfoil sections in perforated wall wind tunnels, a test series performed using a simplified wing fuselage configuration will be cited here. The model used in these tests is given in Fig. 7.14 (see Ref. 5). In order to show the effect of wall interference more clearly, a rather thick fuselage and wing section were selected. The fuselage had a fineness ratio of 4 and the wing a thickness ratio of 10 per cent. The tests were conducted in a small 6 × 6.25 in. test section with perforated walls having an open-area ratio of 22 per cent with the wall setting 1° converged. The model size was selected so that the total blockage of wing and fuselage amounted to 2.47 per cent.

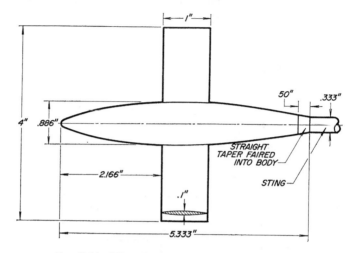

FIG. 7.14. Wing–fuselage model for correlation tests.[5]

The small model was also tested in a large 8 ft closed wind tunnel. Since in this tunnel the model blockage amounted to no more than 0.014 per cent, these data may be considered interference-free except in the immediate vicinity of Mach number one.

The pressure distribution over the fuselage and the wing at zero-degree angle of attack are presented in Fig. 7.15a–d. It is apparent that the agree-

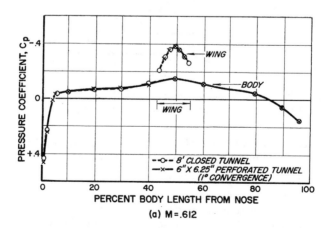

(a) M = .612

FIG. 7.15a.

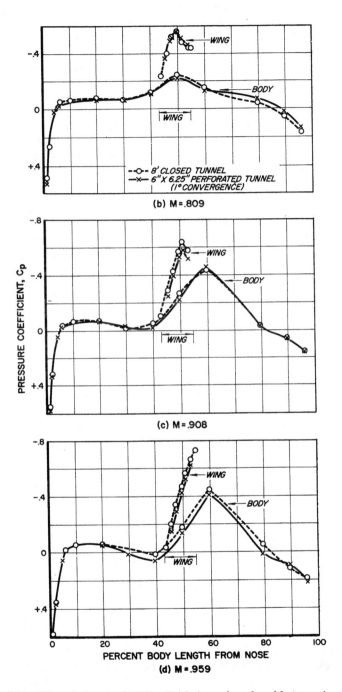

Fig. 7.15. Wing–fuselage model (Fig. 7.14) in perforated model test section (2.47 per cent blockage) and comparison with results of large closed wind tunnel.[5]

ment between the tests in the perforated wall tunnel and in the large closed wind tunnel is very satisfactory. However, at the highest test Mach number of 0.96, a small systematic deviation between the two sets of data occurred which may be explained by a small difference in Mach number. This difference may require a small velocity correction in the perforated wind tunnel in the direction of a somewhat smaller value.

It is particularly remarkable that even at the blockage ratio of 2.47 per cent no indication of a pressure gradient produced by the test-section walls can be found. Such a pressure gradient, the result of interference in perforated test sections, was predicted by the theoretical calculations of Chapter 6. It is apparent, therefore, that the model is either small enough to eliminate the pressure gradient effect or that the boundary layer overshadows this effect so that it is no longer noticeable.

b. Force measurements in 36 × 35 in. test section

A very extensive series of correlation tests are in progress in a number of British transonic wind tunnels of both the slotted and perforated type.[8] For these tests a special correlation model consisting of a fuselage and a 45° swept-back wing with a thickness ratio of 6 per cent was selected (see Fig. 7.16). Various models, some of different sizes but of the same shape and some identical, were built and tested in various wind tunnels. Certain results of these correlation tests, not completed at the present time, are already available and are cited here.

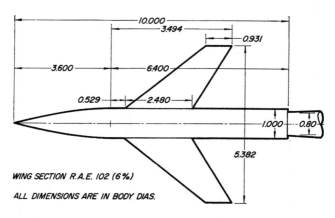

Fig. 7.16. *Wing–fuselage model for correlation tests.*[8]

Models of 0.5 and 0.06 per cent blockage were installed in the full-scale 9 × 8 ft ARA perforated wall wind tunnel, and lift, moment, and drag coefficients were determined. Unfortunately, the Reynolds number during these correlation tests was not always kept constant. For the small model the Reynolds number was an average of 2×10^6. It is not possible at the present time to assess the influence of this Reynolds number difference until more detailed investigations have been conducted.

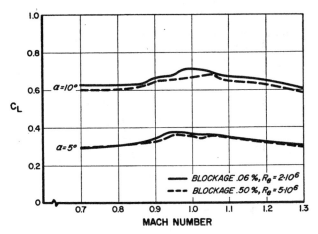

Fig. 7.17a. *Lift coefficient.*

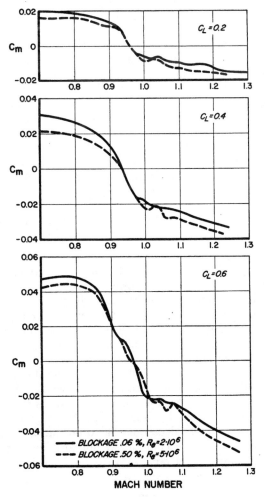

Fig. 7.17b. *Moment coefficient.*

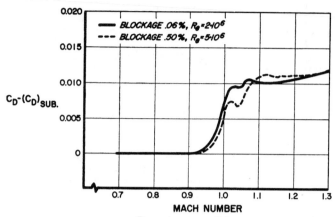

c. *Drag rise at zero lift.*

Fig. 7.17. *Wing–fuselage models of different size (Fig. 7.16) in ARA 9 ft × 8 ft perforated test section with 22.5 per cent open-area ratio.*[8]

Some results are presented in Fig. 7.17a for the lift coefficients, in Fig. 7.17b for the moment coefficients, and in Fig. 7.17c for the transonic drag rise as a function of Mach number. All curves for the small and the large blockage models exhibit the same changes when the Mach number is increased from high subsonic to low supersonic values. On the other hand, definite differences between the results for the various models exist. The waviness of the curves in the Mach number range between Mach numbers 1.0 and 1.1 seems to indicate that some disturbances due to wave reflection occur.

Reference should also be made to the more refined perforated test sections with inclined holes for which systematic measurements have been conducted (see Chapter 10 of this AGARDograph).

7. BODIES OF REVOLUTION AT SONIC SPEED
a. Theoretical considerations

As discussed in detail in previous chapters, the interference effect of tunnel walls cannot be extrapolated reliably to the vicinity of Mach number one from theoretical calculations based on subsonic flow characteristics. In order to obtain some insight into the characteristics of the flow in a slotted or perforated wind tunnel at Mach number one, theoretical investigations, utilizing the transonic similarity rule by von Kármán,[9] were carried out for a body of revolution. The flow field around a body of revolution in free flight at $M = 1$ had been determined previously for specific shapes by Guderley and Yoshihara.[10] By utilizing the Guderley–Yoshihara solution for the flow along the tunnel wall and applying the usual boundary conditions for either slotted or perforated walls, it is possible to determine for each wall geometry the pressure in the plenum chamber and, thus, a Mach number correction for the wind tunnel test. Such calculations based on the von Kármán transonic similarity rule and on the Guderley–Yoshihara theory for bodies of revolution at Mach number one were carried out by Berndt[11] for slotted wall wind tunnels, and later by Page[12] for perforated wall wind tunnels.

It was shown that in neither case was it possible to satisfy the boundary condition at all stations of the wall when the wall geometry was kept constant in the flow direction. For slotted wall wind tunnels, however, a slot geometry can be determined for which the boundary condition is satisfied over a large range of the tunnel wall. It is significant to note that such wall geometry is not identical to the wall geometry determined previously for interference-free flow at subsonic velocities. However, in the example studied by Berndt, the difference between both cases was found to be relatively small. For any other wall geometry, a Mach number correction should be applied which can be estimated from the following equations:

Slotted Wall Wind Tunnel

$$\Delta M = 0.9g \ (r^*/R)^{6/7} \ (r^*/x^*)^{2/7} \qquad \text{(Ref. 11)}$$

For usual configurations, the expression, g, has values between 0·15 and 0·35. r^* is the radius of the model at the station at which sonic speed is reached for the first time at the model surface. Correspondingly, x^* is the coordinate in flow direction starting from the model nose, again at the station where sonic speed is reached first at the surface (R = wind tunnel radius).

Perforated Wind Tunnels

$$\Delta M = -h \ (r^*/R)^{6/7} \ (r^*/x^*)^{2/7} \qquad \text{(Ref. 12)}$$

The numerical value of h for a 20 per cent open perforated wall was calculated to be $h = 0.82$.

These equations show some significant features. First, the sign of the correction is opposite for slotted and perforated walls of conventional geometry. A slotted wall reaches sonic flow conditions at an uncorrected test Mach number which is somewhat smaller than one. On the other hand, a perforated wall reaches sonic speed at an uncorrected test Mach number which is somewhat larger than one.

Furthermore, it is shown that no longer is the maximum cross-sectional area of the model, or the blockage ratio, A_M/A_T, the main parameter of significance, but that rather it is the cross-sectional area of the model at the station where first sonic speed is obtained. It is significant also that the Mach number correction grows considerably less than the blockage ratio at the sonic point; that is specifically, it grows only with $(r^*/R)^{6/7}$. Consequently, the wind tunnel correction is reduced at a much smaller rate than the blockage ratio when the model size is reduced in a given wind tunnel.

Another important conclusion from the calculations in Ref. 11 is that in wind tunnel testing at sonic speed the length of a model is a more critical parameter than the model thickness. From the calculations of Guderley and Yoshihara it is known that only downstream of the reflected limiting Mach wave can a wall interference effect be felt when perfect matching of the boundary conditions is assumed in the subsonic range along the wall. Consequently, because of the smaller curvature of the streamlines around slender models, the same degree of wall interference, or inaccuracy of wind tunnel test data, is obtained for slender bodies in a given wind tunnel when they have a shorter length than blunt models.

The correction factors indicated above for slotted and perforated wind tunnels depend upon the geometry of the models and upon the wind tunnel parameters such as slot geometry, boundary layer, etc. In order to obtain reliable data at sonic speed, extensive test work is required to calibrate the individual wind tunnel for various model shapes. Near sonic speed, a simplification of the calibration task for the large group of possible model shapes is possible when the equivalence rule of Oswatitsch–Keune[13] (transonic area rule) is employed.

It might be mentioned that transonic wind tunnel testing near the sonic speed still offers some unexplored and promising possibilities. Since for a slotted and for a perforated test section, the Mach number correction was found to be of opposite sign, it can be expected that by combining the two wall types, for example by using walls having longitudinal slots with perforated cover plates, the Mach number correction near Mach number one can be greatly reduced, or even eliminated (see Chapter 9, Section 4).

b. Experiments on the validity of Mach number correction at sonic speed

The validity of the Mach number correction at sonic speed was checked in two sets of experiments for a slotted and for a perforated wall wind tunnel. In both cases, a parabolical-arc model of rotational symmetry with a fineness ratio of six was used.

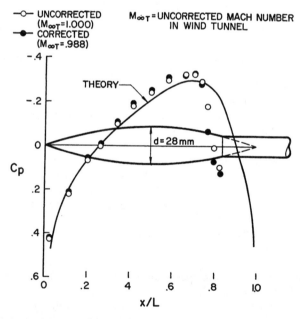

Fig. 7.18. *Pressure distribution along parabolic-arc body of revolution (fineness ratio 6, blockage 0.37%) at sonic speed in a slotted 45 by 48 cm wind tunnel.*[14,15]

Some pertinent results obtained by Drougge[14] are shown in Fig. 7.18. The wind tunnel was equipped with longitudinal slots for which the value of $0.9\, g$ in the correction equation for slotted wall wind tunnels was found to be

$$0.9\, g = 0.26.$$

The theoretical data for this model, indicated also in the figure, were obtained from calculations by Guderley and Yoshihara. The uncorrected wind tunnel data differ from the theoretical values, particularly in the region around and behind the maximum thickness of the model. Application of the Mach number correction, which in this particular case amounted to $\Delta M = 0.012$, was not successful; it shifted the experimental curve in the wrong direction. The deviations between theory and experiments behind station $x/L = 0.75$ can be disregarded since the theory is established for a model without sting support; the deviations can be readily explained by the sting effect.

The same model was also investigated in a perforated wall wind tunnel by Page and Speiter.[15] Significant deviations between theory and uncorrected wind tunnel results are apparent (see Fig. 7.19). The Mach number correc-

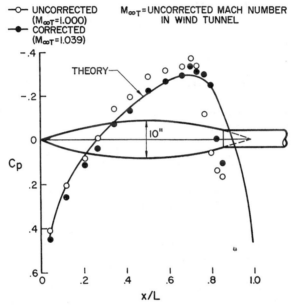

FIG. 7.19. *Pressure distribution along parabolic-arc body of revolution (fineness ratio 6, blockage 0.46%) at sonic speed in Ames 14 foot transonic wind tunnel with perforated walls.*[15]

tion according to the previously given equation results in $\Delta M = -0.039$. This correction is generally in the right direction and brings the pressures around and behind the maximum thickness of the model in good agreement with theory. However, larger deviations are obtained for the forward part of the model.

In summary, it may be stated that at sonic speed the need for a correction of the free-stream Mach number exists even with a wind tunnel geometry which at lower Mach numbers produces practically correction-free data. Transonic similarity considerations can be used to obtain an estimate of the Mach number corrections to be expected near Mach number one. However, more refined calculations and careful wind tunnel calibration tests are required for precision wind tunnel testing in this speed range. Particular attention should be given to the movement of the main shock wave which normally terminates the local supersonic field around models in the transonic speed range. It has been observed in numerous experiments that the wall interference effect frequently results in a delay of the rearward shock movement and thus in a distortion of the flow field and errors in pressure distribution and force and moment measurements.[16] Suitable geometry of transonic wind tunnel walls will reduce the magnitude of such delays. However, whenever the terminating shock position is particularly sensitive to changes of Mach number, boundary layer, etc., special care in the evaluation of wind tunnel test data must be applied.

REFERENCES

[1] GUDERLEY, G. and YOSHIHARA, H. "The Flow over a Wedge Profile at Mach Number 1." *J. Aero. Sci.* **17**, 723–735, 1950.

[2] MAEDER, P. F. "Investigation of Tunnel Boundary Interference on Two-Dimensional Airfoils Near the Speed of Sound." Brown University, WT-6, April, 1951.

[3] HOLDER, D. W., NORTH, R. J. and CHINNECK, A. "Experiments with Slotted and Perforated Walls in a Two-Dimensional High-Speed Tunnel." Aeronautical Research Council TR-14411, R. & M. 2955, November, 1951.

[4] HALLISSY, J. M. "Pressure Measurements on a Body of Revolution in the Langley 16-ft Transonic Tunnel and a Comparison with Free-Fall Data." NACA RM-L51L07a, March, 1952.

[5] CUSHMAN, H. T. "Small-Scale Transonic Interference Studies with Perforated Wall Test Sections, Final Report." UAC-R-95538-16, June, 1953.

[6] GOETHERT, B. H. "Physical Aspects of Three-Dimensional Wave Reflections in Transonic Wind Tunnels at Mach Number 1.20 (Perforated, Slotted and Combined Slotted-Perforated Walls)." AEDC-TR-55-45, March, 1956.

[7] ESTABROOKS, B. B. "Wall Interference Effects on Axisymmetric Bodies in Transonic Wind Tunnels." AEDC-TR-59-12, June, 1959.

[8] O'HARA, F., SQUIRE, L. C. and HAINES, A. B. "An Investigation of Interference Effects on Similar Models of Different Size in Various Transonic Tunnels in the U.K." Royal Aircraft Establishment Technical Note AERO 2606, ARA Wind Tunnel Note 27, February, 1959.

[9] VON KÁRMÁN, Th. "The Similarity of Transonic Flow." *J. Math. Phys.* **26**, No. 3, 182, October, 1947.

[10] GUDERLEY, G. and YOSHIHARA, H. "Axial-symmetric Transonic Flow Patterns" Air Material Command, Wright-Patterson Air Force Base, Ohio, AFTR-5797, September, 1949.

[11] BERNDT, S. "Theoretical Aspects of the Calibration of Transonic Test Sections." The Aeronautical Institute of Sweden, FFA Report 74, Sweden, 1957.

[12] PAGE, W. A. "Experimental Study of the Equivalence of Transonic Flow about Slender Cone-Cylinders of Circular and Elliptical Cross-Section." NACA TN-4233 (Appendix), April, 1958.

[13] OSWATITSCH, K. and KEUNE, F. "The Flow around Bodies of Revolution at Mach No. 1." *Proc. Conf. on High Speed Aerodynamics* pp. 113–131. Polytechnic Institute of Brooklyn, Brooklyn, New York, 1955.

[14] DROUGGE, G. "Some Measurements on Bodies of Revolution at Transonic Speed." *Acts. 9th International Congress of Applied Mechanics*, Vol. II, pp. 70–77, 1957.

[15] PAGE, W. A. and SPIEITER, J. R. "Some Applications of Transonic Flow Theory to Problems of Wind Tunnel Interference." Presented to Wind Tunnel and Model Testing Panel of AGARD, Rhode-St.-Genese, Belgium, March 2–5, 1959.

[16] ROGERS, E. W. E. and HALL, I. M. "Some Experiments with Static Tubes at Transonic Speeds in a Slotted-Wall Wind Tunnel." Aeronautical Research Council, Report A.R.C. 20, 306, July, 1958.

BIBLIOGRAPHY

BATES, G. P. "Preliminary Investigation of 3 in. Slotted Transonic Wind-Tunnel Test Sections". NACA RM L9D18, September, 1949.

KNECHTEL, E. D. "Experimental Investigation at Transonic Speeds of Pressure Distributions over Wedge and Circular-Arc Airfoil Sections and Evaluation of Perforated Wall Interference." NASA TN D-15, August, 1959.

LAURMANN, J. A. and LUKASIEWICZ, J. "Development of a Transonic Slotted Working Section in the NAE 30 in. × 16 in. Wind Tunnel." NAE LR–178, August, 1956.

MAEDER, P. F. "Theoretical Investigation of Subsonic Wall Interference in Rectangular Slotted Test Sections." Brown University, WT–11, September, 1953.

MICHEL, R. and SIRIEIX, M. "Contribution Experimentale à l'Étude du Profil Portant aux Vitesses Transsoniques." ONERA–TN–26, 1955.

MICHEL, R. and SIRIEIX, M. "Repartitions Experimentales du Nombre de Mach Local sur Differents Profils d'Ailes en Encoulement Sonique." ONERA–TM–17, 1959.

NELSON, W. J. and BLOETSCHER, F. "An Experimental Investigation of the Zero-lift Pressure Distribution over a Wedge Airfoil in Closed, Slotted, and Open-Throat Tunnels at Transonic Mach Numbers." NACA RM L52C18, June, 1952.

OHIO STATE UNIVERSITY RESEARCH FOUNDATION. "An Experimental and Analytical Investigation to Determine Important Design Parameters of Wind Tunnels Using Porous Test Section Walls in Sonic and Supersonic Operation." (Sverdrup and Parcel, Inc., Contract AF33(038)–9928), AD 58 838, December, 1952.

PILAND, R. O. "The Zero-Lift Drag of a 60° Delta Wing-body Combination (AGARD Model 2) Obtained from Free-Flight Tests between Mach Numbers of 0.8 and 1.7." NACA TN 3081, April, 1954.

POISSON-QUINTON, PH. "Premiers Resultats d'Essais sur Maquettes Étalons AGARD." Paper presented at the Sixth Session of the AGARD Wind Tunnel Committee, November, 1954.

ROE, F. E. "Some Aspects of Transonic Tunnel Operation in Industry." *J. Roy. Aero. Soc.*, **62**, 16, January, 1958.

SLEEMAN, W. C., KLEVATT, P. L. and LINSLEY, E. L. "Comparison of Transonic Characteristics of Lifting Wings from Experiments in a Small Slotted Tunnel and the Langley High-Speed 7 × 10 ft Tunnel." NACA RM L51F14, November, 1951.

SPIEGEL, J. M. and LAWRENCE, L. F. "A Description of the Ames 2 × 4 ft Transonic Wind Tunnel and Preliminary Evaluation of Wall Interference." NACA RM A55I21, January, 1956.

TAYLOR, H. D. "Progress of Transonic Wind Tunnel Studies of U.A.C." UAC R–95434–8, June, 1951.

THOMPSON, J. R. "Measurements of the Drag and Pressure Distributions on a Body of Revolution Throughout Transition from Subsonic to Supersonic Speeds." NACA RM L9J27, January, 1950.

VINCENTI, W. G., DUGAN, D. W. and PHELPS, E. R. "An Experimental Study of the Lift and Pressure Distribution on a Double-Wedge Profile at Mach Numbers Near Shock Attachments." NACA TN 3225, July, 1954.

WHITCOMB, C. F. and OSBORNE, R. S. "An Experimental Investigation of Boundary Interference on Force and Moment Characteristics of Lifting Models in the Langley 16 and 8 ft Transonic Tunnels." NACA RM L52L29, February, 1953.

WILDER, J. G. "An Experimental Investigation of the Perforated Wall Transonic Wind Tunnel Phase I." Cornell Aeronautical Lab., Inc., AD–706–A–5, August, 1951.

CHAPTER 8

SUPERSONIC WALL INTERFERENCE IN PARTIALLY OPEN TEST SECTIONS FOR TWO-DIMENSIONAL CONFIGURATIONS

1. GENERAL CONSIDERATIONS CONCERNING WAVE REFLECTIONS IN OPEN, CLOSED AND PARTIALLY OPEN TEST SECTIONS

a. Open and closed test sections

In supersonic wind tunnels, models produce compression and expansion waves which, unless special measures are taken, are reflected from the test-section boundaries. If the models are not small enough, the reflected waves will meet the models and will cause discrepancies between the flow in the wind tunnel and the flow in free flight. If the wind tunnel walls are solid boundaries, a shock wave will be reflected with the same sign, that is, as a compression wave. Expansion waves such as those generated around the shoulder of the model shown in Fig. 8.1 are reflected as expansion waves.

It is known as a general principle in supersonic flow that waves are reflected with the same sign and the same absolute flow turning angle (also approximately the same intensity) from a plane solid wall because the flow direction must be maintained parallel to the solid wall (see Fig. 8.2). On the other hand, along an open boundary, the condition of a constant static pressure along the boundary must be met. Therefore a compression wave meeting an open boundary will be reflected as an expansion wave of equal density with the same absolute pressure change. In the case of an expansion wave impinging upon an open boundary, the reverse relationships hold true (Fig. 8.2).

Since solid boundaries and open jet boundaries produce wave reflections having opposite characteristics, there is a possibility of eliminating wave reflections by means of a suitable mixing of open and solid boundaries. Such mixed boundary conditions might be produced by using walls of either the slotted or perforated type.

b. Partially open walls

Several examples of two-dimensional shock-wave reflections without consideration of the wall boundary layer are presented in Fig. 8.3. The inclined flow behind an oblique shock wave meets the perforated wall and produces a pressure drop when it flows through the wall. If this pressure drop is equal to the pressure rise through the main oblique shock wave, the static pressures in the flow and in the plenum chamber are in equilibrium, and the condition of "no reflection" is obtained. If the open-area ratio of the perforated wall is smaller than in the "no reflection" case, a partial reflection in the form of a compression wave occurs since the cross flow through the wall must be reduced to produce pressure equilibrium. On the other hand, if the open-area ratio is greater than in the "no reflection" case, a partial reflection in the form of an expansion wave occurs, and the pressure equilibrium is established by increasing the cross flow through the wall.

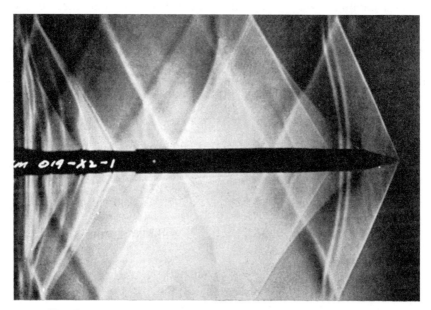

FIG. 8.1. *Schlieren picture showing wave reflection in supersonic test section with solid walls.*[11]

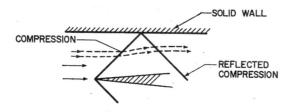

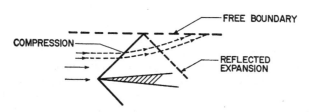

FIG. 8.2. *Wave reflection in supersonic flow on solid and on free-jet boundaries.*

The wave pattern for oblique flow through a partially open wall is shown schematically in Fig. 8.4. A system of expansion and compression waves is produced in which the waves cancel each other gradually and disappear completely at large distances from the wall. It is apparent from this simplified consideration that the flow in close proximity to the walls is non-uniform but at a suitable mixing distance from the wall the non-uniformity decays and the flow assumes a uniform quality. At a given Mach number the thickness of the non-uniform flow layer is proportional to the size of the openings in the wall, so that only a narrow layer of non-uniform flow exists when many small openings, instead of a few large openings, are provided. Since it is obvious that in wind tunnels the layer of non-uniform flow must

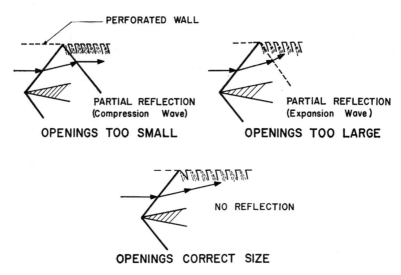

FIG. 8.3. *Shock-wave reflection on partially open walls with various opening ratios.*[12]

not extend to the model, the use of test-section walls with "small-grain" openings is necessary. This requirement holds true for both perforated and slotted test-section walls. In other words, a great number of small individual openings (i.e. circular holes or transverse or longitudinal slots) is required for efficient wave-reflection cancellation in wind tunnels. The permissible size of these openings depends upon the shape of the opening, the thickness of the wall boundary layer, the Mach number, and the flow angle.

The above considerations on the correct open-area ratio of perforated walls are based on the requirement for equilibrium between test-section and plenum chamber pressures behind a shock wave. In general, pressure equilibrium is not obtained with an open-area ratio of 0.50, that is, with a wall having open and closed areas of equal size. Consequently, an oblique shock wave first reflects compression waves which exceed the expansion waves in intensity or expansion waves which exceed the compression waves. This initial unbalance of intensity will then be eliminated by the formation of additional waves behind the main oblique shock wave if the walls have the correct open-area ratio.

The above discussion indicates that the wave cancellation problem is divided into two separate problems: first, the problem of the reflection of the initial shock; and secondly, the problem of establishing the correct

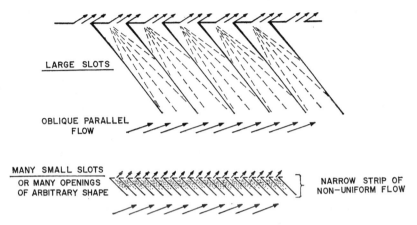

Fig. 8.4. *Wave pattern for supersonic flow crossing a wall with transversal slots.*[12]

pressure equilibrium between the test-section and plenum chamber pressures behind the initial oblique shock through a system of secondary waves.

In addition to the above considerations concerning wave cancellation in inviscous flow, the boundary layer along partially open walls has been found to have a significant influence on the shock cancellation characteristics of the walls. One of the major problems in achieving effective wave cancellation is to find a suitable method of controlling wall boundary layer (see Section 4 of this chapter).

c. Influence of outflow and inflow on shock reflection

In the preceding discussion it is assumed that with partially open walls a definite relationship exists between cross flow and pressure drop. It will now be shown how the pressure drop relationship can be calculated assuming, among other conditions, constant entropy in the flow. In the case of flow out of the test section, the condition of constant entropy can be met reasonably well. In the case of inflow from the plenum chamber into the test section, however, low energy air flows through the wall, and the condition of constant entropy in the cross flow is not met. It is possible, however, to shape partially open walls in such a manner that, in spite of entropy changes, the "pressure drop–cross flow" characteristics for inflow and outflow are approximately equal. Walls with inclined holes will be discussed in some detail in Section 5 of this chapter and in Chapter 9, where experimental results obtained using such walls are presented.

When specially shaped walls are employed which have the characteristic of an equal pressure drop for both outflow and inflow, the analytical calculations for reflection-free walls are again applicable as far as the required characteristic "cross flow–pressure drop" relationship is concerned. However, the open-area ratios determined from isentropic flow relationships must be modified.

2. THEORY OF WAVE CANCELLATIONS IN WIND TUNNELS WITH LONGITUDINAL SLOTS

a. Conditions in the vicinity of the impinging waves

When a two-dimensional shock wave impinges upon a flat wall having longitudinal slots, the impinging shock wave will be reflected as a shock wave from the solid parts of the wall according to the general laws of wave reflection. In a similar manner, the impinging shock wave will be reflected as an expansion wave from the open parts of the wall, that is, from the slots. If no other waves are produced further downstream of the impingement of the initial shock wave, the reflected shock waves and the reflected expansion waves will cancel each other at some distance from the wall where their combined intensity equals zero. In other words, the effect of the reflected shock and expansion waves is nullified at some distance from the wall when the slotted wall has equal areas of solid walls and open slots, that is, when the open-area ratio of the wall is $A_{\text{open}}/A_{\text{total}} = 0.50$. It should be noted that only with a 50 per cent open longitudinally slotted wall is the effect of the primary *reflected* waves thus eliminated at some lateral distance from the wall. However, as will be shown, this 50 per cent open wall will not satisfy the requirement of pressure equilibrium in the flow behind the reflected waves.

In the regions adjacent to the slots of the longitudinally slotted walls, the static pressure for a 50 per cent open wall will equal the undisturbed pressure (the pressure in the plenum chamber). Along the solid portions of the wall behind the reflected shocks, the static pressure will be considerably larger than either the pressure in the plenum chamber or the pressure behind the initial shock wave. Consequently a wave system will develop along the edges of the slots which will tend to equalize the pressures between the regions adjacent to the slots. At a sufficient distance behind the impinging shock wave, the secondary wave system will assume a uniform character along the edges of the individual slots, and there will be no further changes in the flow direction. It is for this uniform flow far behind the initial impinging shock that the condition of pressure equilibrium between the flow in the wind tunnel and the air in the plenum chamber must be satisfied.

b. Conditions downstream of the impinging waves

When a wall with longitudinal slots is exposed to an oblique supersonic flow such as that which occurs far behind a two-dimensional oblique shock wave, the cross flow is basically of the subsonic type. This situation is analogous to the case of inclined flow around slender bodies or small-aspect-ratio wings.[1] The pressure drop through such a slotted wall can therefore be readily calculated from the basic laws of continuity and momentum.

When entropy changes are neglected and the compressible flow equations are linearized, the pressure drop of a flow passing through the wall at a small oblique angle, θ, is (see Section 1 of Appendix to this chapter):

$$\Delta p/q = (1/R^2 - 1)\theta^2$$

The open-area ratio necessary to provide complete cancellation for slotted walls can then be determined:

$$R^2 = \tfrac{1}{2}(M^2 - 1)^{\frac{1}{2}}\theta$$

where R = ratio of open to total area, and
θ = flow inclination angle behind oblique shock.

From these linearized calculations, the open-area ratios for slotted walls were calculated for complete and partial cancellation of shock-wave reflections (Figs. 8.5a and b). It is significant that the correct open-area ratio

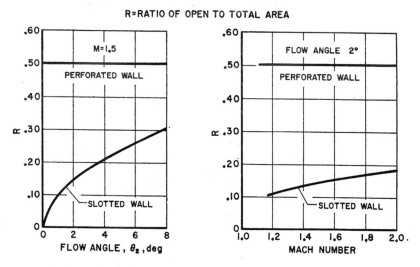

FIG. 8.5a. *Open-area ratio of longitudinally slotted and perforated walls for complete shock cancellation.*[12]

for longitudinally slotted walls changes with both the Mach number and the flow angle of the flow approaching the slotted wall. The longitudinally slotted wall, therefore, is not well suited to cancellation of shock waves in a wind tunnel capable of a variety of Mach numbers because the wall open-area ratio would have to be adjusted continuously in order to meet the "no-reflection" condition. An even more serious difficulty is the fact that it would also be necessary to adjust the wall open-area ratio when the supersonic flow approaching the longitudinally slotted wall changed its direction. Such an adjustment is not easy to accomplish because the flow direction around a complex wind tunnel model changes locally, even at a given constant angle of attack.

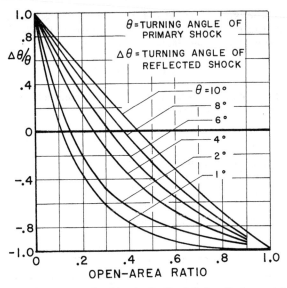

FIG. 8.5b. *Open-area ratio of longitudinally slotted walls for partial shock cancellation at* $M = 1.80$.[2]

3. THEORY OF WAVE CANCELLATION IN PERFORATED WIND TUNNELS

a. Linearized theory

When thin walls with transverse slots (which can be assumed to act in a manner similar to thin walls with perforations) are considered, the flow pattern can be readily determined by the method of characteristics. If entropy changes are neglected and the compressibility equations are linearized, the pressure drop of a supersonic flow passing the wall at a small oblique angle, θ, is (see Section 2 of Appendix to this chapter):

$$\frac{\Delta p}{q} = \frac{2}{(M^2-1)^{\frac{1}{2}}}\left(\frac{1}{R} - 1\right)\theta = K\theta$$

The open-area ratio, R, necessary to provide complete wave cancellation for perforated walls can then be calculated:

$$R = 0.50$$

The value of the open-area ratio of perforated walls thus obtained from linearized theory for the "no-reflection" case is independent of Mach number and shock intensity. Such a characteristic is highly desirable and makes the perforated wall superior to the longitudinally slotted wall because one wall with constant geometry is then capable of satisfying the requirements for reflection-free conditions at different Mach numbers and flow angles (see Fig. 8.5a).

If the open-area ratio of perforated walls is not selected correctly, a partial reflection of an impinging shock wave will occur. In this case, when the same assumptions are made as previously, the pressure rise Δp_r due to partial reflection of the primary shock wave can be determined from the following relationship:

$$\Delta p_r / \Delta p_{sh} = 1 - 2R$$

This relationship is presented graphically in Fig. 8.5c.

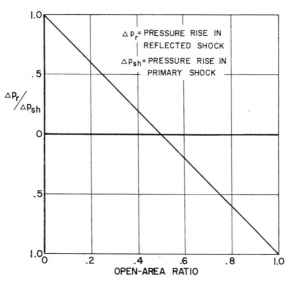

Fig. 8.5c. *Open-area ratio of perforated walls for partial shock cancellation at $M = 1.80$.*

It should be noted that the above equation is valid only for conditions at some distance behind the impinging shock wave. As in the case of longitudinally slotted walls, the type of reflection at the impingement point of the initial shock wave depends upon the character of the wall in that area. If the shock wave should strike a solid portion of the wall, it would be reflected as a shock wave. However, as with longitudinally slotted walls, this initial reflected shock from a perforated wall is supplemented by another wave system which is instrumental in establishing pressure equilibrium between the plenum chamber flow and the tunnel flow downstream of the initial shock. More detailed consideration of the wave pattern shows that the transition from the initial reflected wave to the final stabilized wave system will be accomplished within a distance which will be smaller than the distance between the leading edges of two adjacent transverse wall elements.

b. Deviations from linearized theory

The preceding calculations are based on the assumption that the flow is isentropic and follows the linearized Prandtl–Glauert theory. The factor K in the equation for the pressure drop at the wall,

$$\Delta p / q = K\theta$$

can be written as (see Section 2 of Appendix to this chapter):

$$K = \frac{2}{(M^2-1)^{\frac{1}{2}}}\left(\frac{1}{R} - 1\right)$$

It is known that in real flow this equation does not hold true, particularly in the case of Mach numbers near one, because the linearized theory requires that for $M \to 1$ the factor K would tend to go to infinity. In actual flow the factor K maintains a finite value.

For complete wave cancellation it would be necessary to use a wall with cross-flow characteristics according to:

$$K = 2/(M^2-1)^{\frac{1}{2}}$$

As was shown by analytical investigations, such a condition for K can be met in linearized isentropic flow when the wall open-area ratio R of the perforated wall is equal to 0.50. However, since in real flow the factor K does not vary according to the above law, particularly in the case of Mach numbers near one, the open-area ratio of perforated walls must be reduced significantly near Mach number one in order to produce the necessary large values of K.

4. EXPERIMENTAL RESULTS FOR TWO-DIMENSIONAL SHOCK-WAVE CANCELLATION

a. Longitudinally-slotted wind tunnel walls

A number of systematic experiments have been conducted in several laboratories in order to check the correctness of the equations describing wave reflection from slotted walls.

The staff of the United Aircraft Corporation[2] studied the cancellation of shock reflections on slotted walls having open-area ratios between 10 and 50 per cent. Figure 8.6 presents a number of selected Schlieren photographs from this test series. A two-dimensional plate set at an angle of 3° in supersonic flow at Mach number 1.8 was used to produce a two-dimensional shock front.

It can be seen that for solid walls ($R = 0$) and for slotted walls having a 10 per cent open-area ratio, the reflection of the primary shock wave occurs in the form of a compression reflection. When the open-area ratio is increased to 20 or 30 per cent, the reflected wave becomes an expansion wave, indicating that the wall is too open. According to theory, a slotted wall with an open-area ratio of approximately 21 per cent would be required to cancel a shock wave of 3° turning angle in Mach number 1.8 flow. The experimental results of these tests are, therefore, in qualitative agreement with theory. However, the slotted wall with 20 per cent open-area ratio produced an *expansion* reflection of the impinging shock wave although theory predicted that this open-area ratio would provide an almost reflection-free geometry and would produce only a weak *compression* reflection. Apparently a longitudinally slotted wall in a supersonic wind tunnel acts somewhat more open in real flow than non-viscous theory predicts when the boundary layer along the wall is not considered.

In order to check quantitatively the reflection characteristics of longitudinally slotted walls, a series of systematic experiments with shock reflection were carried out in the WADC Propulsion Laboratory 6 in. supersonic wind tunnel.[3,4] Again a flat plate was used to produce a two-dimensional shock wave which would impinge upon the test-section wall. The intensity of the reflected wave system was then probed by means of detailed measurements along a line at a constant distance from the wall (see Fig. 8.7).

First the pressure distribution through a shock wave and the reflection

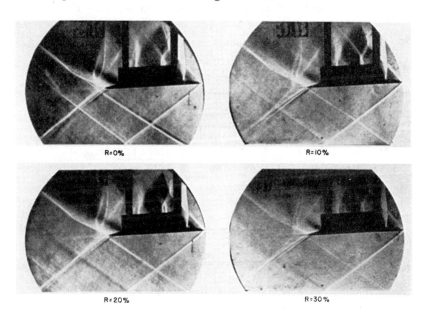

Fig. 8.6. *Reflection of a 3° shock wave on longitudinally slotted walls with different open-area ratios, R, at parallel wall setting, M = 1.80.*[2]

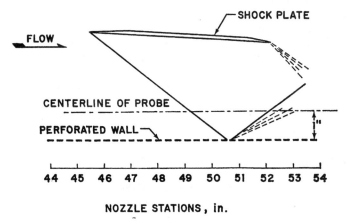

Fig. 8.7. *Test configuration for shock wave reflection.*[12]

from a solid wall were examined. The pressure data for this case are presented in Fig. 8.8. It can be seen that upstream and downstream of the shock fronts the pressure is essentially constant and that the reflected shock has approximately the same intensity as the primary impinging shock.

The same test equipment was then used to investigate the shock reflection from a longitudinally slotted wall with 20 per cent open-area ratio. The slot width was 0.1 in., the distance between slot centers was 0.6 in., and the wall thickness was $\frac{1}{16}$ in. The results obtained for this case are shown in Fig. 8.9. Both surveys, that above the open area and that above the solid

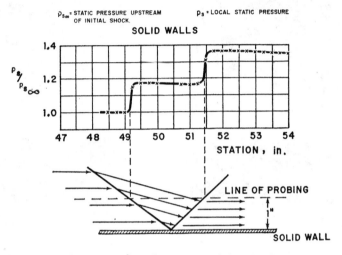

FIG. 8.8. *Reflection of a 3° shock wave on a solid wall at* $M = 1.75$.[12]

area of the wall, are basically very similar. It is also obvious that at a probe distance approximately two times the distance between slot centers, the influence of the open and the solid portions of the wall is no longer noticeable as individual distinct reflections; the open and solid parts of the wall act together as one partially open wall. This selected open-area ratio of 20 per cent should have yielded reflection-free conditions for a 3° shock in Mach number 1.75 flow, according to theory. Although the strong reflections experienced with a solid wall are greatly reduced by the slotted wall configuration, the test results show an irregular pressure distribution behind the initial reflected wave. Apparently a complex system of secondary shock and expansion waves causes unpredictable disturbance conditions, even at a large distance downstream of the initial reflected shock. The cause of these experimentally determined pressure disturbances can be found in viscous flow effects and secondary flow as described in detail by Allen and Spiegel[5] and in Chapter 9.

Thus, despite experimental verification of the fact that slotted walls have the capacity to reduce shock reflection considerably when the open-area ratio of the wall is properly selected, the potential of such a wall in a supersonic wind tunnel is not considered very favorable. This conclusion results from the disturbing influence of the viscosity and secondary flow effects and, most significantly, from the necessity of selecting a different open-area ratio for each shock strength and each Mach number. The number of experimental investigations of the shock reflection characteristics of longitudinally slotted walls is therefore relatively limited.

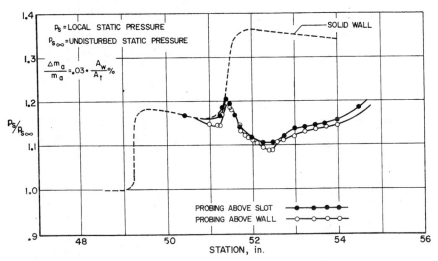

Fig. 8.9. *Reflection of a 3° shock wave on a longitudinally slotted wall with 20 per cent open-area ratio at $M = 1.75$ (pressure surveys at 1-in. distance from wall).*[12]

b. Perforated wind tunnel walls

Qualitative Experimental Results.—As early as 1950 results were reported for experiments conducted in a transonic wind tunnel at the Cornell Aeronautical Laboratory.[6,7] The Schlieren pictures of the tests showed that in the low supersonic Mach number range up to 1.3, no visible reflections could be detected when shock waves produced by a two-dimensional wedge impinged upon perforated wind tunnel walls. The selected Schlieren pictures of this early experimental study shown in Fig. 8.10 indicate that in the critical Mach number range from just above Mach number one to Mach number 1.3 it is possible to cancel the shock reflections completely or to reduce them to an acceptable level using a wall of constant geometry.

The same conclusions can be drawn from experiments conducted by United Aircraft Research Laboratories on the shock reflections produced on perforated plates in supersonic flow in the Mach number range between 1.5 and 2.0 (Ref. 8). The Schlieren pictures obtained during these tests indicate that a 50 per cent perforated wall, which has approximately the correct open-area ratio for reflection-free flow, produced pronounced wall reflections in the form of expansion waves (see Fig. 8.11). More detailed investigations showed that the observed reflection pattern, that is, the shock wave reflected as an expansion wave, is controlled by the thick boundary layer which forms along perforated walls and acts as an open jet boundary. Since the primary condition for wave reflection from open jet boundaries is that the static pressure be maintained, the shock waves are

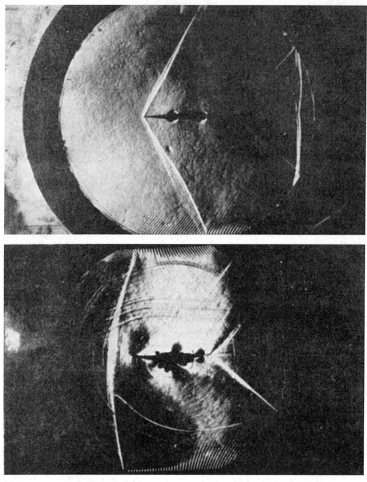

Fig. 8.10. *Shock reflection on perforated wind tunnel walls*, M = *1.15 and 1.30*.[7]

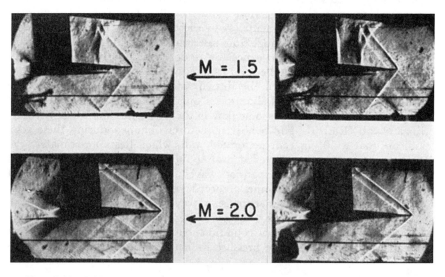

Fig. 8.11. *Schlieren photographs of shock reflections on perforated walls with and without boundary layer removal.*[8]

reflected as expansion waves. This interpretation of the test results led to the concept that the boundary layer, *particularly its subsonic portion*, must be removed or at least sufficiently thinned if a perforated wall were to be effective for wave cancellation. Experimental results presented in Fig. 8.11 verify the fact that with proper boundary layer removal a perforated wall acts as theory predicts: no visible reflections for a perforated wall having the proper open-area ratio can be detected.

Quantitative Experimental Results.—The initial quantitative results reported briefly in the preceding section were supplemented by a number of systematic experimental studies in which the reflected wave system was surveyed in detail at certain distances from the test wall.[9] United Aircraft Corporation showed by Schlieren photos and by systematic pressure probing that at a Mach number of 1.2 a perforated wall with a 23 per cent open-area ratio acts in almost the same manner as an open jet boundary when the boundary layer is not removed (see Fig. 8.12). The boundary layer thickness at a point just upstream of the shock impingement point was also determined, and the fact was verified that perforated walls develop extremely detrimental and thick boundary layers if no auxiliary suction is provided to remove the layer. At the same test conditions, that is, at a shock angle of 4° and a supersonic flow of Mach number 1.2, perforated walls can effectively eliminate shock reflections when boundary layer at the shock impingement point is thinned by means of auxiliary suction (see Fig. 8.12). These United Aircraft experiments show the significance of boundary layer removal from perforated walls and the potential value of such walls in eliminating shock reflections.

Another extensive systematic series of shock reflection tests in supersonic flow was conducted in the 6 in. supersonic wind tunnel of WADC with the same equipment described previously (see Fig. 8.8). For a shock wave of 3° in a supersonic flow of Mach number 1.75, a perforated wall having a 39 per cent open-area ratio reduced the reflected shock strength to a very small value when the boundary layer thickness was sufficiently reduced (see Fig. 8.13). Various boundary layer thicknesses were studied in this test series. With the thickest investigated, that is, with an auxiliary suction of only $\Delta m_a/m_a = 0.02 \cdot A_w/A_t$ per cent) the boundary layer acted much like an open jet boundary. Hence the impinging shock wave was reflected as an expansion wave having nearly the same intensity as the initial shock wave. Behind the shock the boundary layer gradually flowed through the perforated wall into the plenum chamber so that at a certain distance behind the impingement point a thin boundary layer and good reflection cancellation were obtained. When the boundary layer was thinned by auxiliary suction (i.e., when $\Delta m_a/m_a = 0.62 \cdot A_w/A_t$ per cent) the open jet boundary character of the shock wave reflection largely disappeared so that the initial disturbance field behind the primary shock wave was weaker and decayed more quickly until the final state of the flow was reached (Fig. 8.13).

Results similar in principle to those obtained with the 39 per cent open-area wall were also obtained in the WADC series using perforated walls with open-area ratios of 10, 20, and 30 per cent.[4] All pressure surveys from these tests (Figs. 8.13–8.16) show that a local disturbance field produced mainly by the boundary layer develops immediately downstream of the impinging shock wave. In the case of thick boundary layers, that is, when little or no auxiliary suction is provided, the shock wave is reflected as an expansion wave until the boundary layer is bled off through the perforated wall and a thin boundary layer having a constant pressure level is finally established. The thinner the initial boundary layer, that is, the greater the amount of auxiliary suction provided, the smaller is the open boundary layer effect and consequently the disturbance pressure field.

From the results of the WADC boundary layer surveys conducted at a point upstream of the impingement point of the primary shock (see Fig. 8.17), it can be estimated that an axial distance corresponding to four to

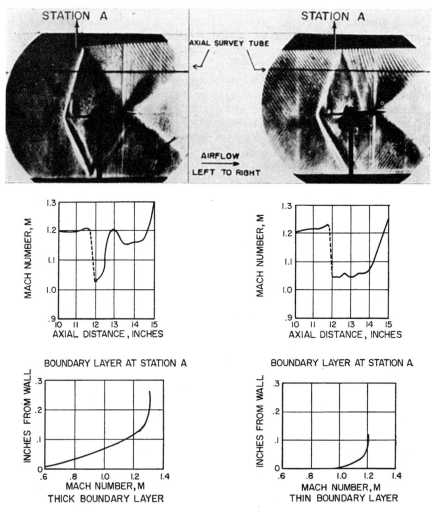

Fig. 8.12. *Reflection of 4° shock wave on perforated wall, 23 per cent open-area ratio, without and with boundary layer removal.*[9]

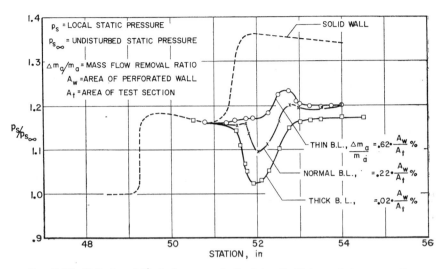

Fig. 8.13. *Reflection of 3° shock wave on perforated wall, 39 per cent open-area ratio, $M = 1.75$ (pressure surveys at 1 in. distance from wall).*[4,12]

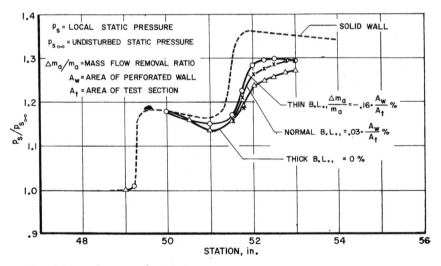

Fig. 8.14. *Reflection of 3° shock wave on perforated wall, 10 per cent open-area ratio, $M = 1.75$ (pressure surveys at 1 in. distance from wall).*[4]

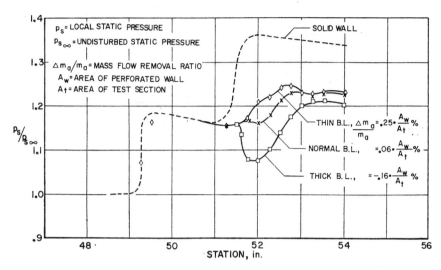

Fig. 8.15. *Reflection of 3° shock wave on perforated wall, 20 per cent open-area ratio, $M = 1.75$ (pressure surveys at 1 in. distance from wall).*[4]

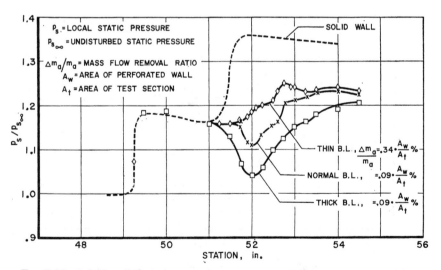

Fig. 8.16. *Reflection of 3° shock wave on perforated wall, 30 per cent open-area ratio, $M = 1.75$ (pressure surveys at 1 in. distance from wall).*[4]

SUPERSONIC WALL INTERFERENCE

five boundary layer thicknesses is required before the boundary layer and the resulting local pressure disturbances reach steady-state conditions. Boundary layer thickness is defined here as the distance from the wall to a point at which 99 per cent of the undisturbed velocity is obtained (see Fig. 8.17).

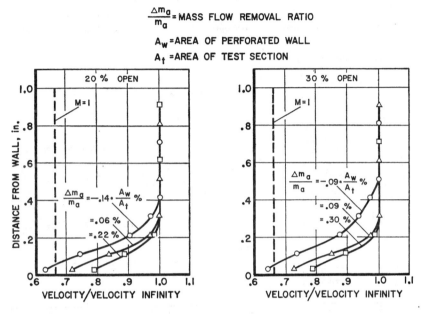

FIG. 8.17. Boundary layer profiles along perforated wall for different auxiliary suction quantities at $M = 1.75$, station 50 in.[12]

Calculations on Boundary Layer Influence on Wave Reflections.—To confirm the concept that even boundary layers with completely *supersonic* velocity profiles can produce local disturbances of considerable intensity, an extremely simplified calculation was carried out for the reflected wave pattern along a perforated wall having a thick boundary layer. A test configuration similar to that of Fig. 8.13 with the suction adjusted to $p_c = p_{s\infty}$ was assumed. The measured boundary layer profile was represented by a simple step function. With this simplification, the wall boundary layer was represented by a layer of uniform flow at a Mach number of 1.2; outside this layer the undisturbed Mach number 1.7 existed (Fig. 8.18a).

It was further assumed that the wall characteristics could be defined by the following relationship:

$$\frac{\Delta p}{(\tfrac{1}{2}\rho v^2)_w} = K\left(\frac{v_y}{v_x}\right)_w$$

where

Δp = pressure drop in flow through wall
$\tfrac{1}{2}\rho v^2 = q$ = dynamic pressure
v_y/v_x = flow inclination angle.

The subscript "w" indicates that the flow parameters were measured in the immediate vicinity of the wall. The constant K was selected in such a way that complete cancellation of the shock reflection would occur in the case of "zero" boundary layer thickness near the wall.

With the above assumptions it was possible to determine the wave pattern using the method of characteristics. To establish equilibrium between the pressures in the boundary layer and the plenum chamber, it was necessary to increase the cross-flow velocity v_{yw} locally in order to compensate for the

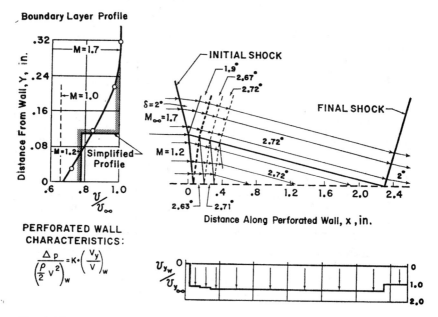

Fig. 8.18a. *Wave pattern for oblique supersonic flow through a perforated wall with idealized boundary layer at $M = 1.70$.*[12]

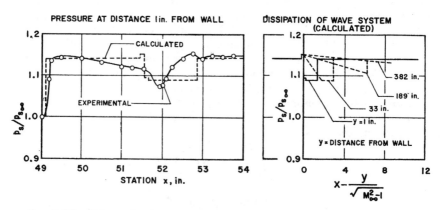

Fig. 8.18b. *Calculated and experimental pressure distribution for shock reflection on a perforated wall at $M = 1.75$.*[12]

reduced mass flow density $(\rho v)_w$. Only after the boundary layer was completely bled off did the final state of pressure equilibrium exist between boundary layer and plenum chamber. The pressure distribution obtained from the above calculations is plotted and compared with experimental data in Fig. 8.18b. Main features of the experimental curve, the initial pronounced pressure drop and the subsequent recovery to the correct pressure level, are properly described by the calculations.

This simplified calculation can also be used to predict the decay of the wave reflection pattern at increasing distances from the wall. Since expansion and compression waves of equal intensity have a tendency to cancel each other, the expectation is that, at large distances from the perforated wall, the pressure non-uniformities will level off and finally disappear completely. As Fig. 8.18b shows, the decay of the initial pressure disturbances is extremely slow for the selected test configuration. It is conceivable that an important parameter for the rapid progress of decay is the initial thickness of the boundary layer.

The above calculations and the experimental results show that the boundary layer has only a local effect. As soon as it is bled off through the wall, the final balance between cross flow through the wall and the pressure drop is established. This final balance is independent of initial boundary layer thickness.

Partial Reflection of Shock Waves from Perforated Walls.—The data reported in the preceding sections were further analyzed with respect to the partial reflection which remains when the open-area ratio of a perforated wall is not correctly selected for reflection cancellation. Some results of the United Aircraft experiments with reflections from perforated walls having different open-area ratios in the Mach number range between 1.3 and 2.0 when the boundary layer has been largely removed (Ref. 8) are presented in Fig. 8.19. This figure shows that by changing the perforated wall open-area

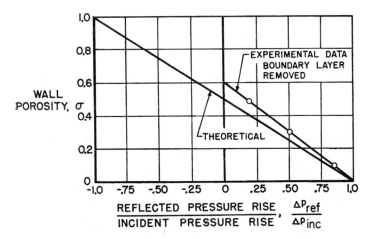

FIG. 8.19. *Partial shock wave reflection on perforated walls with different open-area ratios at $M = 1.3$ to 2.0.*[8]

ratio from completely closed to completely open, the shock-wave reflection should theoretically change from a complete compression wave to a complete expansion wave. The experimental results obtained with open-area ratios of 10 and 50 per cent closely approximate the theoretical prediction for walls with transverse slots. However, somewhat larger open-area ratios are required for the same degree of reduction in the intensity of the reflection wave than are predicted. For example, a 60 per cent open-area ratio was required for complete reflection cancellation instead of the 50 per cent open-area ratio wall with transverse slots predicted by theory.

Some results of the WADC experiments are plotted in a similar manner (Fig. 8.20). Shock intensities corresponding to 3° of flow turning angle at a

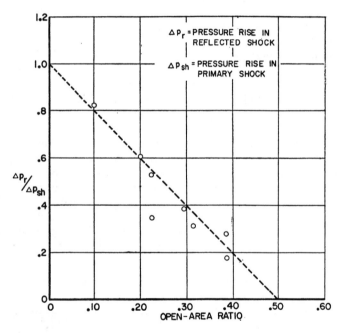

FIG. 8.20. *Partial shock wave reflection on perforated walls with different open-area ratios at M = 1.75 (experiments of WADC)*.[4,13]

Mach number of 1.75 can be extrapolated to produce reflection-free conditions for an open-area ratio of approximately 50 per cent when stabilized conditions behind the primary shock wave are considered.

In summary, it can be stated that perforated walls behave basically as theory predicts when proper measures are taken to reduce the boundary layer thickness. However, because of the presence of the boundary layer, a local disturbance field occurs in the region between the initial impingement point of the primary shock wave and the region where conditions are finally stabilized.

5. EXPERIMENTAL RESULTS FOR TWO-DIMENSIONAL EXPANSION-WAVE CANCELLATION

According to theory an ideal perforated wall should be capable of eliminating reflections of shock waves and expansion waves since, with small flow inclinations, the "pressure-drop vs. cross-flow" characteristics of both the waves and the wall are linear and can be matched by proper geometry of the walls. However, in real flow it is difficult to obtain linear wall characteristics having the same slope for outflow and inflow through the wall due to both the boundary layer effect and the inflow of low energy air from the plenum chamber into the test section. In order to investigate these problems, some experimental studies were conducted in the 12 in. transonic model tunnel of the AEDC.[10] It was found from these tests that by the selection of a perforated wall with suitable geometry (converged to the proper angle), reflections of expansion waves can be eliminated in the same manner as can reflections of compression waves.

Some examples of expansion wave cancellation at Mach number 1.35 are shown in Fig. 8.21. A conventional perforated wall with 22 per cent open-area ratio acts as an open jet boundary when the walls are set parallel, that is, when the boundary layer thickness along the perforated wall is permitted to grow to large values. However, if the boundary layer is thinned sufficiently by converging the walls to 15' and 30', respectively, the expansion wave reflections on these same walls are practically eliminated.

It is important to note that walls which were too open for expansion wave reflection at the parallel wall setting were too closed for proper compression wave cancellation. At the wall setting which gave practically complete cancellation of expansion wave reflections (30' convergence), the reflection of a compression wave was very pronounced. It must be concluded, therefore, that a conventional perforated wall is not capable of eliminating reflections of both expansion and compression waves unless a change is made in the geometry of the wall.

Wave Reflection from Differential-Resistance Walls.—In order to produce reflection cancellation of both expansion and compression waves, it would be necessary to provide a wall which would produce the same pressure drop characteristic for outflow and inflow. A conventional perforated wall with straight holes does not possess this characteristic but, as will be discussed in more detail in Chapter 9, walls with inclined holes can be devised which produce an equal pressure drop through the wall for both outflow and inflow. It is even possible to increase the inflow pressure drop above the values for outflow.

Walls with inclined holes were tested in the transonic model tunnel of the AEDC in order to explore briefly the potentialities of such walls for reflection cancellation of impinging waves of both types.[10] Some typical results are shown in Fig. 8.22. A wall with 60° inclined holes and 6 per cent open-area ratio almost eliminated reflections of expansion waves. The same wall with the same setting also essentially reduced the reflection of compression waves, though it was somewhat too closed for complete cancellation.

To obtain complete cancellation of both expansion and compression waves using the same wall geometry and setting, the differential-resistance character of the wall must still be somewhat amplified.

When results obtained with the conventional wall (see Fig. 8.21) and for the wall with inclined holes (see Fig. 8.22) are compared, it can be concluded that the "inclined-hole" wall is superior to the "conventional-hole" wall if simultaneous cancellation of both compression and expansion waves must

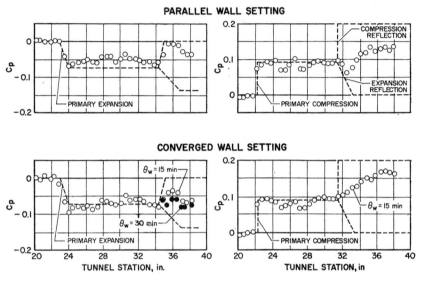

FIG. 8.21. *Reflection of shock and expansion waves on conventional perforated walls, 22 per cent open-area ratio, $M = 1.35$, $C_p = (p - p_\infty)/q_\infty$.*[10]

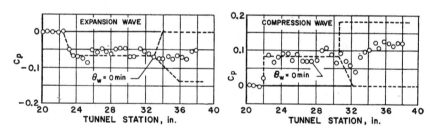

FIG. 8.22. *Reflection of shock and expansion waves on perforated wall with 60° inclined holes and 6 per cent open-area ratio at $M = 1.30$, $C_p = (p - p_\infty)/q_\infty$.*[10]

be accomplished. In transonic wind tunnel testing of airplane models where a variety of different compression and expansion waves usually occur, the wall with inclined holes has a greater potential than the wall with conventional straight-hole perforations.

APPENDIX

DERIVATIONS OF EQUATIONS FOR PARTIALLY OPEN WALLS IN OBLIQUE SUPERSONIC FLOW[11]

1. WALL WITH LONGITUDINAL SLOTS

a. Oblique parallel flow crossing a wall with longitudinal slots

The derivations in the following sections are carried out for small deviations from parallel flow with all deviation terms up to the second order included. As in the case of the wall with transverse slots, the pressure at the back side of the slotted wall is assumed to be adjusted in such a way that the same mass flow per unit area crosses the wall and any plane parallel to the wall in flow area ① (Fig. 8.23). Then with the distance between slot centers being d and the open-area ratio being R, the required continuity equation becomes

$$h_1 d(\rho_1 v_1) = h_3 dR(\rho_3 v_3)$$

and

$$h_1(\rho_1 v_1) = h_3 R(\rho_3 v_3)$$

Referring to Fig. 8.23:

$$h_1 \sin \theta_3 = h_3 \sin \theta_1$$

$$h_1(\theta_3 - \tfrac{1}{6}\theta_3^3) = h_3(\theta_1 - \tfrac{1}{6}\theta_1^3)]$$

Considering only terms of second-order magnitude:

$$h_1/h_3 = \theta_1/\theta_3$$

When this equation is introduced

$$\theta_3 = \frac{1}{R}\theta_1 \left(\frac{\rho_1 v_1}{\rho_3 v_3}\right)$$

The change of the stream density ρv, can be related to the change of the velocity by means of the general energy equation:

$$\rho_3 v_3 = \rho_1 v_1 \left[1 - (M_1^2 - 1)\left(\frac{v_3 - v_1}{v_1}\right) + C_2\left(\frac{v_3 - v_1}{v_1}\right)^2\right]$$

where C_2 is constant.

Another equation is obtained from momentum considerations which, because of the lack of external forces in the x-direction, states that there is no change of the velocity components in the x-direction:

$$v_{1_x} = v_{3_x} \quad \text{or} \quad v_1 \cos \theta_1 = v_3 \cos \theta_3$$

Hence:

$$v_3/v_1 = \frac{1-\frac{1}{2}\theta_1^2}{1-\frac{1}{2}\theta_3^2} = 1-\frac{1}{2}\theta_1^2+\frac{1}{2}\theta_3^2$$

and

$$\frac{v_3-v_1}{v_1} = \frac{1}{2}\theta_1^2+\frac{1}{2}\theta_3^2$$

When the previous equations are combined:

$$\rho_3 v_3 = \rho_1 v_1[1-(M_1^2-1)(-\tfrac{1}{2}\theta_1^2+\tfrac{1}{2}\theta_3^2)+\text{small terms}]$$

and, finally,

$$\theta_3[1-(M_1^2-1)(-\tfrac{1}{2}\theta_1^2+\tfrac{1}{2}\theta_3^2)] = \frac{1}{R}\theta_1$$

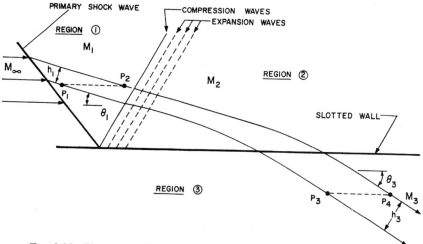

FIG. 8.23. *Flow pattern for cross flow through a wall with longitudinal slots.*[11]

Then, when the terms of the third order are neglected:

$$\theta_3 = \frac{1}{R}\theta_1$$

The pressure change can also be calculated from the velocity change by means of the basic energy equation:

$$\frac{p_3-p_1}{q_1} = -2\left(\frac{v_3-v_1}{v_1}\right)+C_3\left(\frac{v_3-v_1}{v_1}\right)^2$$

with C_3 being a constant.

Hence:

$$\frac{p_3-p_1}{q_1} = (+\theta_1^2-\theta_3^2)+C_3(-\tfrac{1}{2}\theta_1^2+\tfrac{1}{2}\theta_3^2)^2 = \theta_1^2-\theta_3^2 = \theta_1^2\left(1-\frac{1}{R^2}\right)$$

APPENDIX

Unlike the case of the wall with transverse slots, the pressure drop through the wall is represented by a quadratic function of the cross-flow angle θ.

b. Shock-wave cancellation

A shock wave produces a pressure change according to the relationship:

$$\frac{p_1 - p_\infty}{q_1} = \frac{2}{(M_\infty^2 - 1)^{\frac{1}{2}}}\theta_1 + C_4\theta_1^2$$

The correct wall opening ratio R for complete wave cancellation can be determined again from the condition that the pressure on the back side of the slotted wall p_3 is equal to the pressure p_∞ of the undisturbed parallel flow:

$$\frac{p_1 - p_\infty}{q_1} + \frac{p_3 - p_1}{q_1} = 0$$

and, after introducing several previous equations,

$$\frac{2}{(M_\infty^2 - 1)^{\frac{1}{2}}}\theta_1 + C_4\theta_1^2 + \theta_1^2\left(1 - \frac{1}{R^2}\right) = 0$$

$$\frac{1}{R^2} - 1 = \frac{2}{(M_\infty^2 - 1)^{\frac{1}{2}}}\frac{1}{\theta_1} + C_4 \simeq \frac{2}{(M_\infty^2 - 1)^{\frac{1}{2}}}\frac{1}{\theta_1}$$

Then for small values of θ_1 the correct wall open-area ratio R, becomes:

$$R^2 = \tfrac{1}{2}(M_\infty^2 - 1)^{\frac{1}{2}}\theta_1$$

c. Momentum considerations

The preceding calculations are based on the fact that the velocity components in the x-direction, v_{1_x} and v_{3_x}, are equal in flow regions ①, ②, and ③ (Fig. 8.23). This can be readily seen by consideration of the momentum terms for the flow region P_1, P_2, P_3, and P_4, which is bounded by two streamlines. Supersonic waves occur only in the region immediately downstream of the impinging primary shock wave. Further downstream the system is wave-free because of the subsonic character of the flow perpendicular to the wall. In the wave region the pressure terms in the momentum equation are cancelled due to symmetry of the flow. In the region behind the wave system, flow inclination θ_2, as well as flow velocity v_2 and pressure p_2, are equal to the corresponding values in flow region ①. These equalities follow from the requirement for constant unit cross flow through any plane parallel to the wall in regions ① and ② and the assumption of isentropic flow. Consequently the pressure terms in the momentum equation are cancelled completely along the streamlines from P_1 to P_3 and from P_2 to P_4. Hence, also, the horizontal velocity components must be equal.

2. WALL WITH TRANSVERSE SLOTS

a. Oblique parallel flow crossing a wall with transverse slots

The pressure on the back side of the wall (flow area ③) is assumed to be adjusted in such a way that the same mass flow per unit area crosses the wall at any plane parallel to the wall in flow area ① (Fig. 8.24). Then from continuity considerations

$$h_1(\rho_1 v_1) = h_3(\rho_3 v_3)$$

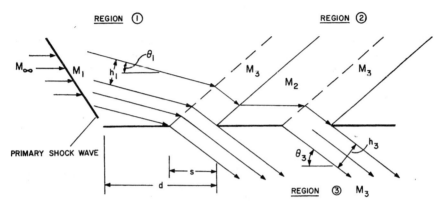

FIG. 8.24. Flow pattern for cross flow through a wall with transversal slots.[11]

If entropy changes are neglected, as is permissible in linear theory, then the well-known supersonic wave relationship exists:

$$\rho_3 v_3 = \rho_1 v_1 [1 - (M_1^2 - 1)^{\frac{1}{2}}(\theta_3 - \theta_1)]$$

Hence:

$$h_1 = h_3 [1 - (M_1^2 - 1)^{\frac{1}{2}}(\theta_3 - \theta_1)]$$

With

$$h_1 = d \sin \theta_1 \simeq d\theta_1$$

and

$$h_3 = s \sin \theta_3 \simeq s\theta_3 = Rd\theta_3$$

the following equation is obtained:

$$\theta_1 = R\theta_3 [1 - (M_1^2 - 1)^{\frac{1}{2}}(\theta_3 - \theta_1)]$$

By solving this equation for $(\theta_3 - \theta_1)$ and developing in a series,

$$\theta_3 - \theta_1 = \frac{(1/R) - 1}{1 - \theta_1 (M_1^2 - 1)^{\frac{1}{2}}} \theta_1$$

or by further linearization:

$$\theta_3 - \theta_1 = \left(\frac{1}{R} - 1\right)\theta_1$$

APPENDIX

In linearized supersonic flow with waves of one kind only, the following relationship between pressure changes and flow direction changes exists:

$$\frac{\Delta p}{2} = \frac{-2}{(M_1^2-1)^{\frac{1}{2}}} \Delta \theta$$

This equation is applied to the flow regions ③ and ①

$$\frac{p_3 - p_1}{q_1} = \frac{-2}{(M_1^2-1)^{\frac{1}{2}}} \left(\frac{1}{R} - 1\right) \theta_1 = K \theta_1$$

b. Shock-wave cancellation

If it is assumed that the oblique parallel flow discussed above is produced by a shock wave in the original parallel flow at the Mach number M (see Fig. 8.24), the pressure change through this shock wave is:

$$\frac{p_1 - p_\infty}{q_1} = \frac{2}{(M_1^2-1)^{\frac{1}{2}}} \theta_1$$

The correct wall opening ratio R can now be determined from the condition that the pressure on the back side of the slotted wall p_3 is equal to the pressure p_∞ of the undisturbed flow:

$$\frac{p_1 - p_\infty}{q_1} + \frac{p_3 - p_1}{q_1} = \frac{p_3 - p_\infty}{q_1} = 0$$

or, after introducing several preceding equations:

$$\frac{2}{(M_1^2-1)^{\frac{1}{2}}} \theta_1 - \frac{2}{(M_1^2-1)^{\frac{1}{2}}} \left(\frac{1}{R} - 1\right) \theta_1 = 0$$

from which

$$R = \tfrac{1}{2}$$

c. Momentum considerations

It is to be noted that, contrary to the case of a longitudinally slotted wall, the velocity components parallel to the initial undisturbed flow are not equal for flow regions ① and ③ (see Fig. 8.24). The momentum equation may be applied to the area bounded by the two streamlines shown in Fig. 8.25 and the oblique lines P_1P_2 and P_3P_4. Since the pressures in each of the areas a–a, b–b, c–c, etc., are constant, the pressure term in the momentum equation is:

$$[(P_1, P_2) \sin \alpha] p_1 - [(P_3 P_4) \sin \alpha] p_3 = C_1(p_1 - p_3)$$

with C_1 being a constant for a given Mach number.

The complete momentum equation is then:

$$m_a(v_{1_x} - v_{3_x}) - C_1(p_3 - p_1) = 0$$

where: $m_a = h_1 \rho_1 v_1$

and finally:

$$(v_{3_x} - v_{1_x}) = \frac{C_1}{m_a}(p_1 - p_3)$$

For a shock-wave reflection the pressure p_1 is larger than the pressure p_3; as a result the velocity component v_{3_x} in the region at the back of the wall is larger than the velocity component v_{1_x} in region ① of the oblique parallel flow behind the initial shock.

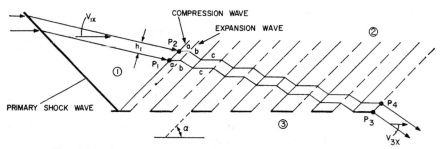

FIG. 8.25. *Streamline and wave pattern for cross flow through a wall with transversal slots.*[11]

d. Partial reflection

The following equations were obtained previously for the pressure drop through the wall (Sections 2a and b of this Appendix and Fig. 8.24).

No Reflections

$$\frac{p_1 - p_3}{q_\infty} = \frac{\Delta p_{sh}}{q_\infty} = K_c \theta_1$$

with

$$K_c = \frac{2}{(M^2 - 1)^{\frac{1}{2}}}$$

Partial Reflection with pressure rise Δp_r, and flow direction change $\Delta\theta$:

$$\frac{(p_1 + \Delta p_r) - p_3}{q_\infty} = \frac{\Delta p_{sh} + \Delta p_r}{q_\infty} = (K_c + \Delta K)(\theta - \Delta\theta)$$

with:

$$K_c + \Delta K = \frac{2}{(M^2 - 1)^{\frac{1}{2}}}\left(\frac{1}{R} - 1\right)$$

By dividing these two equations and after several transformations

$$\frac{\Delta p_r}{\Delta p_{sh}} = \frac{\Delta K}{K_c} - \frac{\Delta\theta}{\theta}\left(1 + \frac{\Delta K}{K_c}\right)$$

APPENDIX

From these linearized shock-wave equations, the following relationship can be derived:

$$\frac{\Delta p_r}{\Delta p_{sh}} = \frac{\Delta \theta}{\theta}$$

By combination of the two equations for $\Delta p_r / \Delta p_{sh}$:

$$\frac{\Delta p_r}{\Delta p_{sh}} = \frac{\Delta K / K_c}{2 + \Delta K / K_c}$$

and, after introducing the values for K_c and $(K_c + \Delta K)$:

$$\frac{\Delta p_r}{\Delta p_{sh}} = 1 - 2R$$

REFERENCES

[1] JONES, R. T. "Properties of Low-Aspect-Ratio Pointed Wings at Speeds below and above the Speed of Sound." NACA TN 1032, March, 1946.

[2] McLAFFERTY, G. H., JR. "Investigation of the Use of Slotted Walls to Cancel Shock Wave Reflections in Supersonic Flow." UAC Report R-15127-3, October 10, 1949.

[3] BUCKLEY, A. B. "Shock Wave Cancellation with Porous Walls at Mach Number 1.75." WADC 53WC-26222, March, 1953.

[4] BUCKLEY, A. B. and ZONARS, DEMETRIUS. "Shock Wave Cancellation Properties of Perforated and Slotted Walls at Mach Number 1.75." WADC Technical Report No. 55-287, July, 1955.

[5] ALLEN, H. J. and SPIEGEL, J. M. "Transonic Wind Tunnel Development at the NACA." S.M.F. Fund Paper No. FF-12, Institute of Aeronautical Sciences publication, 1954.

[6] GOODMAN, T. R. "The Porous Wall Wind Tunnel—Part I: One-dimensional Supersonic Flow Analysis." CAL Report No. AD-594-A-2, October, 1950.

[7] GOODMAN, T. R. "The Porous Wall Wind Tunnel—Part III: The Reflection and Absorption of Shock Waves at Supersonic Speeds." CAL Report No. AD-706-A-1, November, 1950.

[8] TAYLOR, H. D. and MANONI, L. R. "Shock Wave Cancellation Characteristics of Two-dimensional Transonic Test Sections." Arnold Engineering Development Center, July, 1956.

[9] CUSHMAN, H. T. "Small Scale Transonic Interference Studies with Perforated Wall Test Sections." UAC Report R-95538-16, June, 1953.

[10] MARSHALL, JOHN C. "Two-dimensional Wave Cancellation Characteristics of Several Perforated Test Section Wall Liners." AEDC-TM-57-35, June, 1957.

[11] GOETHERT, B. H. "Physical Aspects of Wind Tunnel Wall Interference at Low Supersonic Mach Numbers." Paper I, 129, IX International Congress of Applied Mechanics, Brussels, September 5-13, 1956.

[12] GOETHERT, B. H. "Flow Establishment and Wall Interference in Transonic Wind Tunnels." AEDC-TR-54-44, June, 1954.

[13] McLAFFERTY, G. H., JR. "Notes on the Cancellation of Shock and Expansion Waves in a Perforated-Wall Tunnel at Supersonic Speeds." UAC Report M-13644-2, August 11, 1954.

[14] NELSON, WILLIAM J. and BLOETSCHER, Frederick. "Preliminary Investigation of Porous Walls as a Means of Reducing Tunnel Boundary Effects at Low Supersonic Mach Numbers." NACA RM L50D27, August 23, 1950.

BIBLIOGRAPHY

CARROLL, JAMES B. and RICE, JANET B. "Analytical and Experimental Studies of Wall Interference in Perforated Wind Tunnel." WADC Technical Report No. 56-240, June, 1956.

DAVIS, DON D., JR. and WOOD, GEORGE P. "Preliminary Investigation of Reflections of Oblique Waves from a Porous Wall." NACA RM L50G19a, November 9, 1950.

DUBOSE, H. C. "Experimental and Theoretical Studies on Three-dimensional Wave Reflection in Transonic Test Sections. Part II: Theoretical Investigation of the Supersonic Flow Field about a Two-dimensional Body and Several Three-dimensional Bodies at Zero Angle of Attack." AEDC TN-55-43, March, 1956.

ECKHAUS, W. "On the Theory of Shock Reflection on Walls with Slots." National Aeronautical Research Institute, Holland, N.L.L. Report F.167, July 25, 1955.

GOLDBAUM, G. C. "Research Work on Supersonic Wind Tunnels to Investigate the Use of Perforated Walls." WADC TR-55-185, March, 1956.

HAMILTON, C. V., PARKER, G. H. and RAMM, H. "Experimental Determination of Auxiliary Suction Requirements in the PWT Supersonic Model Tunnel at Mach Numbers 2.0 and 2.6." AEDC-TN-57-2, March, 1957.

MCLAFFERTY, G. H., JR. "Additional Notes on the Cancellation of Reflections from Shock Waves in a Perforated-Wall Wind Tunnel at Supersonic Speeds." UAC Report M-13644-4, November 10, 1954.

NELSON, WILLIAM J. and BLOETSCHER, FREDERICK. "An Experimental Investigation of the Zero-Lift Pressure Distribution over a Wedge Airfoil in Closed, Slotted and Open-Throat Tunnels at Transonic Mach Numbers." NACA RM L52C18, June, 1952.

PINDZOLA, M. "Shock and Expansion Wave Cancellation Studies in a Two-dimensional Porous Wall Transonic Tunnel." UAC Report R-25473-6, September 11, 1951.

SELLERS, T. B., DAVIS, D. D., JR. and STOKES, G. M. "An Experimental Investigation of the Transonic Flow-Generation and Shockwave-Reflection Characteristics of a Two-Dimensional Wind Tunnel with 24 per cent Open, Deep, Multi-slotted Walls." NACA RM L53J28, December 10, 1953.

SPIEGEL, J. M. and TUNNELL, P. J. "An Analysis of Shock-Wave Cancellation and Reflection for Porous Walls which Obey an Exponential Mass-Flow Pressure-Difference Relation." NACA TN 3223, August, 1954.

TAYLOR, H. D. "Progress of Transonic Wind Tunnel Studies at U.A.C." UAC Report R-95434-8, June, 1951.

WILDER, JOHN G., JR. "An Experimental Investigation of the Perforated Wall Transonic Wind Tunnel, Phase II." CAL Report No. AD-706-A-5, August, 1951.

WILDER, John G., JR. "An Experimental Investigation of the Perforated Wall Transonic Wind Tunnel, Phase II." CAL Report No. AD-706-A-6, January, 1952.

WOOD, GEORGE P. "Reflection of Shock Waves from Slotted Walls at Mach Number 1.62." NACA RM L52E27, July 21, 1952.

CHAPTER 9

SUPERSONIC WALL INTERFERENCE IN PARTIALLY OPEN TEST SECTIONS FOR MODELS OF ROTATIONAL SYMMETRY AND COMPARISON WITH TWO-DIMENSIONAL WALL INTERFERENCE

It has been demonstrated in the preceding chapters that, in the case of two-dimensional waves, test-section walls can be developed which satisfactorily eliminate reflections (see Chapter 8). However, it can be shown theoretically that a perforated wall with constant geometry is not capable of completely eliminating wave reflections over the entire model region in the case o three-dimensional wave fields. The physical reasons for the difference in the behavior of two-dimensional and three-dimensional shock reflections will be discussed in the following sections.

1. BOUNDARY CONDITIONS ALONG WIND TUNNEL WALLS: TWO-DIMENSIONAL AND THREE-DIMENSIONAL MODELS

a. Two-dimensional wedge-flat-plate model

To obtain insight into the order of magnitude of flow inclinations and pressure disturbances along wind tunnel walls, the flow disturbance of typical simple two-dimensional models in supersonic flow at $M = 1.5$ was calculated. The model sizes were selected to give test-section blockage ratios of 2, 4, and 6 per cent (see Refs. 1 and 2). Figure 9.1 shows that with these simple wedge-flat-plate models the pressure rise is concentrated in the initial shock wave. Behind the initial wave, the pressure is gradually reduced by means of expansion waves originating at the shoulder of the models.

b. Three-dimensional cone-cylinder model

A schematic showing the compression and expansion wave pattern around cone-cylinder models in free flight is presented in Fig. 9.2. In contrast to the wave field for a two-dimensional wedge-flat-plate body, this wave system consists not only of an initial shock and a following system of expansion waves initiated at the shoulder of the body; in addition, a secondary wave system covering the entire three-dimensional flow field is produced to satisfy the continuity requirements for wave fronts proceeding in a three-dimensional flow field. The pressure distribution calculated along a plane which would correspond to the ideal wind tunnel wall for a 2 per cent blockage model would then have only a small portion of the compression occurring in the initial conical shock (point A in Fig. 9.2). The major portion of the compression would occur in the system of secondary compression waves downstream of this initial shock. Compression along the ideal tunnel wall is then abruptly interrupted because the expansion wave system causes the pressure to drop rapidly to values below the pressure level of the undisturbed flow (region B-C). Behind this expansion wave system, the pressure gradually returns to the level of the undisturbed pressure.

The relationship between pressure disturbance $\Delta p/q_\infty$ and flow inclination θ is presented in Fig. 9.3. The magnitude of the wall disturbances is seen to be considerably greater with two-dimensional than with three-dimensional models. In the two-dimensional case, the 6 per cent blockage model reaches

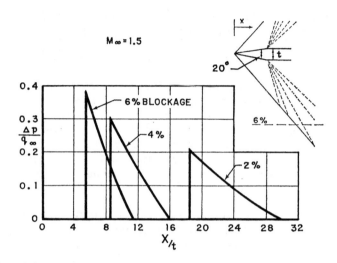

Fig. 9.1. *Calculated pressure distributions along the test-section wall for "wedge-flat-plate" models in supersonic flow.*[3]

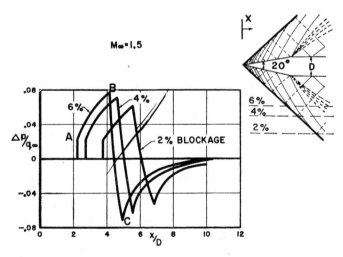

Fig. 9.2. *Calculated pressure distribution along the test-section wall for cone-cylinder models in supersonic flow.*[3]

a maximum flow inclination of approximately 9° at the wall; a three-dimensional model with the same blockage ratio reaches a maximum of only 3°. The graph also indicates that, with the two-dimensional model, a wall having linear characteristics matches the model disturbance characteristics very well; therefore only small wall-interference effects should be expected.

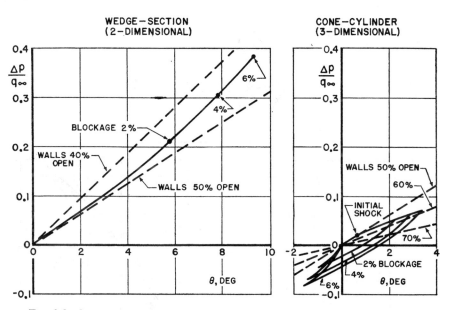

FIG. 9.3. *Comparison between flow disturbance along the test-section wall at $M = 1.5$ for two-dimensional and three-dimensional models in supersonic flow.*[1]

For cone-cylinder models, the characteristic curve $\Delta p/q = f(\theta_w)$, that is, pressure disturbance as function of flow inclination, can no longer be represented by a straight line as is approximately true in the two-dimensional case (linear theory). Some noteworthy features of the characteristic curve are as follows (see Fig. 9.4):

1. The compression waves upstream of the expansion wave system (region A–B) increase the pressure and the flow inclination angle so that the flow angles at a given pressure rise are considerably larger than predicted by the Prandtl–Meyer curve.
2. In the expansion wave region (region B–C), the pressure drops rapidly below the value of the undisturbed pressure. At the point at which the pressure reaches the undisturbed value, that is, at $\Delta p/q = 0$, the flow still has an inclination of 0.7° in a direction away from the body. At zero flow inclination, $\theta_w = 0$ (flow parallel to the cone-cylinder axis), the disturbance pressure is negative, that is, the local pressure is smaller than the static pressure of the undisturbed flow. As the flow progresses downstream into the negative pressure region, it attains flow inclinations directed toward the body (inflow) and reaches absolute magnitudes only slightly smaller than those in the outflow region (point C).

3. In the inflow region, particularly in the region C–O, the pressure deviates even more from linear characteristics since considerably lower static pressures occur than are predicted by linear theory.

It is obvious in the case of the three-dimensional wave system under consideration that boundary conditions for an ideal disturbance-free tunnel wall cannot be obtained with a perforated wall having linear characteristics

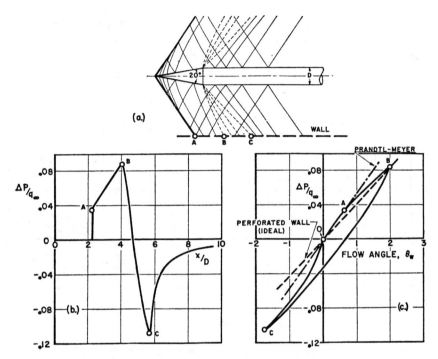

FIG. 9.4. *Theoretical disturbance distribution and wave system of cone-cylinder model with 2 per cent blockage at $M = 1.20$.*[1]

c. Models equipped with various contours

The most remarkable feature in testing a cone-cylinder model in a partially open test section, and, as will be shown later, the main source of interference, is associated with the concentrated expansion wave system initiated at the shoulder of the model. If the sharp-edged shoulder of the cone-cylinder model is rounded off, however, these expansion waves are distributed over a larger area, and consequently the matching problem can be expected to be less severe. To determine the degree of improvement which might be anticipated, the flow field around a cone-cylinder model having a rounded shoulder and that around a smoothly-curved model, the NACA RM-10 model, were calculated theoretically to determine the flow disturbance characteristics produced along an ideal wind tunnel wall. The results of these calculations[1,2] are presented in Fig. 9.5.

The rounding of the shoulder of the cone-cylinder model was assumed to extend over a region equal to approximately one-third of the cone length. Even this relatively slight rounding of the shoulder resulted in a considerable reduction of the flow disturbances (both pressure and flow inclination) at the plane of the wall. In particular, the magnitude of the maximum inflow angle was reduced to nearly one-third that of the corresponding cone cylinder. However, the maximum outflow angle was reduced only slightly.

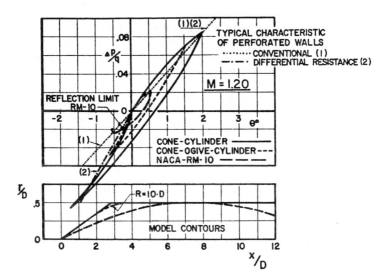

FIG. 9.5. *Theoretical disturbance distribution along the test-section wall for different bodies of revolution with 2 per cent blockage at $M = 1.20$.*[1]

The disturbance characteristics of this cone-ogive model are thus much closer to the characteristic cross flow of a test-section wall having linear characteristics than are those of the cone-cylinder model. In the extreme case of the continuously-curved NACA RM-10, the mismatch between model disturbances and linear wall characteristics was reduced even further.

2. MODEL DISTURBANCES COMPARED WITH CROSS-FLOW CHARACTERISTICS OF PERFORATED TEST SECTIONS (CONVENTIONAL AND DIFFERENTIAL-RESISTANCE TYPE)

a. Two-dimensional wedge-flat-plate model

Wind tunnel models produce distinct types of disturbances, both in the local flow direction and in the local static pressure along the wind tunnel walls. If the walls are properly designed, they can support this difference between the pressures along the inside and outside of the test section produced by the local oblique directional flow calculated previously (see Figs. 9.1 and 9.2). To a first approximation it may be assumed that at each point along the wall only the local conditions, not the influence of the adjacent areas upstream or downstream, need be considered. With the two-dimensional wedge-flat-plate model these local conditions can be readily supported by a partially open wall having linear characteristics (see Fig. 9.3). It should be noted that for a given model the local pressure disturbance is related to the local directional flow disturbance by a linear function. Consequently, not only can the condition for one particular point, that is, one value of pressure disturbance associated with one value of flow inclination, be satisfied, but a whole series of such conditions can be met, as is necessary for interference-free testing of a model at any given value of model setting and Mach number.

As shown in more detail in Chapter 8, a perforated wall with an open-area ratio of approximately 50 per cent will satisfy to a first approximation the requirements for interference-free testing of a wedge-flat-plate model or any other two-dimensional model in supersonic non-viscous flow.

b. Cone-cylinder models

When the disturbance characteristics of a cone-cylinder model in supersonic flow (see Fig. 9.3) are considered, it is immediately apparent that no wall having a definite relationship between cross flow and pressure drop can match the requirements of the model-produced disturbances. Therefore, in principle it is not possible to design a partially open wall having fixed geometry which will be capable of matching locally the "cross flow-pressure drop" characteristic of cone-cylinder models. At best it may be hoped that a suitable compromise can be accomplished which will satisfactorily reduce the wave reflections from the wind tunnel walls.

It appears possible that in the compression region (region A–B of Fig. 9.4) the correct boundary conditions can be approximated by a wall which has linear pressure-drop characteristics similar to those of a perforated wall. In the expansion region, however, such a wall is not capable of producing the necessary boundary conditions because large deviations from linearized flow occur. Moreover, in the region where the pressure disturbance is zero, the correct boundary condition should be one of outflow or parallel flow, depending upon the wall station under consideration. Such opposite characteristics are obviously impossible to attain with a perforated wall of constant geometry since the flow field around a cone-cylinder model in a conventional perforated test section would be distorted because inflow into the test section would occur in the region where outflow is required. The result would be strong compression disturbances originating in this region.

A part of the inflow difficulties described above could be overcome if a wall could be designed which resists flow into the test section more than it does flow out of the test section. With a differential-resistance wall having a characteristic according to O–B in the outflow region and according to O–C in the inflow region, it appears possible to achieve simultaneous matching of the outflow and inflow over large areas (Fig. 9.4). Even in such a case, though, some mismatch would remain in the expansion wave region (Fig. 9.4, region B–C) because even a differential-resistance wall could not support a negative pressure in the test section while the flow was proceeding outward against a higher pressure in the surrounding plenum chamber.

It should be realized, however, that the cone-cylinder model under consideration produces the most severe matching problem possible because of the concentrated expansion wave region which originates at the shoulder of the model.

c. Models of rotational symmetry having various contours

The main source of interference in the testing of cone-cylinder models in perforated test sections is associated with the concentrated expansion wave system originating at the shoulder of such models. It can therefore be expected that models with rounded shoulders (the cone-ogive-cylinder model or, even more, the continuously-curved NACA RM-10 model) will produce disturbances along a test-section wall which can be more closely matched by the cross-flow characteristics of a typical perforated wall.

When the characteristic disturbance curves of three models having different contours are compared (Fig. 9.5), it becomes plain that the same wall can provide an adequate matching of the flow in the compression wave regions if the wall open-area ratio is properly selected. In the expansion wave regions, and particularly in the inflow regions (that is, in the regions having negative values of θ_w), a perforated wall with linear characteristics will gradually produce less mismatch as the sharp shoulder of the cone-cylinder model is rounded into the cone-ogive and then into the continuously-curved NACA model. If a differential-resistance-type wall were employed, the matching would be improved still more since one wall characteristic could be selected for the compression wave regions and another for the expansion wave regions.

This latter statement is not entirely true for the RM-10 model with continuous curvature because, in this case, the improvement in matching is restricted to the upstream portion of the inflow region. In the downstream part of this region, the matching is actually better with a conventional perforated wall having equal resistance to both outflow and inflow. Fortunately the mismatch downstream is not very critical and causes no interference on the model since shock waves, if deflected from the wall in this region, do not intersect the model but pass behind it (see reflection limit indicated in Fig. 9.5). As a result, it may be stated that even for the continuously-curved RM-10 model a suitably designed differential-resistance wall would reduce the wall interference effect as compared with that of a conventional perforated wall.

In summary, the theoretical investigation of bodies of revolution having various contours shows that the elimination of wave reflections from wind tunnel walls with linear characteristics becomes particularly difficult when a system of concentrated expansion waves exists, as, for instance, around three-dimensional models having sharp corners or small radii of curvature. Gradually curved contours can be expected to produce less severe wave reflection problems. In the case of the three-dimensional models investigated, neither conventional nor differential-resistance-type perforated walls can provide the required wall characteristics in those regions where outflow from the test section occurs in spite of the fact that the static pressure in he test section is lower than in the surrounding plenum chamber. However, in these cases, the replacement of the conventional wall with linear characteristics by a suitably perforated differential-resistance-type wall would considerably improve the possibility of obtaining interference-free data from wind tunnel tests.

3. MODEL DISTURBANCE CHARACTERISTICS COMPARED WITH THE CROSS-FLOW CHARACTERISTICS OF COMBINED SLOTTED–PERFORATED TEST-SECTION WALLS

a. Comparison of perforated and slotted walls for testing of three-dimensional models

Perforated Walls.—It was shown in the preceding section that in testing three-dimensional models a perforated wall of either the conventional or the differential-resistance type cannot adequately match the disturbance characteristics of the model in those regions along the wall where the test-section pressure is lower than the plenum chamber pressure, but the flow is directed outward into the plenum chamber. This can be readily recognized when the boundary condition for perforated walls is considered (see Chapter 5):

$$\Delta v_x + K \Delta v_n = 0$$

or:

$$\Delta p/q = K\theta_w$$

where Δp = difference between local static pressure in test-section and plenum chamber pressure
q = dynamic pressure of test-section flow
K = characteristic wall constant
θ_w = local flow direction at wall.

The above equation states that in regions having a negative local static pressure difference, that is, when the pressure in the test section is lower than the plenum chamber pressure, the test-section flow must be directed from the plenum chamber into the test section in order to produce a pressure difference having the proper sign. In the case of the interference-free models, however, the model disturbance characteristics are exactly the reverse of the basic characteristics of a perforated wall. This deficiency occurs not only in supersonic flow where it is generally restricted to relatively small regions of mismatch; in subsonic flow the discrepancies extend widely since the subsonic flow upstream of the maximum model thickness must be directed outward from (and behind the maximum thickness inward into) the wind tunnel. Consequently the perforated wall causes basic distortions in the flow pattern in subsonic flow, as shown for instance, in Ref. 3 and as discussed in detail in Chapter 5. This characteristic is known to be a basic disadvantage of perforated wind tunnel walls in subsonic testing. However, due to the fact that the distortion in subsonic flow is usually small, perforated wind tunnel walls have been used satisfactorily in subsonic tunnels, particularly in tests of complete airplane models.

Longitudinally Slotted Walls.—The boundary conditions occurring with longitudinally slotted walls are basically different from those occurring with the perforated wall. In addition to the pressure change caused by the cross-flow velocity component through the slotted wall (see Chapter 11), an additional pressure difference occurs which is dependent upon the curvature of the flow approaching the slotted wall. The additional term for this pressure change can be expressed in the following form when only linear terms are considered (see Chapter 5):

$$\Delta p/q = \frac{2}{\pi} \log\left(\sin\frac{R\pi}{2}\right) \partial \theta_w / \partial(x/h)$$

where R = open-area ratio of wall
$\partial \theta_w / \partial x$ = flow curvature
h = distances between slot centers.

This equation presents the only pressure-drop term within the validity of the linearized theory for non-viscous flow in the vicinity of a slotted wall. It may be applied to both subsonic and supersonic flow using the same assumptions since the theory for longitudinal slots can be handled much like that for slender bodies at small angles of attack.[4]

The additional pressure difference due to flow curvature near the slotted wall is caused by centrifugal forces which are magnified by the slot effect in the manner discussed in detail in Chapter 5. The significance of the curvature effect upon the pressure-drop characteristics of a wall arises from the fact that, with sufficiently large curvature, this effect may become predominant. When such is the case, a longitudinally slotted wall can support cross flow in a direction opposite to that indicated by the difference between the test-section and the plenum chamber pressure. Since a wall with these characteristics is required for the cancellation of expansion waves emanating from the shoulders of a body of revolution such as a cone-cylinder model, the slotted wall offers some significant advantages.

The additional pressure build-up occurring in curved flow approaching a slotted wall is directly proportional to the distance between slot centers when the open-area ratio is kept constant. Thus to maintain the desired constant value of build-up when a larger number of slots is used, it is necessary to reduce the open-area ratio of the wall at the same rate that the distance between slot centerlines is reduced.

Longitudinally Slotted Walls with Protruding Slats.—Another interesting slotted test-section configuration has been studied theoretically. In this configuration the solid wall portions (slats) protrude into the test section, and the slots are equipped with solid or perforated sidewalls which guide the flow. The centrifugal pressure gradients are amplified even more than in a simple slotted test section, and the total pressure build-up due to flow curvature is increased:

$$\Delta p/q = \frac{\partial \theta_w}{\partial(x/h)} \left[\frac{2}{\pi} \log\left(\sin\frac{R\pi}{2}\right) - \frac{2}{R}\frac{l}{h}\right]$$

where l = height of slot walls.

Since the pressure build-up across a wall of this type is increased by increasing the slat height, the slots in such a wall can be made wider without changing their effectiveness.

It might also be noted that improved pressure build-up may be accomplished by a configuration in which the lateral guides of the slots are made to protrude into the flow. Because such guides can be made adjustable and possibly controllable, it appears possible to develop a wall with characteristics that can be easily varied to conform to the boundary conditions existing along a given wall.

The choice of the wall configuration for effective shock cancellation in a transonic wind tunnel involves a compromise between the pressure disturbance conditions of the model and those of the wall. The additional parameter provided by the slot height in the case of longitudinally slotted walls strengthens the possibility of obtaining a satisfactory compromise and makes this configuration quite attractive; the conventional type of slotted walls previously discussed does not possess this flexibility.

As shown in the preceding discussions, a slotted wall can support outflow from the test section, even when the pressure inside the test section is smaller than the pressure in the plenum chamber. Therefore a perforated test section equipped with longitudinal slots can be expected to eliminate some of the difficulties occurring with both the conventional perforated test-section wall and the conventional longitudinally slotted test-section wall.

b. Various bodies of revolution in test sections equipped with the combined perforated-slotted-type walls

It is possible, of course, to produce a perforated-slotted test section which combines the characteristics of both by covering the openings of a slotted test section with perforated cover sheets. To determine whether or not this configuration had promise, a theoretical investigation was made of the flow around several three-dimensional bodies in test sections of the combined type.

For this study the flow around a cone-cylinder model was calculated for a plane corresponding to the tunnel wall (see Fig. 9.4). To the pressure occurring in free flight was added the pressure which can be supported by a longitudinally slotted wall in curved flow. The resulting disturbance pressure is the pressure left to be supported by the perforated cover plates. Since perforated walls have essentially linear characteristics, a satisfactory match can be obtained if this disturbance pressure is also linear.

The results of the calculation for the 2 per cent blockage cone-cylinder model in a test section having sixteen longitudinal slots and an open-area ratio of 36 per cent are presented in Fig. 9.6. According to the equations given in Section 3a of this chapter, approximately the same results can be obtained with walls having 32 or 64 slots and open-area ratios of 18 and 5 per cent, respectively. For the purpose of these calculations, the regions with abrupt changes of pressure (see Fig. 9.4, points A, B, and C) were smoothed out over a distance corresponding to the distance between the wall slots. It can be seen that the slot effect considerably reduces the previously noted difficulty with respect to the support of negative pressures inside the tunnel in outflow regions. Also the characteristics of the remaining wall pressure disturbances match the characteristics of an ideal perforated wall fairly well over the entire region of the initial compression and expansion waves. In the region behind the expansion wave system, however, significant differences between the desired characteristics and those of a conventional perforated wall still remain. These remaining differences can be reduced considerably by means of properly matched differential-resistance cover plates (see Section 2c of this Chapter).

The effect of the slots on the disturbance flow characteristics of the continuously-curved NACA RM-10 model was also investigated using wall configurations with sixteen and eight slots and an open-area ratio of 5 per cent. The results, presented in Fig. 9.7, indicate that with a suitable com-

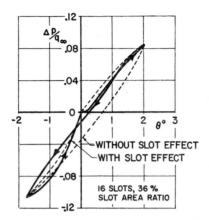

FIG. 9.6. *Flow disturbance along the walls of a test section with and without longitudinal slots for a cone-cylinder model with 2 per cent blockage (20° cone angle) at $M = 1.20$.*[1]

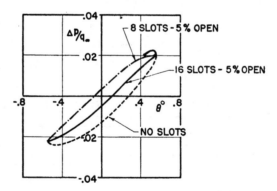

FIG. 9.7. *Flow disturbance along the walls of a test section with and without longitudinal slots for NACA RM-10 Model with 2 per cent blockage at $M = 1.20$.*[1]

bination of perforated and slotted walls the disturbance pressure along the tunnel wall produced by this model can be satisfactorily matched. In order to make the matching perfect it would be necessary either to reduce the total open-area ratio to 1.5 per cent while sixteen slots were retained or to keep the open-area ratio of 5 per cent and reduce the number of slots to eleven.

In summary, the theoretical investigation of bodies of revolution having various contours shows that combined perforated-slotted walls offer promise for the testing of smoothly-curved models. However, some shock reflection difficulties remain in testing models which produce regions of strong compression waves since a large portion of the test section must be solid.

4. WIND TUNNEL TEST RESULTS OBTAINED FOR CONE–CYLINDER MODELS IN VARIOUS PARTIALLY OPEN TEST SECTIONS AT A MACH NUMBER OF 1.20

a. Conventional perforated walls with different open-area ratios

A systematic series of cone-cylinder models having different cone angles and blockage ratios was investigated over the Mach number range between 0.75 and 1.25 in the transonic model tunnel of the AEDC using various perforated test sections.[5] In these tests, significant wall interaction difficulties were observed in the supersonic speed range which could be traced back to the mismatch between the wall characteristics of a conventional perforated wall and the characteristics required for effective cancellation of compression and expansion waves emanating from the model.

Pressure Distribution at Model Surface.—Figure 9.8 presents the experimental pressure distribution for a 2 per cent blockage cone-cylinder model in a perforated test section having a 12 per cent open area. Hole diameter and

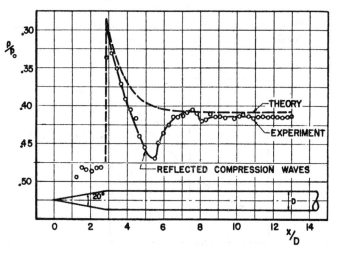

FIG. 9.8. *Pressure distribution along surface of cone-cylinder model in perforated test sections with 12 per cent open-area ratio, 0° wall setting, 2 per cent blockage, at $M = 1.20$.*[1]

wall thickness were each $\frac{1}{16}$ in., and the wall setting was parallel to the centerline of the test section. It can be seen that the compression waves originating from the conical flow field of the model were reflected as compression waves. Consideration of the initial and the reflected wave systems and comparison with theory (see Sections 2 and 3 of this chapter) indicate that the wall with 12 per cent open area is too closed.

Perforated test sections having 22 and 33 per cent open areas and the same hole size, wall thickness, and parallel wall setting, were therefore investigated. Figure 9.9 shows that the 22 per cent open wall very nearly eliminated wave reflections in the compression wave region ($x/D = 3.6$–5.6); the 33 per cent open wall was definitely too open since the compression waves were reflected as expansion waves in this region. However, further analysis of Fig. 9.9 indicates that in the case of the 22 and 33 per cent open

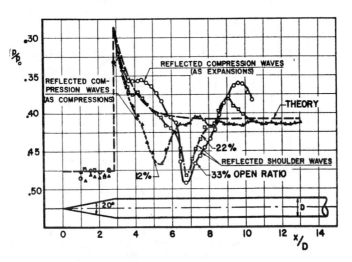

FIG. 9.9. *Pressure distribution along surface of cone-cylinder model in perforated test section with different open-area ratios, $0°$ wall setting, 2 per cent blockage at $M = 1.20$.*[1]

walls, a much more serious disturbance region occurs behind the compression wave region of the model. Reflection waves from this region can be expected to impinge on the model approximately 5.6 diameters behind the nose. These disturbances can be traced to the expansion waves originating from the shoulder of the model and are even greater in magnitude than the disturbances observed in the compression wave region. The nature of these expansion wave disturbances implies that the walls with 22 and 33 per cent open area reflect the expansion waves as compression waves or, as the magnitude of the observed compression disturbances indicates, that a concentrated inflow into the test section produces a flow inclination considerably larger than that produced by interference-free flow.

The test results obtained with a cone-cylinder model in perforated test sections having different open-area ratios at a parallel wall setting show that it is possible to select a wall geometry which satisfies the condition of no-reflection in the compression wave region. It is also possible to select a wall geometry which cancels the expansion wave reflections. However, none of the conventional perforated walls will satisfy the no-reflection requirement of wind tunnel testing in both the compression and expansion regions.

Basically the same conclusion results from the experiments with perforated walls having 12 and 33 per cent open areas when the walls are diverged to $+30'$ and $+40'$, respectively (see Fig. 9.10).

It must be concluded, therefore, that conventional perforated walls are not capable of producing satisfactory reflection-free data with the selected cone-cylinder model.

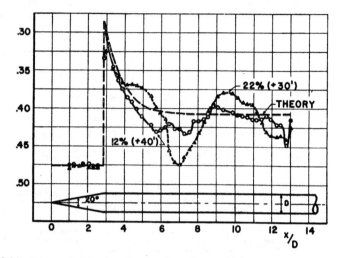

FIG. 9.10. *Pressure distribution along surface of cone-cylinder model in perforated test section with different open-area ratios, 30' and 40' diverged wall setting, 2 per cent blockage, at $M = 1.20$.*[1]

Survey of Reflected Wave System.—To gain specific insight into the origin of the observed disturbances, the flow field was surveyed at various distances from the model; in this way the disturbances were traced to their origin. In Fig. 9.11, some typical survey results are presented for the case of a 2 per cent blockage cone-cylinder model in a perforated test section having 22 per cent open walls. The initial shock wave produced only a slight disturbance on the surface of the model (point 1b). Furthermore, it can be clearly seen that the entire compression wave field between the initial shock wave and the expansion wave system from the shoulder was properly absorbed by the walls since the experimental and theoretical curves coincide very closely in this region. The major difficulty in testing a cone-cylinder model in the selected perforated test section, therefore, did not arise because of unsatisfactory absorption of the compression wave system but because of the difficulty of eliminating reflections produced by a strong expansion wave system originating at the shoulder of the model.

Difficulties similar to those described in the preceding section also occurred in the testing of a smaller $\frac{1}{2}$ per cent blockage cone-cylinder model in a 33 per cent open perforated test section (see Fig. 9.12). Because of the large open-area ratio of the wall, the compression wave system was reflected as an expansion wave system, as is proved by the pressure distribution between points 1b and 2b at the surface of the model. Also, in the region influenced by reflections of the expansion wave, that is, between points 2b and 3b, very significant disturbances occurred which were similar to reflections from walls that are too open. As before, these disturbances could be traced back to an excessively large inflow in the expansion wave region along the tunnel wall.

The surveys presented above are typical and confirm the conclusions drawn in the previous section, namely, that conventional perforated walls are not capable of simultaneously eliminating the reflections of the compression waves and the expansion waves produced in the extreme case of a three-dimensional cone-cylinder model.

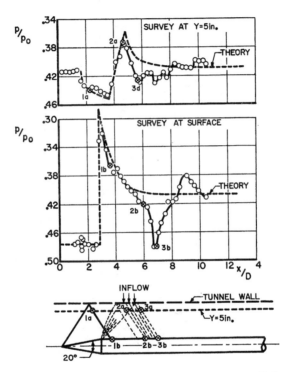

FIG. 9.11. *Wave system and pressure survey results for a cone-cylinder model with 2 per cent blockage in conventional perforated test sections with 22 per cent open-area ratio at* $M = 1.20$.[1]

b. Perforated walls of the differential-resistance type

Characteristics of a Differential-Resistance Wall.—The guiding idea utilized for the development of a differential-resistance wall is as follows. It is obvious that the resistance to inflow into a test section can be increased if the flow can be directed against the main test-section flow. The inflow will then have to overcome at least a part of the dynamic pressure of that flow. This end can be obtained rather simply with a sufficiently thick perforated wall by inclining the holes obliquely against the main flow direction (see Fig. 9.13).

181

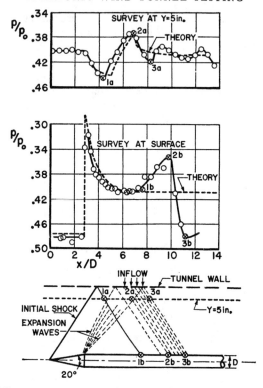

Fig. 9.12. *Wave system and pressure survey results for a cone-cylinder model with $\frac{1}{2}$ per cent blockage in conventional perforated test sections with 33 per cent open-area ratio at $M = 1.20$.*[1]

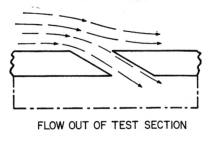

FLOW OUT OF TEST SECTION

FLOW INTO TEST SECTION

Fig. 9.13. *Streamline pattern for inflow and outflow through a wall with inclined holes.*

Some typical test results* for holes inclined at angles of 60° are shown in Fig. 9.14 (Ref. 1). When such a wall has a 6 per cent open-area ratio, the outflow region matches the characteristics of a conventional wall with 22 per cent open area, and the resistance of the inflow into the test section is considerably increased, as intended.

Also, in the case of a given partially open wall, the reference point for the empty test section can be shifted along the characteristic cross-flow curve by changing the wall alignment from a parallel to a converged or diverged setting. It is with respect to this reference point for the empty test section that the pressure and flow direction disturbances produced by the models must be related (see points A and B in Figs. 9.14a and b). The choice

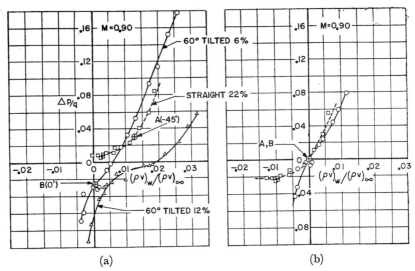

FIG. 9.14. *Cross-flow characteristics of conventional perforated and differential-resistance-type walls at $M = 0.90$.*[1]

of reference points A and B indicated that the differential-resistance wall should be set at 0°; the conventional wall should be set at a convergence angle of 45'. In considering the model disturbances, therefore, the characteristics of these two walls must be compared on the basis of Fig. 9.14b in which the two reference points are brought into coincidence.

The differential-resistance-type wall matches much more closely the characteristics required for wave cancellation in the case of cone-cylinder models in both compression and expansion regions. However, even such a wall cannot completely meet the theoretical cross-flow characteristics of a cone-cylinder model since it cannot support a lower pressure in the test section in the outflow region.

* The experimental results shown in Fig. 9.14 were obtained at a Mach number of 0.90 instead of at the desired Mach number of 1.20. At the time of this test series, the AEDC transonic model tunnel was not equipped to obtain the necessary data in the supersonic speed range.

Cone-Cylinder Model Test Results.—A series of investigations was conducted to verify the improved testing characteristics of differential-resistance-type wind tunnel walls having 60° inclined holes and a 6 per cent nominal open-area ratio. Some typical results[1,5] for a 2 per cent blockage cone-cylinder model obtained in the AEDC transonic wind tunnel with this new test section are presented in Fig. 9.15. The experimental test data follow the

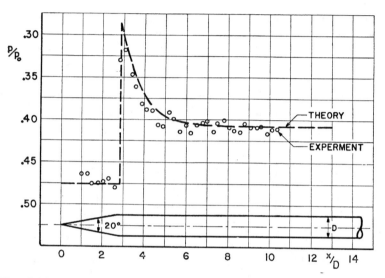

Fig. 9.15. *Pressure distribution along surface of cone-cylinder model in perforated test section with hole axis inclined 60°, 6 per cent open-area ratio, 2 per cent blockage, and 0° wall setting, at $M = 1.20$.*[1]

theoretical curve closely; differences are no greater than $1-1\frac{1}{2}$ per cent of the total pressure and may be explained to some extent by the non-uniformity of the empty test-section flow. The improvement in the test results with the differential-resistance-type wall becomes obvious when the results are compared with data obtained using the conventional perforated test section (see Fig. 9.8). Reflected waves are considerably reduced, not only in the compression wave region, but also in the expansion wave region.

Although smoothly curved models should be less critical than cone-cylinder contours, further investigation is needed to determine whether the same good results obtained with the new type wall can also be obtained with models which do not have the extreme shape of the cone cylinder.

c. Longitudinally slotted test sections with and without perforated cover plates

To obtain insight into the wall interference characteristics of typical test sections having longitudinal slots, some test results obtained with the same cone-cylinder model (20° cone angle) in various slotted test sections will be discussed in the following sections.

Conventional Tests.—The tests were conducted at the AEDC in a typical transonic model tunnel test section having sixteen uncovered longitudinal slots and 11 per cent open area. Some results of the cone-cylinder investigations[1] with this test section are presented in Fig. 9.16. The comparison of the experimental pressure distributions along the surface with theoretical free-flight distributions indicates that the agreement between experiment and theory is even poorer than in the case of the corresponding experiments

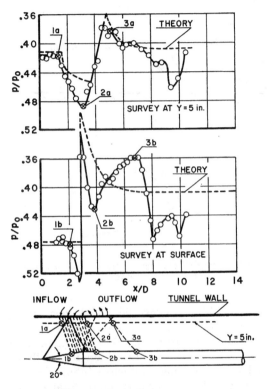

FIG. 9.16. *Wave systems and pressure survey results for a cone-cylinder model with 2 per cent blockage in a longitudinally slotted test section (sixteen slots in parallel walls with 11 per cent open-area ratio at $M = 1.20$.*[1]

made using conventional perforated test sections. It is particularly noteworthy that a large discrepancy between experiment and theory begins at a point on the surface of the cone (point 1b) which is *considerably ahead of the point where the initial conical shock would be expected to be reflected against the surface of the model.*

The disturbance wave pattern which produced these discrepancies upstream of the initial conical shock shows that a secondary flow occurred in the slots (discussed by Allen and Spiegel in Ref. 6). This secondary flow field can be attributed to a layer of subsonic flow close to the walls and in the slots. Although in supersonic flow a sudden pressure rise may result from impinging shock waves, such a sudden rise is not possible in subsonic flow. Instead, particularly in the subsonic flow region, the pressure increase caused by impinging shock waves is propagated upstream in the flow. Consequently a new boundary is established for the supersonic flow, and a secondary wave system occurs (see Fig. 9.17). Besides the disturbances

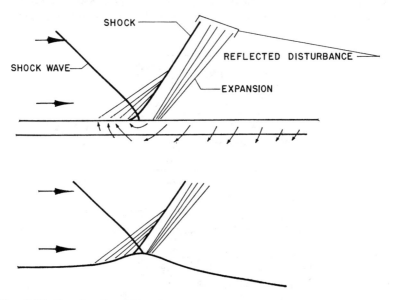

Fig. 9.17. *Secondary flow field and wave system for shock-wave reflection at a slotted wall.*[16]

produced directly by this secondary flow field, very strong expansion and compression waves occur which impinge on the model further downstream (Fig. 9.16) and result in serious deviations from the theoretical flow field.

It can be concluded that a conventional longitudinally slotted test section is even less capable of producing satisfactory interference-free results for the selected test conditions than a conventional perforated test section.

Slots with Perforated Cover Plates.—One major difficulty in testing cone-cylinder models in slotted test sections was found to be connected with the fact that a strong interference effect of the walls occurs upstream of the model-initiated shock waves. To find a remedy for this difficulty, further tests were carried out in which perforated cover sheets (such as those employed in the WADC 10 ft wind tunnel[7]) were placed over the longitudinal slots. These cover plates are similar in principle to the corrugated slot plates employed in the test by Allen and Spiegel.[6] Their purpose is to reduce the secondary flow in the slots and to thereby prevent the upstream propagation of disturbances. Since it has been verified that perforated walls reduce the axial momentum of the flow on the outside of the walls to a negligibly small magnitude,[8] the cover plate may be expected to operate similarly.

Wind tunnel tests of the 2 per cent blockage cone-cylinder model were conducted in a slotted test section having sixteen longitudinal slots which were covered by perforated sheets with an open area of approximately 33 per cent. Hole diameter and thickness of the perforated sheets were each $\frac{1}{16}$ in. The area ratio of the slots was 30 per cent, which, with the 33 per cent open cover sheets, produced a geometric open-area ratio of 10 per cent. This open-area ratio is approximately equal to the open-area ratio of the conventional slotted wall discussed in the preceding sections (see Fig. 9.16). Typical pressure distributions from these tests are presented in Fig. 9.18. No serious disturbance due to secondary slot flow can be detected. The pressure distribution coincides well with the theoretical pressure distribution upstream of the point where the initial conical shock reflection strikes the model. However, the test results indicate that in the compression wave reflection region, $x/D = 3.6$–5.6, the selected wind tunnel wall is too closed; the compression waves produced by the model are reflected as compression waves. In the region behind the expansion wave reflection, $x/D > 6.5$, the experimental pressure distribution coincides satisfactorily with the theoretical distribution.

These tests verified the fact that a slotted test section with perforated cover sheets over the slots is capable of effectively eliminating the secondary flow disturbances previously discussed. However, the walls must be opened more than for the conventional slotted configuration in order to eliminate reflections in the compression wave region. To produce a test section having the proper open area for successful absorption of compression waves, it was necessary to increase the slot width to as much as 53 per cent of the wall area and to diverge two walls of the test section to $+30'$.

The preceding results of cone-cylinder tests in slotted test sections show that, as was true in the case of conventional perforated test sections, wall configurations consisting of longitudinal slots with perforated cover plates can be devised which will eliminate the wall interference in either the compression or the expansion wave region. However, no satisfactory slotted-wall configuration having *conventional perforated cover plates* over the slots has been found which can *simultaneously* eliminate the wave reflections in both the compression and the expansion wave regions.

Slotted Test Sections with Perforated Cover Plates of the Differential-Resistance Type.—Differential-resistance-type cover plates over longitudinal slots should be beneficial because differential resistance to the cross flow over longitudinally slotted walls is required to eliminate the wall interference in both the compression and expansion wave regions. In order to verify this expectation, a series of tests was conducted in the AEDC transonic model tunnel using a wall configuration having twelve slots (open-area ratio of 20 per cent without cover plates) and cover plates with holes inclined at 60° angles and an open-area ratio of 22 per cent. With such a wall configuration, both a noticeable slot flow curvature effect and a strong differential-resistance wall effect should be produced. Typical results of this test series are presented in Fig. 9.19. The experimental results coincide much better with theory than do the results obtained using conventional slots without cover plates. It should be noted that the test Mach number was somewhat smaller than

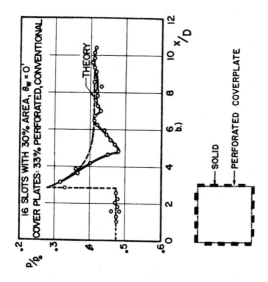
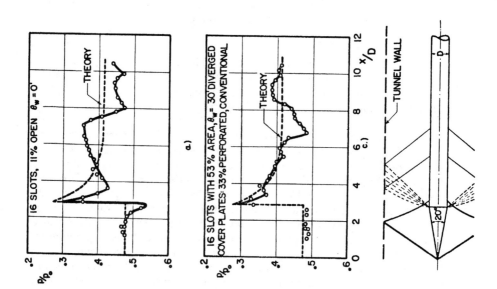

Fig. 9.18. *Pressure distribution along surface of cone-cylinder model with 2 per cent blockage in various slotted test sections with and without perforated cover plates at* $M = 1.20$.[1]

1.20, as is obvious from the generally high pressures along the model surface (see Fig. 9.19).

The test with the slotted walls and cover plates of the differential-resistance type indicates that it is possible to obtain fairly reliable test results for cone-cylinder models in the Mach number range near 1.20. However, in the selected case of the cone-cylinder model, the concentrated shock waves were of small intensity, that is, the initial conical shock produced at the tunnel wall was no more than one-fifth of the entire pressure rise produced by the model. With other models which produce strong shock

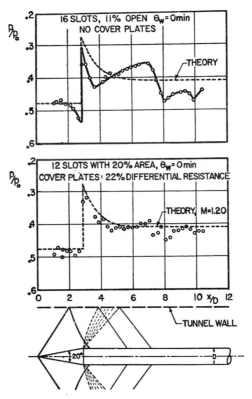

FIG. 9.19. *Pressure distribution along surface of cone-cylinder model in conventional slotted wall test sections and in combination slotted test sections with differential-resistance-type cover plates at $M = 1.20$.*[1]

waves the conditions may be more severe. In these cases reflections from the large area of solid wall between the slots will meet the model before the intensity is sufficiently reduced by the counteracting effect of the expansion waves originating at the slots.

A comparison between the results obtained with cone-cylinder models in perforated test sections and in longitudinally slotted sections with perforated cover plates indicates that more reliable test results were obtained near Mach number 1.20 when perforated walls of the differential-resistance type were employed.

5. RESULTS OF WIND TUNNEL TESTING WITH THREE-DIMENSIONAL MODELS IN PERFORATED TEST SECTIONS AT VARIOUS MACH NUMBERS AND BLOCKAGE RATIOS

In the preceding section the emphasis was placed on test results obtained with 2 per cent blockage cone-cylinder models in perforated test sections at Mach number 1.20. Some additional typical test results obtained using cone-cylinder and cone-ogive-cylinder models in perforated test sections with varying blockage and open-area ratios of the wall will be presented and discussed in this section. In all these tests the wall geometry was selected according to the differential-resistance principle. *In particular*, holes with 60° inclination and a ratio of hole size to wall thickness of 1.0 were used to produce the differential-resistance character.

a. Cone-cylinder model with 2 per cent blockage at subsonic and supersonic Mach numbers

The same 2 per cent blockage cone-cylinder model discussed in the preceding section was utilized to obtain test results over the entire transonic Mach number range, that is, from high subsonic to supersonic Mach numbers of 1.4. The test results were first obtained in the AEDC transonic model tunnel (12 in. test-section height) in a test section having differential-resistance walls with a 6 per cent open-area ratio. Then the same models were transferred to the full-scale 16 ft transonic wind tunnel of the AEDC, and again test data were obtained over the identical Mach number range.

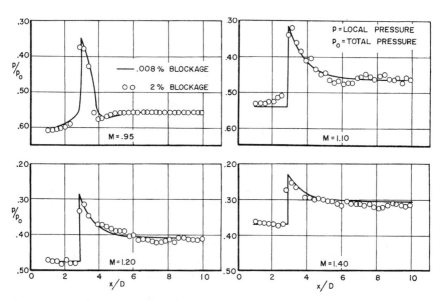

Fig. 9.20. *Pressure distribution of cone-cylinder model in perforated test section for various Mach numbers (6 per cent open-area ratio, 60° inclined holes, $d = \frac{1}{4}$ in., $t = \frac{1}{4}$ in., parallel wall setting) (D = cylinder diameter, x = axial distance behind apex) (unpublished data by Gardenier and Estabrooks obtained in the AEDC transonic model tunnel).*

Since in the latter case, the blockage ratio dropped to the extremely small value of 0.008 per cent, the data resulting can be considered to be practically interference-free.

Pressure distribution curves for the 2 per cent blockage model are compared with the interference-free distributions shown in Fig. 9.20 for Mach numbers 0.95, 1.1, 1.2, and 1.4. The Mach number range around Mach number one is not included in this figure. Agreement between the data from the 2 per cent blockage model and from the interference-free test of the 20° cone-cylinder model is satisfactory. Though the influence of wave reflections can be recognized at all supersonic Mach numbers, their magnitude is reasonably small.

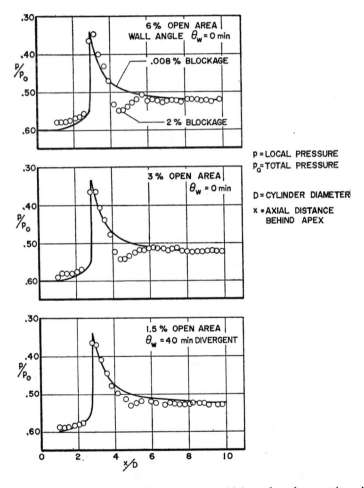

Fig. 9.21. *Pressure distribution of cone-cylinder model in perforated test section with various open-area ratios for $M = 1.00$ (60° inclined holes, $d = \frac{1}{4}$ in., $t = \frac{1}{4}$ in.) (unpublished data by Gardenier and Estabrooks obtained in the AEDC transonic model tunnel).*

b. Cone-cylinder model with 2 per cent blockage in differential-resistance test sections with different open-area ratios near Mach number one

Some test results obtained with the 2 per cent blockage cone-cylinder model in the test section with 6 per cent open walls discussed previously are presented in Figs. 9.21 and 9.22 at the most sensitive Mach numbers of 1.00 and 1.05. In these cases the test results exhibit large disturbances.

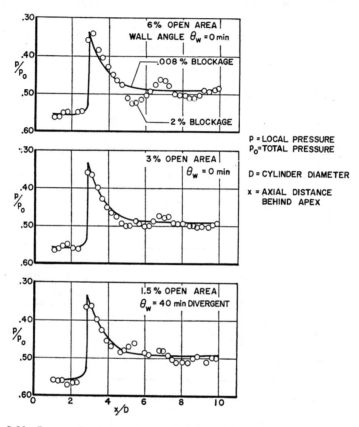

FIG. 9.22. *Pressure distribution of cone-cylinder model in perforated test section with various open-area ratios for $M = 1.05$ ($60°$ inclined holes, $d = \frac{1}{4}$ in., $t = \frac{1}{4}$ in.) (unpublished data by Gardenier and Estabrooks obtained in the AEDC transonic model tunnel).*

As pointed out in Chapter 8, it can be expected that in the Mach number range around one the open-area ratio of perforated walls must be reduced in order to accomplish suitable matching between shock cross-flow characteristics on the one hand, and wall cross-flow characteristics on the other hand. In order to check this conclusion the same model was investigated in perforated test sections having the same hole geometry but with the open-area ratio reduced from 6 per cent to 3 and 1.5 per cent, respectively. Typical results for the smaller open-area ratios are also presented in Figs. 9.21 and 9.22. As expected, the less open walls produced smaller wall interferences in the Mach number range under consideration than did the 6 per cent open wall. At Mach number one, a wall with 1.5 per cent open-area ratio and a wall setting of 40′ divergence gave the least interference; at a Mach number of 1.05, the open-area ratio of the wall should be increased to approximately 3 per cent with the wall setting parallel.

In view of the test results over the entire transonic Mach number range, it is apparent that a 6 per cent differential-resistance wall would be capable of producing reasonably interference-free data in the subsonic Mach number range up to 0.95. As Mach number one is approached, the wall should be closed to approximately a 1.5 per cent or less open area. When

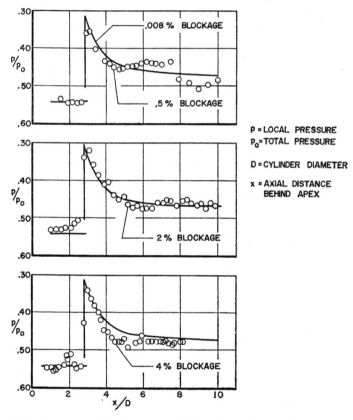

FIG. 9.23. *Pressure distribution of cone-cylinder models of different blockage ratios in perforated test sections at $M = 1.10$ (6 per cent open-area ratio, $60°$ inclined holes, $d = \frac{1}{4}$ in., $t = \frac{1}{4}$ in.) (unpublished data by Gardenier and Estabrooks obtained in the AEDC transonic model tunnel).*

the Mach number is further increased to Mach 1.05, the wall should be opened again to 3 per cent open area, and the opening should be increased up to 6 per cent for Mach numbers of 1.1 and above. It should be remembered, however, that such variation of the open-area ratio of a perforated wall would be required mainly when the emphasis of the wind tunnel testing was placed on obtaining correct *pressure distributions* over the model. If the emphasis is placed on *force or moment measurements* of complete models, reasonably good data even in the most critical range of Mach number one and above, can frequently be obtained with walls having the same geometry because a compensating effect is produced over the model surface. Some of those measurements are given in Section 5e of this chapter.

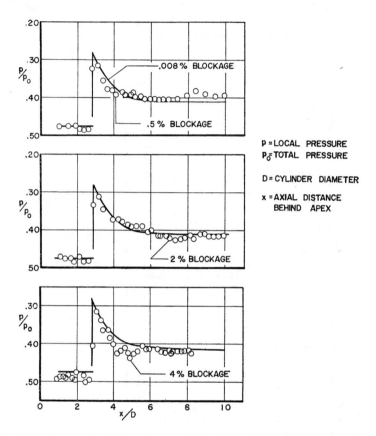

FIG. 9.24. *Pressure distribution of cone-cylinder models of different blockage ratios in perforated test sections at $M = 1.20$ (6 per cent open-area ratio, $60°$ inclined holes, $d = \frac{1}{4}$ in., $t = \frac{1}{4}$ in.) (unpublished data by Gardenier and Estabrooks obtained in the AEDC transonic model tunnel).*

c. Cone-cylinder models with different blockage ratios

Cone-cylinder models with the same $20°$ cone angle and blockage ratios of $\frac{1}{2}$, 2, and 4 per cent were tested in the AEDC transonic model tunnel. The same wall geometry was utilized for these tests, that is, a 6 per cent open wall with $60°$ inclined holes. The test results presented in Figs. 9.23 and 9.24 compare the data for these models with those for the 0.008 per cent blockage model which can be considered to be practically interference-free. It is remarkable that even the large 4 per cent blockage model still yields reasonably good pressure distribution data. This is more surprising in view of the fact that the test Mach numbers for the 4 per cent blockage model were obviously somewhat lower than the nominal values of Mach numbers 1.1 and 1.20.

In contrast the data for the ½ per cent blockage model show noticeable deviations from the interference-free data. This surprising result is not explained completely and requires further detailed probing before the cause of the phenomena can be determined. It might be that the selected wall geometry is most suitable for testing of relatively large models which produce strong disturbance wave systems. For the weaker waves which are produced by the small ½ per cent blockage model, a wall of a different geometry might be more advantageous. In support of this point, it should also be remembered that the wall utilized in the test series was developed by using a 2 per cent blockage model. Naturally therefore the 2 per cent blockage model produced the best correlation with interefrence-free data in this test section.

d. Pressure distribution measurements for cone-ogive-cylinder models in perforated test sections

The cone-cylinder models utilized in the experimental investigation discussed previously develop a strong concentrated expansion wave system around their shoulders. This concentrated expansion wave system causes many difficulties in matching model and wall characteristics. In some phases of the AEDC testing, therefore, the shoulder of these models was rounded, and the resulting cone-ogive-cylinder models were investigated through the same Mach number range. The cone angle of these models was again 20° total, and the radius of the ogive was ten times the diameter of the model cylinder.

Models of different sizes were built which produced 2 per cent blockage in the small transonic model tunnel and ½ per cent blockage in the full-scale transonic wind tunnel of the AEDC. Some unpublished data from this test series, conducted by Gardenier and Estabrooks, are shown in Fig. 9.25. In this case, as in the case of the cone-cylinder model tests, the test results up to Mach number 0.95 and above 1.2 indicate satisfactorily small wall interference effects. In the low supersonic Mach number range, however, definite wall interference effects are noticeable, particularly at Mach numbers 1.0, 1.05, and 1.1. Again, it should be possible to develop a wall geometry which produces small wall interference in the lower supersonic Mach number range, as was the case for the cone-cylinder models. This possibility has not been verified, however, since too few systematic measurements of continuously-curved bodies of revolution have been conducted in transonic wind tunnels to permit a comparison with interference-free data.

e. Force measurements on cone-cylinder models with different blockages

As mentioned before, the pressure distribution measurements along the surface of models reveal most clearly the interference effects produced by reflections from the wind tunnel wall. Since force measurements produce an integrated effect over the entire surface of the models, many of the pressure

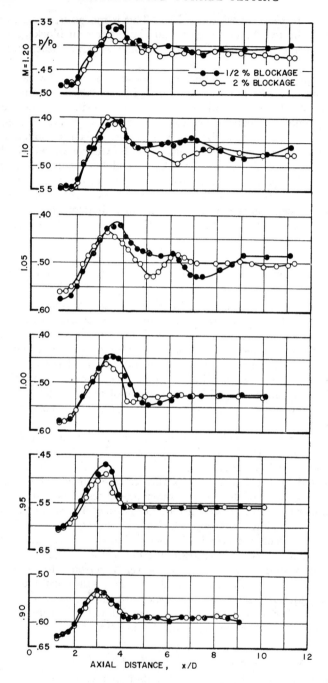

Fig. 9.25. *Pressure distribution for cone-ogive-cylinder models in perforated test section (6 per cent open-area ratio, $60°$ inclined holes, $d = \frac{1}{4}$ in., $t = \frac{1}{4}$ in.) (unpublished data by Gardenier and Estabrooks obtained in the AEDC transonic model tunnel).*

disturbances will disappear because of mutual compensation of the various disturbance regions. Consequently it can be expected that, in many tests, force and moment measurements will produce reliable data, even when the pressure distribution measurements reveal noticeable interference effects.

As an example of this phenomena, some normal force and moment coefficient test results for cone-cylinder models in the Mach number range from 0.6 to 1.6 are shown for a model with 1 per cent blockage and a model with 0.01 per cent blockage. The 1 per cent blockage tests were conducted in the AEDC transonic model tunnel; the 0.01 per cent blockage tests were conducted with another cone-cylinder model in the full-scale transonic wind tunnel of the AEDC. The walls in both cases were of the differential resistance type, 6 per cent open, with inclined holes of 60°.

The slopes of the normal force and moment coefficients of both models, determined for the angle-of-attack range near $\alpha = 0°$ are shown in Fig. 9.26.

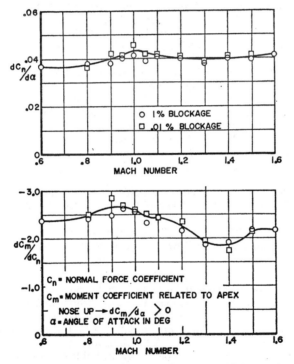

FIG. 9.26. *Force measurements of cone-cylinder model with $L/D = 10$ in. in perforated test section around angle of attack $= 0°$ (6 per cent open walls, 60° inclined holes, $d = \frac{1}{4}$ in., $t = \frac{1}{4}$ in.) (unpublished data by Gardenier and Estabrooks obtained in the AEDC transonic model tunnel).*

The correlation between the two sets of data is very satisfactory considering the fact that two different models in two different wind tunnels were used for these tests.

In summarizing the test results obtained from various models of axial symmetry in transonic test sections, it can be concluded that reasonably accurate pressure distributions, and, in particular, force and moment data, can be obtained with 1 per cent blockage models in transonic test sections through the entire subsonic, sonic, and supersonic Mach number range with the desired differential-resistance-type walls. However, when precise pressure distribution measurements are required in the immediate vicinity of Mach number one and slightly above, it will be necessary to study the interference flow pattern more closely. Additional preliminary tests with walls of a small open-area ratio and possibly different wall geometry might be required in order to cover this range satisfactorily (see also Chapter 10).

REFERENCES

[1] GOETHERT, B. H. "Physical Aspects of Three-dimensional Wave Reflections in Transonic Wind Tunnels at Mach Number 1.20 (Perforated, Slotted and Combined Slotted–Perforated Walls)." AEDC–TR–55–45, March, 1956.

[2] GOETHERT, B. H. "Physical Aspects of Wind Tunnel Wall Interference at Low Supersonic Mach Numbers." Paper I, 129, IX International Congress of Applied Mechanics, Brussels, September 5–13, 1956.

[3] GOETHERT, B. H. "Flow Establishment and Wall Interference in Transonic Wind Tunnels." AEDC–TR–54–44, June, 1954.

[4] JONES, R. T. "Properties of Low-Aspect-Ratio Pointed Wings at Speeds Below and Above the Speed of Sound." NACA TN–1032, March, 1956.

[5] GRAY, J. DON and GARDENIER, HUGH E. "Experimental and Theoretical Studies of Three-dimensional Wave Reflection in Transonic Test Sections, Part I: Wind-Tunnel Tests on Wall Interference of Axisymmetric Bodies at Transonic Mach Numbers." AEDC–TN–55–42, March, 1956.

[6] ALLEN, H. J. and SPIEGEL, J. M. "Transonic Wind Tunnel Development at the NACA." S.M.F. Fund Paper No. FF–12, Institute of Aeronautical Sciences publication, 1954.

[7] GOETHERT, B. H. "Development of the New Test Sections with Movable Side Walls of the Wright 10 ft Wind Tunnel Part I." WADC TR 52–296, November, 1952.

[8] CHEW, WILLIAM L. "Total Head Profiles Near the Plenum and Test Section Surfaces of Perforated Walls." AEDC TN–55–59, December, 1955.

BIBLIOGRAPHY

DuBOSE, H. C. "Experimental and Theoretical Studies on Three-dimensional Wave Reflection in Transonic Test Sections. Part II: Theoretical Investigation of the Supersonic Flow Field about a Two-dimensional Body and Several Three-dimensional Bodies at Zero Angle of Attack." AEDC–TN–55–43, March, 1956.

GLOVER, LOUIS S. and VOLZ, WILLIAM C. "Investigation of a Cascade-Slotted Wind Tunnel Test Section. Part I: Effect of a Single Airfoil Row on Wave Reflections." DTMB Aero Report 911 Part I, Washington, 1956.

GOETHERT, B. H. "Properties of Test Section Walls with Longitudinal Slots in Curved Flow for Subsonic and Supersonic Velocities (Theoretical Investigations)." AEDC–TN–55–56, August, 1957.

HALLISSY, JOSEPH M., JR. "Pressure Measurements on a Body of Revolution in the Langley 16 ft Transonic Tunnel and a Comparison with Free-Fall Data." NACA RM L51L07a, March, 1952.

HAMILTON, C. V., PARKER, C. H. and RAMM, H. "Experimental Determination of Auxiliary Suction Requirements in the PWT Supersonic Model Tunnel at Mach Numbers 2.0 and 2.6." AEDC–TN–57–2, March, 1957.

LAURMANN, J. A. and LUKASIEWICZ, J. "Development of a Transonic Slotted Working Section in the NAE 30 in. × 16 in. Wind Tunnel." National Aeronautical Establishment, Canada LR–178, August, 1956.

McLAFFERTY, G. H., JR. "Notes on the Cancellation of Shock and Expansion Waves in a Perforated-Wall Tunnel at Supersonic Speeds." UAC Report No. M–13644–2, August 11, 1954.

RITCHIE, V. S. and PEARSON, A. O. "Calibration of the Slotted Test Section of the Langley 8 ft Transonic Tunnel and Preliminary Experimental Investigation of Boundary-Reflected Disturbances." NACA RM L51K14, July 7, 1952.

SOLOMON, GEORGE E. "Transonic Flow Past Cone Cylinders." NACA TN 3213, September, 1954.

SPIEGEL, J. M., TUNNELL, P. J. and WILSON, W. S. "Measurements of the Effects of Wall Outflow and Porosity on Wave Attenuation in a Transonic Wind Tunnel with Perforated Walls." NACA–TN–4360, August, 1958.

STAFF OF THE COMPUTING SECTION, Center of Analysis, under the Direction of Zdenek Kopal, Massachusetts Institute of Technology, Dept. of Electrical Engineering. "Tables of Supersonic Flow of Air Around Cones." Cambridge, Mass., MIT Technical Report No. 1, 1947.

THOMPSON, JIM ROGERS. "Measurements of the Drag and Pressure Distributions on a Body of Revolution Throughout Transition from Subsonic to Supersonic Speeds." NACA RM L9J27, January 16, 1950.

WILDER, JOHN G., JR. "An Experimental Investigation of the Perforated Wall Transonic Wind Tunnel, Phase II." CAL Report No. AD–706–A–6, January, 1952.

CHAPTER 10

TYPICAL TEST RESULTS FROM COMPLETE AIRPLANE MODELS IN TRANSONIC WIND TUNNELS

The preceding chapter discusses methods for transonic testing of individual airplane components in different types of transonic test sections. However, reliable testing to determine the aerodynamic characteristics of new aircraft is possible only when the conclusions from suitable test configurations, arrived at on the basis of individual component tests, also hold true for complete airplane tests. Therefore the critical evaluation of complete airplane test results is the final check of the testing possibilities of transonic wind tunnels.

1. INFLUENCE OF FLOW NON-UNIFORMITIES

In Chapter 9, it was demonstrated that force and moment measurements for bodies of revolution frequently produce satisfactory results, even when the pressure distribution along the surface of the models indicates considerable disturbance due to incomplete cancellation of wave reflections or to other causes (see Fig. 9.26). This result can be readily understood since forces and moments of complete models are produced by the integrated local pressures along the model surface; there is usually therefore a compensating effect to such disturbances.

In some transonic test sections, however, flow non-uniformities occur which are not produced by the test model itself. They already exist in the empty test section. Such non-uniformities may be caused by wall irregularities, by incorrect distribution of the open area of slotted and perforated walls, or by boundary layer effects.

A brief systematic study was conducted by the NACA to determine experimentally to what extent a typical disturbance in an otherwise satisfactorily uniform test-section flow influences the aerodynamic characteristics of a complete airplane model.[1] The tests were performed in the NACA 8 ft wind tunnel where a circular test section with solid walls was installed. The Mach number distribution along an axial line at a distance 1 in. from the centerline of the tunnel is shown in Fig. 10.1. It was generally uniform except for a pronounced dip which reduced the Mach number locally from the average value of Mach 1.20 to approximately 1.15. This deficiency of $\Delta M = 0.05$ extended over a range of ± 1 in. radial distance from the tunnel centerline and decayed gradually at larger distances. For example, at a distance of 7 in. from the centerline the disturbance was reduced to $\Delta M = 0.022$. Also the Mach number dip extended obliquely out from the center-

TYPICAL TEST RESULTS FROM COMPLETE AIRPLANE MODELS

line along approximately the Mach 1.20 lines. No detailed measurements of the flow inclination were made; however, it may be assumed that flow inclination disturbances were small.

A complete airplane model with a span of approximately 18 in. was placed in the test section at different axial locations. During one test series

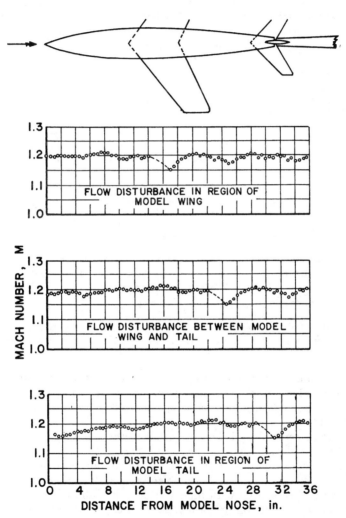

FIG. 10.1. *Mach number distribution for NACA 8 ft wind tunnel 1 in. from tunnel centerline at three different axial locations.*[1]

the main wing was exposed to the Mach number disturbance. Then the model was positioned further forward so that the disturbance was first positioned between wing and tail and, finally, in the tail area (see Fig. 10.1). At the three locations indicated, the lift, moment and drag coefficients of the complete model were measured (Fig. 10.2). It is apparent from the test results that a flow non-uniformity with a Mach number decrement of $\Delta M = 0.05$ need not produce significant irregularities in the model aerodynamic parameters greater than the accuracy of normal wind tunnel testing. This result is particularly noteworthy in view of the fact that the disturbance extended outward along the Mach 1.20 lines, and in addition, the aerodynamic centerline of the wing was swept back at an angle which also corresponded roughly to the Mach lines. Because of this coincidence maximum interference flow disturbance and wing characteristics would be expected to occur in this Mach number range.

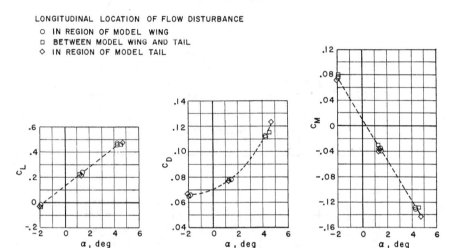

Fig. 10.2. *Lift, drag and pitching moment coefficients of an airplane model for three positions of flow disturbances along the length of the model for $M = 1.2$.*[2]

The conclusion may therefore be drawn that flow non-uniformities of the type found in the brief test series described above can be tolerated without seriously compromising the validity of the aerodynamic test parameters of complete models.

It must be remembered, however, that in this test series only a disturbance of the local Mach number occurred; the local flow direction, although not determined experimentally, was probably quite small. If a Mach number or flow-inclination non-uniformity extends over an axial distance which is small compared with the chord length of the wing or tail surface, the influence of such non-uniformity may be expected to be small. On the other hand, if the axial extent of a disturbance region exceeds the chord length of the wing or tail, the measured forces on the wing and tail may be significantly impaired (see also Ref. 2). In judging the permissibility of flow disturbances in wind tunnel testing, not only must the disturbance magnitude be considered but also, and more significantly, the extent of the disturbance related to the characteristic length of the wing and tail of the complete model.

2. COMPARISON OF FORCE MEASUREMENTS FOR A FUSELAGE–WING MODEL IN SLOTTED AND PERFORATED TEST SECTIONS

In order to investigate the relative merits of different transonic test sections a simple "fuselage–wing" model with 1.0 per cent blockage was investigated in the AEDC transonic model tunnel using both longitudinally slotted and perforated test sections.[3,4] The same model was also tested in the WADC 10 ft transonic wind tunnel where the blockage was only 0.014 per cent. Data obtained in this latter tunnel may be assumed to be practically interference-free, except possibly in a narrow Mach number range around Mach number one. The model shown in Fig. 10.3 had a

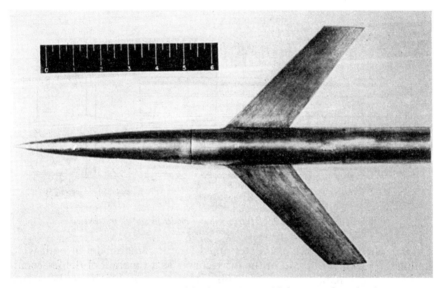

FIG. 10.3. *Photograph of fuselage–wing model for transonic testing.*[4]

fuselage with a cylindrical afterbody and a swept-back wing. The installation in the transonic model tunnel is shown in Fig. 10.4. Force measurements were obtained for drag and lift in the angle-of-attack range from $-2°$ to $+3\frac{1}{2}°$. The moments could not be determined owing to a malfunction of the moment measuring elements of the balance.

a. Testing in slotted test sections

The slotted test sections used in the test series at WADC and AEDC had sixteen slots and a total open-area ratio of 11 per cent. The slot was tapered at the upstream end of the test section in order to provide uniform flow through the Mach number range. Supersonic flow was obtained using a sonic nozzle ahead of the test section and plenum chamber section.[4]

Lift coefficients for the model as a function of angle of attack are shown in Fig. 10.5 for various Mach numbers between Mach 0.80 and 1.10. Obviously the results from the transonic testing of the 1.10 per cent blockage model were not greatly affected by placing the top and bottom walls at 30′ divergent, parallel, or 30′ convergent settings. The setting of only two walls was changed. The lift coefficients obtained in the 10 ft WADC wind tunnel coincided well with the model tunnel data. It can thus be concluded that reliable lift *coefficient measurements* can be obtained for small angles of attack ($\pm 4°$) using wing–fuselage models with blockage ratios of 1.10 per cent in longitudinally slotted test sections.

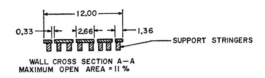

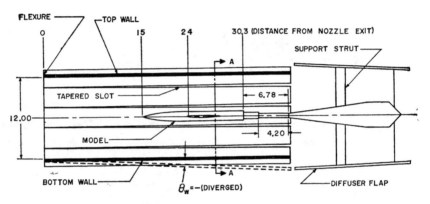

Fig. 10.4. *Installation of fuselage–wing model in slotted test section.*[4]

Drag coefficients were also determined for the same model in both wind tunnels. The general trend of the drag curves as a function of the lift coefficient is in reasonable agreement through the transonic Mach number range although there is a more noticeable shift in the curves (see Fig. 10.5). The drag coefficients for zero lift were measured more closely over the same Mach number range; both the total measured drag and the fore drag of the model are shown in Fig. 10.6. It is apparent that in the supersonic Mach number range between 1.04 and 1.20 significant irregularities occurred which were obviously caused by wave reflections from the slotted test-section walls. A change of the wall setting from 30′ divergent to 30′ convergent did not produce significantly different results.

b. *Testing in perforated test section*

The same fuselage–wing model was also tested in a perforated test section having an 11.8 per cent open-area ratio. The walls, $\frac{1}{16}$ in. thick and provided with straight holes $\frac{1}{16}$ in. in diameter, were selected as suitable on the basis of preliminary tests conducted in the same model tunnel using similar models and a variety of geometry walls. Wall settings of 30′ diverged, parallel, and 30′ and 45′ converged were used.[3]

TYPICAL TEST RESULTS FROM COMPLETE AIRPLANE MODELS

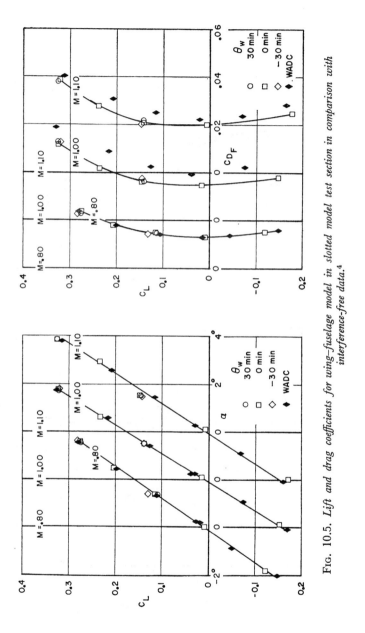

FIG. 10.5. *Lift and drag coefficients for wing–fuselage model in slotted model test section in comparison with interference-free data.*[4]

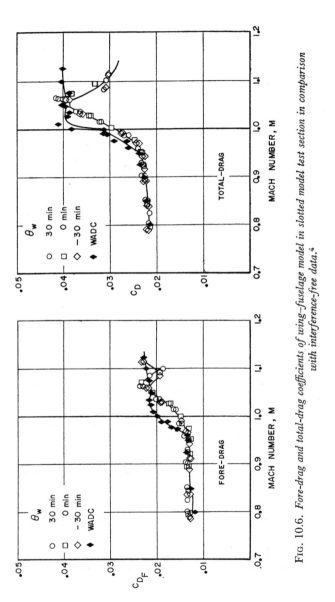

Fig. 10.6. Fore-drag and total-drag coefficients of wing–fuselage model in slotted model test section in comparison with interference-free data.[4]

TYPICAL TEST RESULTS FROM COMPLETE AIRPLANE MODELS

Lift coefficient as a function of angle of attack at Mach numbers 0.80, 1.00 and 1.10 are shown in Fig. 10.7. At both convergent wall settings the agreement between the WADC interference-free data and the model tunnel data was very good. The divergent setting, however, produced a noticeably smaller lift-curve slope than did the WADC tunnel. This discrepancy was greatly reduced when the walls were placed at parallel setting. Good agreement between model tunnel and large tunnel measurements was again exhibited by the drag test results when the walls were placed at a convergent setting (see Fig. 10.7). The influence of wall setting can be mainly attributed to differences in wall boundary layer formation as discussed later.

Drag results with the model for zero lift are shown in Fig. 10.8. The fore-drag coefficients obtained in the model tunnel are in excellent agreement with the WADC data. The influence of the wall setting is negligibly small and within the accuracy of the test. It must be remembered, however, that the fore drag is not a very sensitive parameter since it is the base drag which is mainly influenced by shock reflection.

The total drag coefficients showed a large influence from the reflected waves when the wall was placed at the divergent setting and also, to a smaller degree, at the parallel wall setting. For these conditions the boundary layer along the wall can be expected to be comparatively thick, and, in agreement with previous tests on two-dimensional shock reflection characteristics (see Chapter 8), intense shock reflections occurred. At the converged wall settings of $-30'$ and $-45'$, that is, at wall settings which produce thin boundary layers, shock reflections were greatly reduced so that the total drag coefficients of the model tunnel coincided satisfactorily with the practically interference-free data obtained in the large WADC wind tunnel. However, the drag coefficient determined at Mach number 1.02 in the WADC tunnel is obviously influenced by wave reflection from the wall (see Fig. 10.8). In the model tunnel, wave reflection at this Mach number was effectively eliminated.

In the *subsonic speed range*, the drag values for the converged settings of the perforated model tunnel walls are somewhat larger than the corresponding values obtained from the WADC wind tunnel. At subsonic Mach numbers the best agreement was obtained when the walls were set parallel; at supersonic Mach numbers best agreement was obtained when the walls were converged $-30'$ or $-45'$. The exact reason for this discrepancy is not fully known. It might be an indication of the axial pressure gradient which is established by the interference flow around a model in perforated test sections and is accentuated by wall convergence because of the reduction of the wall boundary layer thickness. Possibly also a difference in the turbulence of Reynolds number level of the two tunnels might have caused this difference in the measurements of base drag.

c. Comparison between slotted and perforated test-section results

The tests of the wing–fuselage model in the transonic model tunnel with both longitudinally slotted and perforated test sections show that reliable data can be obtained with both types of test sections in the subsonic and sonic Mach number range when the geometry of the test-section walls is

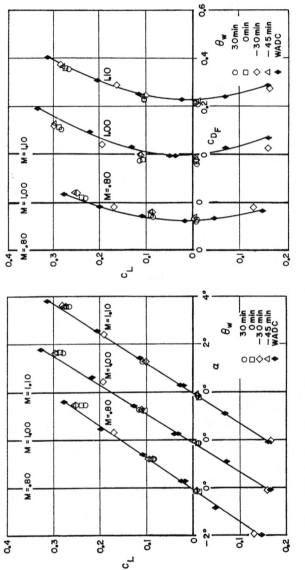

Fig. 10.7. *Lift and drag coefficients of wing–fuselage model in perforated test section in comparison with interference-free data.*[3]

properly chosen. In the *supersonic* Mach number range, however, the longitudinally slotted test section was found to be incapable of effectively eliminating the wave reflections from the tunnel wall so no reliable drag data could be obtained. On the other hand, with perforated test sections, wave reflection disturbances could be effectively minimized when the walls were placed at converged settings of $-30'$ or $-45'$, in order to keep the wall boundary layer thin.

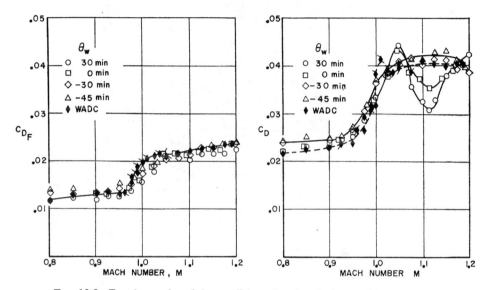

FIG. 10.8. *Fore-drag and total-drag coefficients for wing–fuselage model in perforated test section in comparison with interference-free data.*[3]

On the basis of the comparative tests with a 1.10 per cent blockage model and as a result of a large number of other tests, it can be stated that the greater potential for testing complete airplane models through the transonic Mach number range lies with suitably shaped perforated test sections, particularly when special emphasis is to be placed on the supersonic Mach number range.

3. TESTS IN SLOTTED TEST SECTIONS

a. Tapered unswept wing in slotted model tunnel

The NACA conducted a series of transonic wind tunnel tests with several semi-span wings in a small model tunnel having a cross section of 6.25 × 4.50 in. (Ref. 5). These wings were mounted on one of the solid sidewalls of the tunnel; the upper and lower walls were equipped with eighteen slots per wall, producing open-area ratios of 20 and 12.5 per cent, respectively. (Open-area ratio is related to the area of the upper and lower wall only—see Fig. 10.9.) The semi-span of the wing was 4.24 in., that is, 68 per cent of the width of the tunnel; blockage thus amounted to 1.9 per cent of the tunnel cross section at zero angle of attack. The aerodynamic parameters of the wing were determined through the transonic Mach number range in

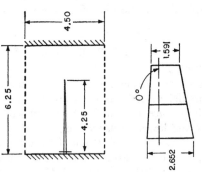

WING DATA (DIMENSIONS IN INCH)
ASPECT RATIO 4.0
AIRFOIL SECTION (PARALLEL TO FREE STREAM) NACA 65A006

FIG. 10.9. *Small model wing in NACA slotted test section.*[5]

the small slotted model tunnel and then in the large NACA 7 × 10 ft tunnel to permit comparison. Data obtained in the large 7 × 10 ft tunnel may be assumed to be interference-free.

Some of the results relating to the lift and pitching moment characteristics obtained during this test series are presented in Figs. 10.10a and b. The 20 per cent open walls were too open and produced lift curve slopes much like open wind tunnel data. The 12.5 per cent open walls, however, produced results which compared well with the data obtained in the large 7 × 10 ft wind tunnel.

The slopes of the lift and moment curves determined at zero lift are presented for several tunnel configurations in Figs. 10.11a and b. The lift curve graph also compares data obtained from the closed and open wind tunnels to which no corrections for wind tunnel boundary effects were applied. Again the data from the 12.5 per cent open slotted wind tunnel coincide well with the data from the 7 × 10 ft wind tunnel. In particular, the large boundary effects which are evident in the data obtained using the closed and open wind tunnels are greatly reduced in both slotted wind tunnels. Only in the Mach number range slightly above Mach number one, do significant deviations occur between the interference-free 7 × 10 ft wind tunnel data and the data obtained in the small 12.5 per cent open slotted wind tunnel. The observed agreement between the various test results is particularly remarkable when effects of the wall boundary layer and the existence of a gap around the semi-wing are taken into consideration.

The curves for the pitching moment slopes (Fig. 10.11b) for both slotted wind tunnels largely coincide. However, noticeable deviations occur between the data obtained in the slotted wind tunnels and the data obtained in the interference-free 7 × 10 ft wind tunnel since the model wing was obviously too large for the small model test section.

In the same NACA test series a smaller wing having a similar plan and with the blockage ratio reduced from 1.9 per cent to 0.7 per cent was also tested in both the previously-defined slotted model wind tunnels and in the full-scale 7 × 10 ft wind tunnel. The results from the model and full-scale tunnel tests coincide closely.[5]

In these same tests the trend of the *drag rise due to lift* was in good agreement for the test series in the three tunnels. However, the results were not conclusive for the absolute level of the drag coefficient through the Mach number range, conceivably because of the small size of the models used.

The results of these early NACA tests indicate that the slotted wind tunnel is indeed a very effective means of eliminating choking and reducing boundary interference effects in the transonic Mach number range. Even with large wings having a span 68 per cent of the tunnel width, the test results obtained match the trend of the interference-free data reasonably well as far as lift and pitching moments are concerned.

b. Wing–fuselage model in full-scale slotted test section

A typical fuselage–wing model with a 45° swept-back wing was investigated through the transonic speed range in both the NACA 8 ft transonic wind tunnel and the NACA 16 ft wind tunnel at Langley Field.[6] Two

TRANSONIC WIND TUNNEL TESTING

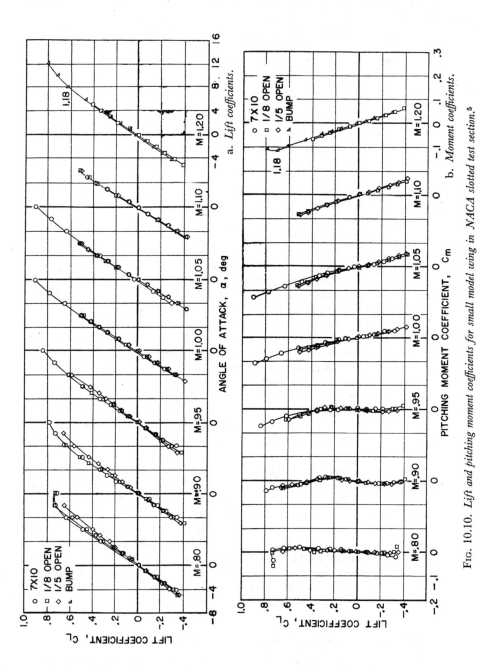

Fig. 10.10. Lift and pitching moment coefficients for small model wing in NACA slotted test section.[5]

models were built, the small one having a fuselage length of 32.6 in. and a 24 in. wing span and the larger one having a fuselage length of 100 in. and a 73 in. wing span (Fig. 10.12). The blockage of the small model amounted to 0.2 per cent in the 8 ft tunnel and 0.04 per cent in the 16 ft wind tunnel. The blockage of the large model in the 16 ft wind tunnel was 0.38 per cent. Both the 8 ft and the 16 ft wind tunnel walls were of the longitudinally slotted type. The 8 ft wind tunnel had twelve slots for an open-area ratio of 11 per cent (see Ref. 7). The 16 ft transonic wind tunnel had eight slots for an open-area ratio of 12.5 per cent (see Ref. 8). The Reynolds number of the small model was approximately 5.8×10^6 based on the mean aerodynamic chord of the model wings.

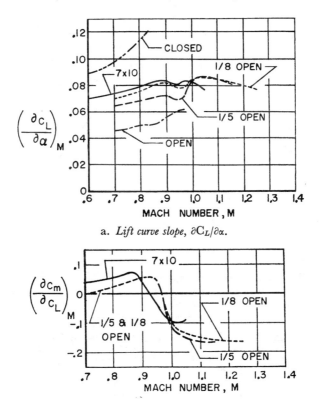

a. *Lift curve slope, $\partial C_L/\partial \alpha$.*

b. *Moment coefficient slope, $\partial C_m/\partial C_L$.*

Fig. 10.11. *Slopes of lift and pitching moment curves as function of Mach number for model wing in slotted test section at zero lift coefficient.*[5]

Some typical measurements of the aerodynamic characteristics of the model configurations are presented in Figs. 10.13–10.15. The correlation at all Mach numbers within the test range is very good. Not merely the general trends of the aerodynamic parameter curves but also the quantitative values are reasonably well matched. Only in the case of the pitching moment characteristics as a function of the lift coefficient are some differences

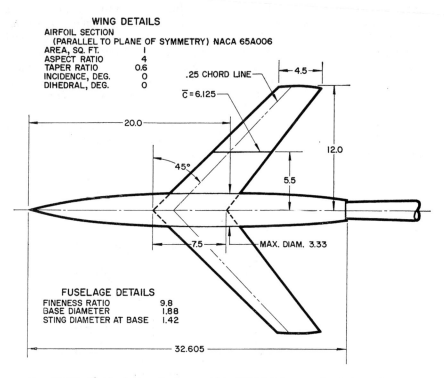

Fig. 10.12. *Sketch of wing–fuselage model in NACA 8 ft transonic wind tunnel tests.*[6]

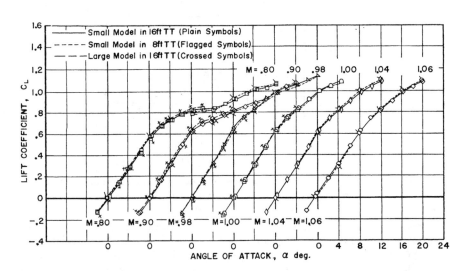

Fig. 10.13. *Lift coefficients as function of angle of attack for NACA transonic wing–fuselage model.*[6]

between the measurements of the large and small models noticeable in the Mach number range near Mach number one.

A cross plot of the drag coefficients for the same models at 0° and 4° angle of attack is presented in Fig. 10.16. Agreement among all three measurements at zero angle of attack is very good; although some deviations occur in the subsonic Mach number range below 0.9 at 4° angle of attack. However, in the transonic speed range the levels again coincide satisfactorily at this angle of attack. These observed drag deviations may be attributed to small differences in lift coefficient occurring during the various test series. In the supersonic Mach number range shock interference with the drag measurements is clearly indicated by the waviness of the drag curves in this speed range. Unfortunately, in the large 16 ft tunnel the measurements were taken only to Mach number 1.07; the test range in the 8 ft wind tunnel extended to Mach 1.13. Thus no direct comparison is possible in the supersonic Mach number range between Mach 1.07 and 1.13.

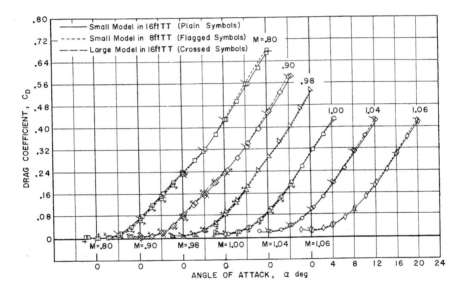

Fig. 10.14. *Drag coefficient as function of angle of attack for NACA transonic wing–fuselage model.*[6]

The results of the test series just described indicate that wall interference in tests of single-wing and tailless airplane configurations in the transonic Mach number range is reasonably small with respect to the quantitative lift and moment characteristics and the general trend of the drag curves. This conclusion is valid as long as the blockage ratio is kept below 0.38 per cent and the "wing–tunnel width" ratio is lower than 40 per cent. However, even at such low blockage area ratios, some supersonic interference effects are clearly noticeable in the measured drag curves.

It should be remembered that the models used in these test series were not equipped with horizontal tails and, consequently, their sensitivity to

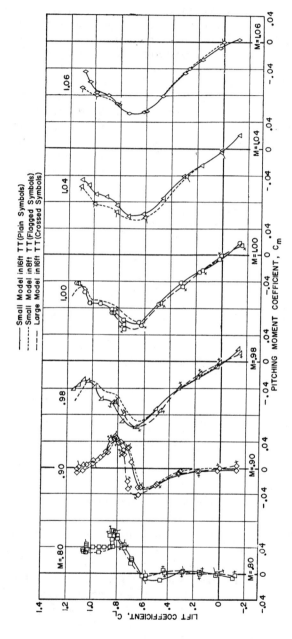

Fig. 10.15. *Pitching moment coefficient as function of the lift coefficients for NACA transonic wing–fuselage model.*[6]

flow curvature effects was greatly reduced. At the present time there are no published results from any systematic test series in slotted wind tunnels which would permit the definition of wall interference from complete airplane models having horizontal tail surfaces and different blockage ratios.

4. TESTS IN PERFORATED TEST SECTIONS

a. "Wing–fuselage" AGARD model "B" in model and full-scale wind tunnels

The AGARD model B, which consists of a fuselage with cylindrical center section and a delta wing, was tested at various blockage ratios in the transonic model tunnel and in the full-scale 16 ft transonic wind tunnel of the AEDC.[9,10] Sketches of the models are shown in Fig. 10.17. One model was built to produce $2\frac{1}{2}$ per cent blockage in the transonic model tunnel

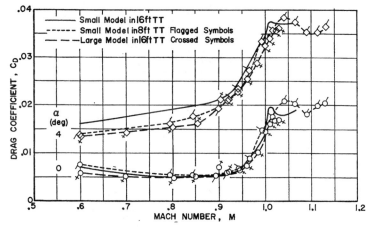

FIG. 10.16. *Drag coefficient as function of Mach number for 0° and 4° angle of attack for NACA transonic wing–fuselage model.*[6]

and 0.01 per cent blockage in the full-scale tunnel. Another larger model had a blockage of 1.15 per cent in the full-scale tunnel. The Reynolds numbers for these comparison tests were $Re = 0.9 \cdot 10^6$ for the small model in the 1 ft model tunnel, $Re = 0.5 \cdot 10^6$ for the small model with 0.01% blockage in the full-scale 16 ft tunnel, and $Re = 4.2 \cdot 10^6$ for the large model with 1.15% blockage in the 16 ft tunnel. (Reynolds numbers are based on the model fuselage diameter.) Both the model tunnel and the full-scale tunnel were equipped with differential-resistance perforated walls having an open-area ratio of 6 per cent (Ref. 11, and Chapter 8).

Some results from these tests for lift, pitching moment, and drag at several Mach numbers are presented in Figs. 10.18–10.20. Agreement among all three model tests is within acceptable limits.

The drag coefficient and the slopes of the lift curve and the pitching moment curves, $\partial C_L/\partial \alpha$ and $\partial C_m/\partial C_L$ for $C_L = 0$, are presented in Fig. 10.21 as a function of Mach number. In general the agreement is reasonably good. The slight variations are within the accuracy of the testing. This fact is particularly remarkable when the differences in Reynolds numbers mentioned above are taken into consideration.

b. *"Wing–fuselage, horizontal tail" AGARD model "C" in model and full-scale wind tunnels*

The same models which were tested without a horizontal tail (described in the preceding section—AGARD model B) were later modified. An elongation of the cylindrical fuselage and a horizontal tail were added (see Fig. 10.17). By this modification, the AGARD models C become more sensitive to flow curvature, and, hence, more suitable for checking the wall interference in transonic wind tunnels. The two models built for this test series had the same blockage ratios as the AGARD model B. (Model tunnel: $2\frac{1}{2}$ per cent blockage; full-scale tunnel: 0.01 and 1.15 per cent blockage.)

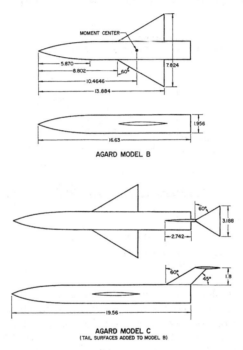

Fig. 10.17. *AGARD calibration models B and modified C (airfoil surfaces— symmetrical 4% circular arc)*.[10]

Typical measurements of the aerodynamic characteristics of AGARD models C are presented in Figs. 10.22–10.24. The curves for the fore-drag coefficients are in close agreement for all three model blockages investigated, but the lift curves, and especially the moment curves, show some differences that appear to grow with the relative size of the models.

The increased wall interference with increased model blockage is also noticeable on the curves showing the lift and pitching moment slopes and the drag coefficients for zero lift (see Fig. 10.25). Some of the differences may be the result of different boundary layer development along the models in the two wind tunnels. An indication of such a boundary layer influence seems to be noticeable in the pitching moment results for subsonic Mach

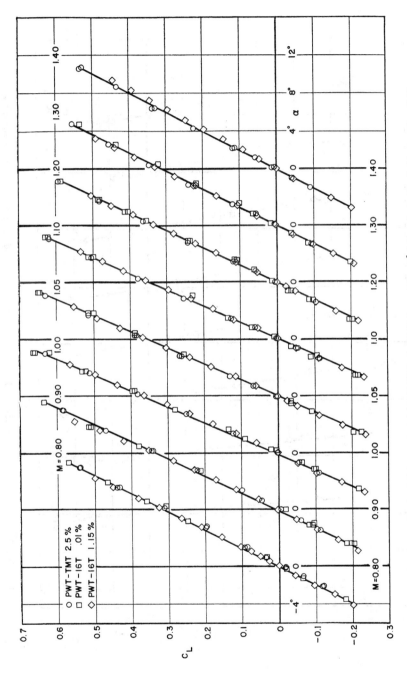

FIG. 10.18. Lift coefficient curves for AGARD model B in model and full-scale perforated test sections.[9,10]

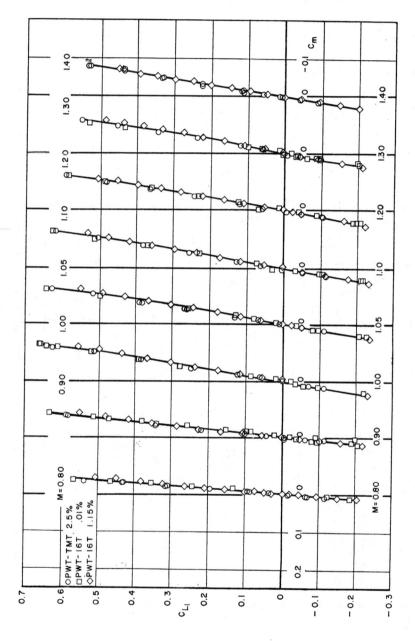

Fig. 10.19. Pitching moment curves for AGARD model B in model and full-scale perforated test sections.[9,10]

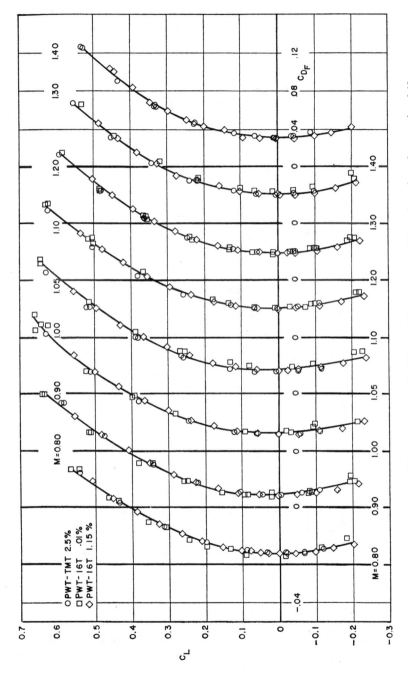

Fig. 10.20. Fore-drag coefficient curves for AGARD model B in model and full-scale perforated test sections.[9,10]

numbers (Fig. 10.23): the curves for pitching moment as a function of lift coefficient exhibit a noticeable S-shape in the case of the small model in the 1 ft tunnel; the curves for this small model and for the large model in the full-scale tunnel do not show a similar S-shape. Such S-shapes in pitching moment curves are a typical indication of large boundary layer influence on the aerodynamic parameters.

The measurements made with AGARD models C do not indicate strong wave reflection interference. It is instead apparent from the gradual and smooth type of the deviation among the pitching moment curves that widely distributed wall interference effects, not wave reflections, are probably responsible for the observed discrepancies.

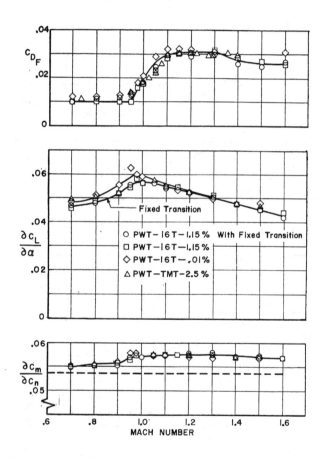

FIG. 10.21. *Fore-drag coefficients and slopes of lift and pitching moment curves as function of Mach number for AGARD model B in model and full-scale test sections.*[9,10]

In general, however, the model with 1.15 per cent blockage exhibited lift, pitching moment, and drag coefficients which coincide reasonably well with the measurements obtained with the interference-free 0.01 per cent blockage models.

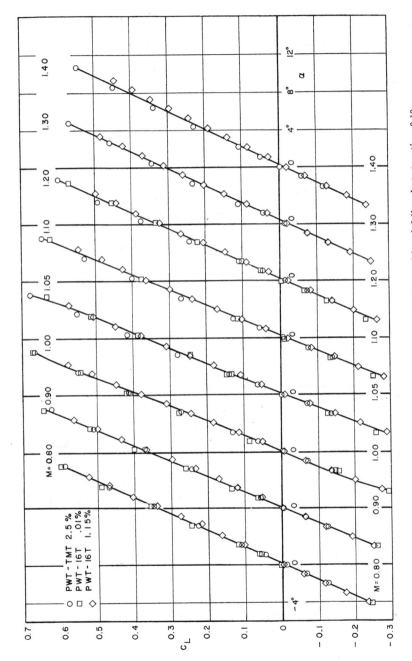

FIG. 10.22. Lift coefficient curves for AGARD model C in model and full-scale test sections.[9,10]

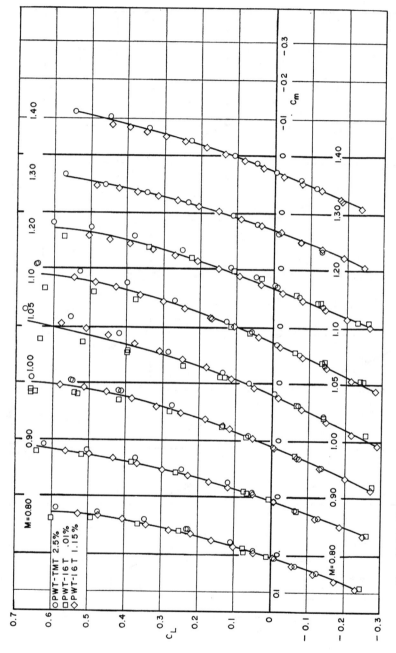

Fig. 10.23. Pitching moment curves for AGARD model C in model and full-scale perforated test sections.[9,10]

TYPICAL TEST RESULTS FROM COMPLETE AIRPLANE MODELS

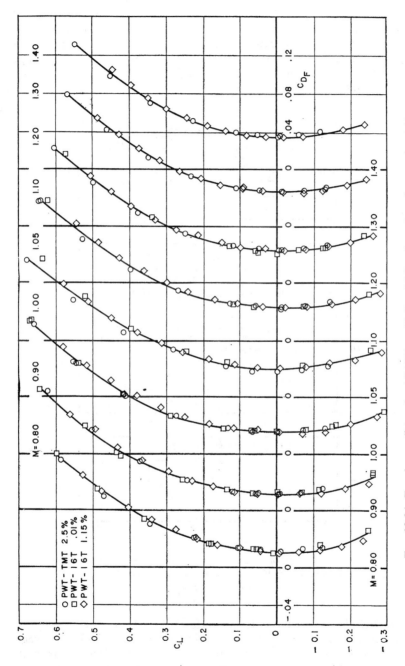

FIG. 10.24. Fore-drag coefficient for AGARD model C and full-scale perforated test sections.[9,10]

c. Airplane model with unswept wing and horizontal tail surface in different wind tunnels

Two different sized models of a transonic airplane were investigated through the transonic speed range in two perforated wind tunnels at the Cornell Aeronautical Laboratory (Ref. 12). These models had straight unswept wings and horizontal tails which made them sensitive to changes of flow curvature. One model was built with a 6 in. wing span, the second with an 18 in. span.

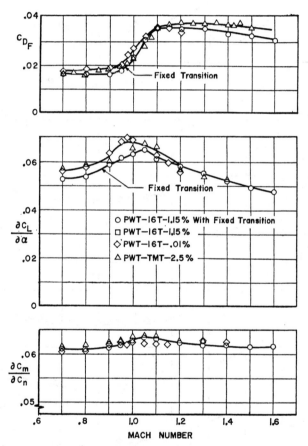

Fig. 10.25. *Fore-drag coefficient and slopes of lift and pitching moment curves as function of Mach number for AGARD model C in model and full-scale perforated test sections at zero lift.*[9,10]

The model airplanes were tested in perforated wall wind tunnels 3×4 ft and 1×1 ft in cross section, respectively. The small model had a blockage of 0.79 per cent and a ratio of wing span to tunnel width of 0.50 in the small wind tunnel and a blockage of 0.07 per cent and a wing span to tunnel width ratio of 0.17 in the large wind tunnel. The larger 18 in. wing span model, tested only in the 3 × 4 ft transonic tunnel, had a blockage ratio of 0.59 per cent and a ratio of wing span to tunnel width of 0.50.

Some aerodynamic parameters determined in this test series are presented in Figs. 10.26–10.28. The general shape of the lift, pitching moment, and drag curves is the same for the small model tests in both the 1 ft and the 4 ft tunnels. There are, however, some parallel shifts of the moment and

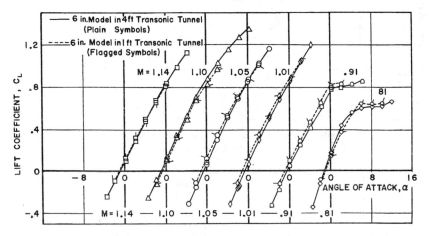

Fig. 10.26. *Lift coefficient curves for complete airplane model with horizontal tail in two perforated transonic wind tunnels of Cornell Aeronautical Laboratory.*[12]

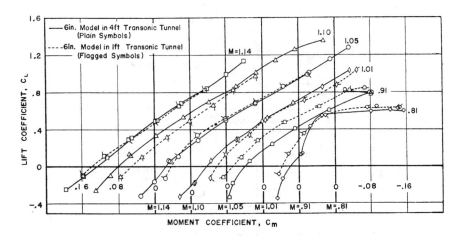

Fig. 10.27. *Lift coefficient as function of pitching moment coefficient for complete airplane model with horizontal tail in perforated transonic wind tunnels of Cornell Aeronautical Laboratory.*[12]

drag curves which can be only partially explained by slight differences in the set Mach number of the tests.

The larger 18 in. wing span model was later tested in both the 4 ft transonic wind tunnel and a 12 ft subsonic wind tunnel. In the latter case

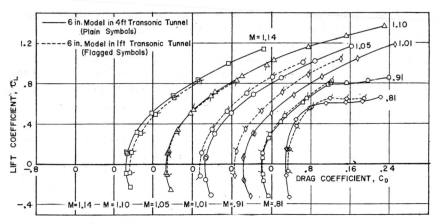

Fig. 10.28. *Drag coefficient for complete airplane model with horizontal tail in perforated transonic wind tunnels of Cornell Aeronautical Laboratory.*[12]

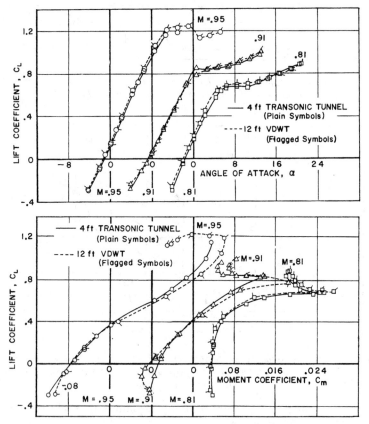

Fig. 10.29. *Lift and pitching moment coefficients at subsonic Mach numbers for complete airplane model with horizontal tail in perforated transonic wind tunnel and subsonic wind tunnel of Cornell Aeronautical Laboratory.*[12]

the wall interference was negligibly small so that a direct comparison of the transonic wind tunnel data with interference-free data was obtained. Some results are shown in Figs. 10.29 and 10.30. The agreement among the curves for lift, pitching moment, and drag coefficients is excellent. Not only the general trends of the curves, particularly in the region close to stall, but also the quantitative results, are in good agreement.

The lift curve slope, $\partial C_L/\partial \alpha$, and the drag coefficient at zero lift for the three configurations are presented in Fig. 10.31 as a function of Mach number. For this comparison the test results for the lift curve slope of the small model in the 4 ft wind tunnel were omitted for Mach numbers below 0.90 because of some unexplainable irregularities which occurred during the testing. The general trends and the quantitative results for the lift curve slopes are in good agreement. The drag coefficients for all three wings also

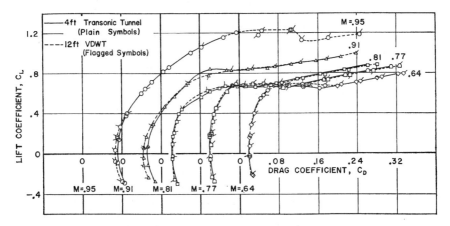

Fig. 10.30. *Drag coefficient as function of lift coefficient at subsonic Mach numbers for complete airplane model with horizontal tail in perforated transonic wind tunnel and subsonic wind tunnel of Cornell Aeronautical Laboratory.*[12]

coincide reasonably well. However, in the low supersonic range some wall interference is noticeable, obviously a result of incomplete cancellation of wave reflections at the tunnel wall.

The agreement among the tests with the various complete airplane models is particularly significant because the unswept wings of these models were expected to present a considerably more severe problem of wave reflection cancellation over a larger range of Mach number than would models with swept-back or delta wing configurations.

5. COMPARISON BETWEEN WIND TUNNEL AND FLIGHT TEST RESULTS

A direct comparison between the aerodynamic characteristics of an airplane model which were measured in a transonic wind tunnel and those of the full-scale airplane which were measured during flight is possible in only a few instances.

In the case of a typical supersonic airplane with the wing swept back approximately 35° and the tail surface likewise swept back horizontally and vertically, the Cornell Aeronautical Laboratory could obtain such a comparison.[12] The model which had a wing span of 18 in. was tested in the Cornell 3 × 4 ft transonic wind tunnel with perforated walls. Some of the aerodynamic characteristics measured during both the wind tunnel and the flight test are discussed in the following paragraphs.

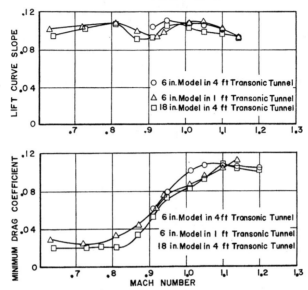

Fig. 10.31. *Lift curve slope and drag coefficient for zero lift as function of Mach number for complete airplane model with horizontal tail in perforated transonic wind tunnels of Cornell Aeronautical Laboratory.*[12]

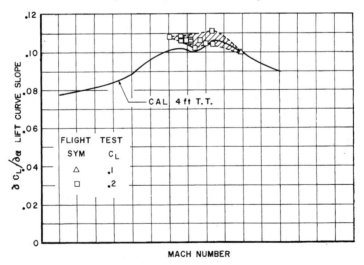

Fig. 10.32a. *Lift curve slope.*

TYPICAL TEST RESULTS FROM COMPLETE AIRPLANE MODELS

Agreement in the lift curve slope $\partial C_L/\partial \alpha$ is very good (Fig. 10.32). The pitching moments show the same slope change above Mach 0.90; however, in the flight tests the slope $\partial C_m/\partial \alpha$ levels off at somewhat smaller values than it does in the wind tunnel test. Flight data are in good agreement with the wind tunnel data for neutral point location $(\partial C_m/\partial C_L)$ in the Mach number range slightly above Mach number one and seem to fall between the data obtained from the Cornell 4 ft wind tunnel tests and those obtained from the 8 ft Langley Field transonic wind tunnel. At the two highest Mach

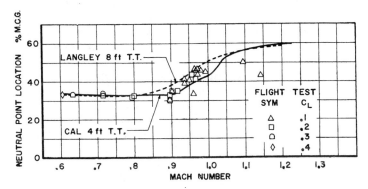

b. *Neutral point location.*

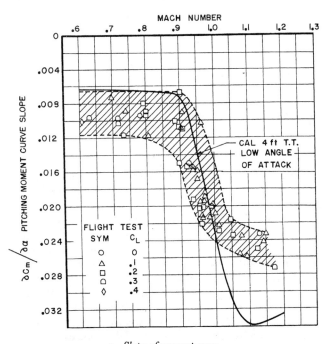

c. *Slope of moment curve.*

FIG. 10.32 Longitudinal stability parameters for supersonic airplane with horizontal tail at transonic Mach numbers for perforated wind tunnel and flight tests.[12]

numbers, 1.1 and 1.15, of the test series; however, the flight test points show a pronounced deviation from the curves obtained in both wind tunnels though the two wind tunnel curves themselves are in excellent agreement.

In Fig. 10.33a some typical measurements of the yawing stability parameters of the airplane are shown. In the wind tunnel tests the yawing moment stability parameter, $\partial C_n/\partial \beta$, becomes much smaller as the Mach number approaches one; at higher Mach numbers a step recovery to higher values was exhibited. The flight test data do not show the pronounced reduction when the Mach number approaches one; however, they do indicate the general tendency towards larger values at Mach numbers slightly above one. A definite explanation of this discrepancy is not available. The differences could be due to the Reynolds number differences between the model tests (mean aerodynamic chord of only 4.7 in.) and the full-scale flight tests. This explanation seems to be supported by transonic data from another wind tunnel in which the Reynolds number was somewhat higher than in the Cornell tests though it was still lower than that of the flight test. These latter tests indicate directional stability characteristics which fall between the flight test and the Cornell wind tunnel test results.

The general trend of the roll-stability parameter, $\partial C_L/\partial \beta$, agrees reasonably well (Fig. 10.33b), particularly beyond Mach number one, where the lateral stability values decrease rapidly in both the wind tunnel and the flight tests. This trend is especially evident in the flight test results at a lift coefficient of 0.2. However, the graph also demonstrates the great difficulty of obtaining measurements with a high degree of accuracy during flight testing. The scatter of the flight test points is considerable and definitely much greater than the scatter occurring with the better controlled conditions of wind tunnel testing.

The side force stability parameter, $\partial C_y/\partial \beta$, shown in Fig. 10.33c, exhibits no violent changes through the transonic Mach number range. The trends of both the wind tunnel data and the scattered data from the flight tests show reasonably good agreement.

6. CONCLUDING REMARKS

In view of the results obtained in transonic wind tunnel and flight testing, it is possible to state that even with wind tunnel models having an 18 in. wing span and an aerodynamic chord of approximately 4 in. it was possible to predict the general trends and the critical changes of the aerodynamic parameters for the complete airplane as it passed from high subsonic Mach numbers through sonic speeds to supersonic Mach numbers. This fact is particularly remarkable when it is remembered that not only did a large difference in Reynolds number exist between the wind tunnel and the flight tests, but that the flight test results could also be greatly influenced by aero-elasticity effects. By using larger models and more refined methods (i.e. by applying corrections for differences in Reynolds number, surface roughness, aero-elasticity, etc.), it appears highly possible to predict reliably the characteristics of the full-scale airplane through the transonic speed range as a result of wind tunnel model testing.

It is suggested that as a general rule for transonic wind tunnel tests the

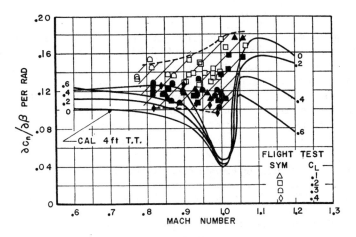

a. *Yawing moment slope.*

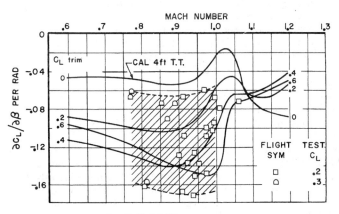

b. *Roll stability parameter.*

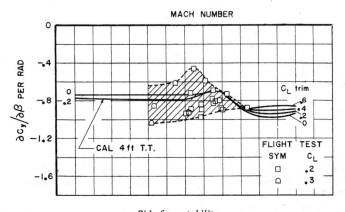

c. *Side force stability.*

FIG. 10.33. *Directional stability parameters for supersonic airplane with horizontal tail at transonic Mach numbers from perforated wind tunnel and flight tests.*[12]

models should not exceed a blockage ratio of 1 per cent. For precision measurements the blockage ratio should be kept as small as one-half of 1 per cent with a wing span not exceeding one-half the tunnel width. In addition, the model length is often a critical parameter, particularly when the tail surface effectiveness is to be determined. The influence of tail surface position with respect to the location and width of possible reflected wave bands needs careful consideration in each individual case.

REFERENCES

[1] RITCHIE, V. S. "Effects of Certain Flow Non-uniformities on Lift, Drag and Pitching Moment for a Transonic-Airplane Model Investigated at a Mach Number of 1.2 in a Nozzle of Circular Cross Section." NACA RM L9E20a, August 31, 1949.
[2] MORRIS, D. E. and WINTER, K. G. "Requirements for Uniformity of Flow in Supersonic Wind Tunnels." AGARD Memorandum AG 17/P7, 1954.
[3] GRAY, J. DON. "Transonic Interference Effects upon Lift and Drag Force Measurements in a Perforated Test Section." AEDC–TR–54–27, Novmber, 1954.
[4] GRAY, J. DON. "Transonic Interference Effects upon Lift and Drag Measurements in a Slotted Test Section." AEDC–TR–54–47, March, 1955.
[5] SLEEMAN, W. C., KLEVATT, P. L. and LINSLEY, E. L. "Comparison of Transonic Characteristics of Lifting Wings from Experiments in a Small Slotted Tunnel and the Langley High-Speed 7 ft × 10 ft Tunnel." NACA RM L51F14, November, 1951.
[6] WHITCOMB, C. F. and OSBORNE, R. S. "An Experimental Investigation of Boundary Interference on Force and Moment Characteristics of Lifting Models in the Langley 16 ft and 8 ft Transonic Tunnels." NACA RM L52L29, February 24, 1953.
[7] RITCHIE, V. S. and PEARSON, A. O. "Calibration of the Slotted Test Section of the Langley 8 ft Transonic Tunnel and Preliminary Experimental Investigation of Boundary-Reflected Disturbances." NACA RM L51K14, July 7, 1952.
[8] WARD, V. G., WHITCOMB, C. F. and PEARSON, M. D. "Air Flow and Power Characteristics of the Langley 16 ft Transonic Tunnel with Slotted Test Section." NACA RM L52E01, July, 1952.
[9] MILILLO, J. R. and CHEVALIER, H. L. "Test Results of the AGARD Calibration Model B and Modified AGARD Model C in the AEDC Transonic Model Tunnel." AEDC–TN–57–6, May, 1957.
[10] MILILLO, J. R. "Transonic Tests of an AGARD Model B and Modified Model C at 0.01 per cent Blockage." AEDC–TN–58–48, August, 1958.
[11] DICK, R. S. "Calibration of the 16 ft Transonic Circuit of the Propulsion Wind Tunnel with an Aerodynamic Test Cart Having 6 per cent Open-Inclined-Hole Walls." AEDC–TN–58–90, November, 1958.
[12] BIRD, K. D. "Model Configuration Tests in the C.A.L. 4 ft Transonic Tunnel: Correlation of Results from Different Scale Models over Wind Tunnels and Free Flight." Arnold Engineering Development Center, July, 1956.

BIBLIOGRAPHY

ALLEN, EDWIN C. "Wind Tunnel Investigations of Transonic Test Sections—Phase III: Tests of a Slotted-Wall Test Section in Conjunction with a Sonic Nozzle." AEDC–TR–54–53, February, 1955.

BROMM, A. F., JR. "Investigation of Lift, Drag and Pitching Moment of a 60° Delta Wing–Body Combination (AGARD Calibration Model B) in the Langley 9 in. Supersonic Tunnel." NACA TN 3300, September, 1954.

DICK, R. S. "Calibration of the Propulsion Wind Tunnel 16 ft Transonic Circuit with a Full Test Cart Having Inclined Hole Perforated Walls." AEDC-TN-58-97, January, 1959.

DICK, R. S. "Tests in the PWT 16 ft Transonic Circuit of an AGARD Model B and a Modified Model C at 1.15 per cent Blockage." AEDC TN-59-32, April, 1959.

O'HARA, F., SQUIRE, L. C. R.A.E. and HAINES, A. B. "An Investigation of Interference Effects on Similar Models of Different Size in Various Transonic Tunnels of the U.K." Aeronautical Research Association, RAE-TN-AERO-2606, February, 1959.

PILAND, R. O. "The Zero-Lift Drag of a 60° Delta Wing–Body Combination (AGARD Model 2) Obtained from Free-Flight Tests between Mach Numbers of 0.8 and 1.7." NACA–TN–3081, April, 1954.

POISSON-QUINTON, PH. "Premiers Résultats d'Essais sur Maquettes Étalons AGARD." Paper presented at the Sixth Session of the AGARD Wind Tunnel Committee, November, 1954.

"Specifications for AGARD Wind Tunnel Calibration Models." AGARD Memorandum AG–4/M3, August, 1955.

SPIEGEL, J. M. and LAWRENCE, L. F. "A Description of the Ames 2 ft × 4 ft Transonic Wind Tunnel and Preliminary Evaluation of Wall Interference." NACA RMA 55I21, January 20, 1956.

CHAPTER 11

CROSS-FLOW CHARACTERISTICS OF PARTIALLY OPEN WALL ELEMENTS

1. PHYSICAL ASPECTS OF THE FLOW THROUGH PARTIALLY OPEN WALLS

Numerous theoretical and experimental investigations have been carried out to determine the pressure drop characteristics of transonic wall elements exposed to flow into and out of the test section. Before these various studies are discussed, a brief review of the wall characteristics required for interference-free testing will be presented.

a. Requirements for transonic test-section walls

As pointed out in the preceding sections, the wall interference in transonic wind tunnels can be eliminated or minimized when specific predetermined characteristics of the test-section walls are established. One main requirement, naturally, is that characteristics of the wall be reliably predictable, either theoretically or experimentally, and that such a wall consistently maintain these characteristics. It was assumed in most calculations of wall interference that the walls produce a linear relationship between cross-flow velocity and pressure drop. This requirement must be satisfied, not only for outflow from the test section but also for flow into the test section. Since in the latter case low-energy air accumulated in the plenum chamber flows into the test section, it is possible to approach this goal in the case of inflow, only by means of special configurations such as walls with differential-resistance geometry.

Also the wall disturbance waves which are generated by the edges of each opening must decay to a sufficiently low level before they strike the test model. This requirement usually results in the "small-grain" wall openings discussed in detail later in this section.

Finally, it is important, particularly in large wind tunnels, that the boundary layer growth along the walls, that is, the viscosity effect, be kept reasonably small. This growth determines decisively the power requirements for a wind tunnel, and, in a more direct manner, the amount of auxiliary suction required to eliminate undesirably thick boundary layers along the test-section walls.

The various requirements for suitable transonic wind tunnel walls are so complex that only by using the orientation of theoretical calculations in conjunction with extensive experiments has it been possible to develop suitable wall shapes for use in transonic wind tunnels.

b. Basic characteristics of different types of walls

Longitudinally Slotted Wall.—In the absence of friction, the pressure drop through a longitudinally slotted wall exposed to straight oblique flow follows a quadratic relationship (see Chapter 8). This result is easily understood since only the velocity component perpendicular to the wall produces a pressure change similar to the flow through an orifice. It has also been shown that with a slotted wall a pressure change is produced which depends upon the curvature of the flow approaching the wall (see Chapter 5). Besides these two pressure change components, an additional pressure drop is caused by viscous forces. Experiments have shown that the viscosity effect is predominant when the flow angle is small with respect to the wall and when no appreciable flow curvature exists.

The total pressure drop through a wall with longitudinal slots can be represented by the equation:

$$\frac{\Delta p}{q} = K_1 \theta + K_2 \theta^2 + K_3 \frac{\partial \theta}{\partial (x/h)}$$

where h = distance between slot centers. With typical longitudinally slotted walls, the constants in the above equation have approximate magnitudes of $K_1 = 0.50$ and $K_2 = 1.0$; thus the linear term predominates for wall angles up to 10°.

In the vicinity of impinging shocks pure longitudinally slotted walls develop a strong flow recirculation that is very detrimental to effective wave cancellation. By means of corrugated sheets placed in each slot or by perforated cover plates,[1] this recirculation can be largely eliminated; combined "perforated–slotted" walls have been found to produce satisfactory results in transonic testing.

Porous Walls.—Walls made of a porous material have cross-flow characteristics of a linear type. In this case the pressure drop is produced mostly by friction in the narrow channels, and the dynamic effects are practically eliminated. It has been shown that the pressure drop for such walls is directly proportional to the mass flow through the wall:

$$\Delta p = K m_{ay}$$

where m_{ay} = mass flow per unit area through the porous wall. With some transformation, the pressure drop can also be expressed in the following manner:

$$\Delta p / q_\infty = (K_1 / v_\infty) \theta$$

This equation indicates that, as desired, porous walls produce a pressure drop coefficient proportional to the flow inclination θ. However, as indicated, the pressure drop coefficient also depends upon the velocity of the undisturbed flow in the test section. This very undesirable characteristic postulates that, for each velocity v_∞ in the test section, a different geometry of the porous wall is necessary to maintain a constant ratio of $\Delta p/q_\infty$. In addition, practical difficulties could arise from the fact that the narrow channels in the porous walls might have a tendency to clog up and would then require frequent cleaning.

As a consequence of these disadvantages, not many test results for wind tunnel walls of this type have been published.

Thin Perforated Test-Section Walls.—When a perforated wall is placed at a large oblique angle in *parallel flow*, a pressure drop is produced which closely follows a quadratic law. As Pindzola (Sverdrup and Parcel, Inc.) showed, the pressure drop through a perforated wall is given by the equation:

$$\Delta p/q_\infty = K_p \sin^2\theta$$

where K_p = pressure drop coefficient $\Delta p/q_\infty$ when the plate is positioned perpendicular to the flow ($\theta = 90°$).

Experimental data obtained with a perforated wall having approximately 22 per cent open-area ratio and $\frac{1}{2}$ in. diameter holes in 0.040 in. thick walls are shown in Fig. 11.2. The quadratic law holds between the limits presented, that is, from 15° to 90°. However, in the case of smaller angles of attack—which are of primary interest in transonic wind tunnel testing—these test data are not conclusive, and more refined experiments are required for the angle-of-attack range which is of particular interest in wind tunnel testing.

When oblique flow approaches a perforated wall at small angles, it can be expected that the solid wall elements will act as wing elements and will produce lift. This characteristic becomes obvious when an idealized perforated wall is replaced by a wall with numerous transverse slots. Such a wall behaves exactly as the lattice which has been extensively investigated for use with compressor and turbine developments. At small angles of attack such lattices are known to produce a lift (equivalent to a pressure drop through the wall) which is proportional to the angle of attack of the lattice with respect to the approaching flow. Also, at large angles of attack such lattices produce an additional pressure drop which is proportional to the square of the cross-flow velocity. The total pressure drop through a perforated wall can then be represented by:

$$\Delta p/q_\infty = K\theta + K_1\theta^2$$

Usually this linear term is sufficient to represent the behavior of perforated walls in transonic wind tunnel testing.

According to the lattice theory the linear relationship of the cross-flow coefficient can be expected to apply only in cases of small viscosity effects. When viscosity effects become large—for instance, when the ratio of boundary layer thickness to the area of the walls becomes large—significant deviations from the linear law can be expected. These deviations are equivalent to the behavior of single wings at low Reynolds number where thick boundary layers occur along the wing surfaces.

Thick Perforated Walls.—When the thickness of perforated walls is large in comparison with the diameter of the wall openings, the individual wall elements can no longer act as the individual wings of a lattice, and therefore considerable deviation from the observed laws occurs (see Fig. 11.1). Besides impairing the lift-producing circulation of each wall element, thick walls tend to guide the cross flow like channels. Then in the absence of viscosity effects, the walls may produce a pressure rise instead of a pressure drop because these individual channels may act as diffusers. This effect can be expected particularly when the inlet to each individual hole is rounded so that flow separation at the inlet is avoided. Usually no pressure recovery

CROSS-FLOW CHARACTERISTICS OF WALL ELEMENTS

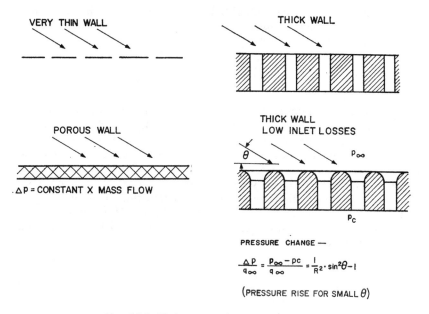

Fig. 11.1 *Various types of partially open walls.*

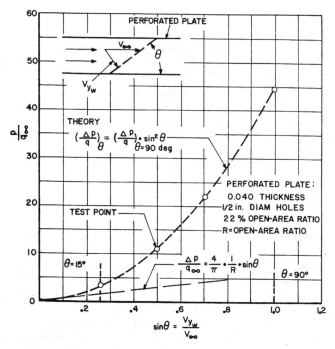

Fig. 11.2. *Pressure drop of perforated wall through large range of flow inclination angles (ref. Pindzola, Sverdrup and Parcel, Inc.).*

occurs at the exit of the individual holes as a result of the abrupt change of cross section. For such cases, when viscosity effects are neglected, the pressure change can be written:

$$\frac{\Delta p}{q_\infty} = \frac{1}{R^2}\sin^2\theta - 1$$

where a minus sign indicates that the pressure change is a pressure rise.

It is apparent from this equation that with thick perforated walls the pressure drop characteristic will not be as steep as in the case of thin walls with otherwise equal geometry. Furthermore, under certain circumstances, irregularities may be expected which might lead to a pressure rise instead of to the normal pressure drop which occurs when flow passes such a wall. The limits for the transition from "aerodynamically thin" walls to acceptably thick walls cannot be readily determined by theory alone. In a number of systematic experiments, however, the influence of wall thickness on transonic testing has been thoroughly explored.

2. THEORETICAL CALCULATIONS FOR THE PRESSURE DROP THROUGH PERFORATED WALLS

Several theoretical studies of the pressure drop of cross flow through perforated walls have been conducted for both subsonic and supersonic flow.

In most of these calculations the assumptions are made that the flow is non-viscous and that potential flow exists on both sides of the perforated wall; that is, at the test-section side and the plenum chamber side. In the case of outflow from the test section into the plenum chamber, these assumptions should not cause the calculated pressure distribution along the test-section side of the walls to differ significantly from the actual.

In the case of inflow from the plenum chamber into the test section, the assumptions are not applicable. Most significantly, the total pressure of the flow from the plenum chamber is equal to the mean static pressure of the plenum chamber, which is considerably smaller than the total pressure of the test-section flow. No adequate theoretical calculations have been published to define quantitatively the pressure drop through a wall for inflow conditions.

a. Thin wall with single transverse slot in incompressible flow

The flow field and the disturbance velocities have been calculated in Ref. 2 for a wall having a single transverse slot. In these extremely simplified calculations it is demonstrated that the essential boundary conditions for a thin transverse slotted wall are similar to those of a flat plate with lift. When the disturbance velocity potential of the single slot is multiplied by the imaginary number $i = \sqrt{-1}$, the disturbance potential of a flat plate in oblique flow is obtained (see Fig. 11.3). In the case of the flat plate, the condition of the circulation corresponding to the lift of the wing must then be introduced. In the case of the single slot, the condition of smooth flow at the upstream edge of the slot is introduced, and thus the circulation is defined (Kutta condition). The disturbance velocity potential of the single slot is then obtained:

$$\phi(\xi) = \frac{2}{\pi} v_h (\xi^{-1} + \log \xi)$$

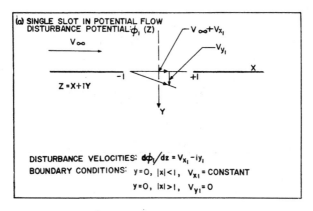

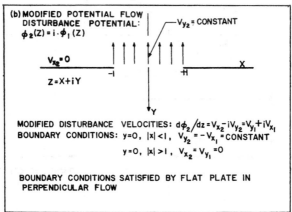

FIG. 11.3. *Potential flow relationships for single transverse slot.*[2]

Here ξ is a complex variable connected with $z = x + iy$ in the real plane by $z = \tfrac{1}{2}(\xi + 1/\xi)$. The mean vertical velocity over the slot area v_h is determined in such a manner that the continuity equation is satisfied, that is:

$$v_y dx = v_h w$$

where

$$w = \text{slot width}$$

From this equation the distribution of the disturbance velocities Δv_y and Δv_x along the wall and across the slot, and the pressure in the plenum

chamber, can be easily determined. It is found that:
Slot Area ($y = 0$)

$$\Delta y_x / v_\infty = \frac{2}{\pi} \frac{v_h}{v_\infty} = -\frac{1}{2} \frac{\Delta p}{q_\infty}$$

$$\Delta v / v_\infty = \frac{2}{\pi} \frac{v_h}{v_\infty} \left(\frac{1+x}{1-x}\right)^{\frac{1}{2}}$$

Solid Wall ($y = 0$)

$$\Delta v / v_\infty = \frac{2}{\pi} \frac{v_h}{v_\infty} \left(\frac{x+1}{x-1}\right)^{\frac{1}{2}} - 1$$

$$\Delta v_y = 0$$

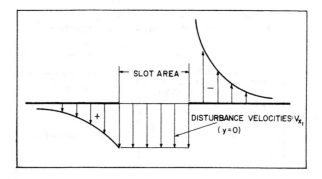

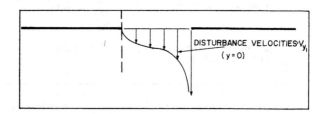

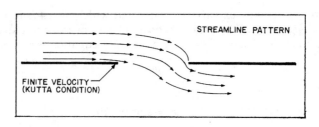

Fig. 11.4. *Disturbance velocity distribution and streamline pattern for wall with single transverse slot.*[2]

and for the cross-flow pressure drop (positive sign for pressure drop):

$$\Delta p / q_\infty = \frac{4}{\pi} \frac{v_h}{v_\infty}$$

The disturbance velocities along the wall were determined from these equations; their distribution is shown in Fig.11.4. As required, the upstream edge of the slot acts like the trailing edge of a wing, that is, the flow is parallel to the wall surface in this area. The downstream edge of the slot acts like the leading edge of a wing exposed to oblique parallel flow. According to this physical flow picture, increased axial velocities occur along the wall upstream of the slot, and reduced axial velocities occur downstream of the slot. Along the plenum chamber side of the wall, the static pressure is constant and corresponds in magnitude to the constant disturbance velocity Δv_x over the slot width.

b. Thin wall with single transverse slot in compressible subsonic flow

The results obtained for a single transverse slot in incompressible flow can be made applicable to compressible subsonic flow by means of the Prandtl–Glauert approximation for flow with small perturbations. According to the generalized correlation rule,[3] the disturbance velocities in the x direction and the disturbance pressures are increased by the factor $1/(1-M^2)$; the disturbance velocities in the y direction are increased by $1/(1-M^2)^{\frac{1}{2}}$. Consequently the basic relationship for the pressure drop through a wall having a single slot can be represented by:

$$\frac{\Delta p}{q_\infty} = \frac{1}{(1-M^2)^{\frac{1}{2}}} \frac{4}{\pi} \frac{v_h}{v_\infty}$$

It can be seen that the pressure drop through a transverse slot increases with the Prandtl factor $1/(1-M^2)^{\frac{1}{2}}$. Naturally, as in other linearized compressible flow problems, this relationship does not hold true in the Mach number one range. Also, since viscosity effects are not considered, the slot width must be large in comparison with the local boundary layer thickness if good correlation between theory and experiment is to be obtained.

c. Thin walls with series of transverse slots in subsonic flow

If the distance between individual slots is large in comparison with the slot widths, as is the case with small open-area ratios, the same relationship derived previously for the single slot will hold true since the interaction between the individual slots is negligibly small. In other words, the pressure drop through a wall with several lateral slots may be presented as:

$$\frac{\Delta p}{q_\infty} = \frac{1}{(1-M^2)^{\frac{1}{2}}} \frac{4}{\pi} \frac{v_h}{v_\infty}$$

where v_h is again the mean vertical velocity in the open slot area. Then if the velocity v_y and the flow inclination θ at large distances from the wall are introduced:

$$\frac{\Delta p}{q_\infty} = \frac{1}{(1-M^2)^{\frac{1}{2}}} \frac{4}{\pi} \frac{1}{R} \theta$$

If, however, the open-area ratio between the individual slots can no longer be considered small, as is the case with walls having large open-area ratios, the interaction between the individual slots becomes appreciable, and the above simple relationships no longer hold true. A thin wall with numerous transverse slots may be treated as a lattice, and each wall element between two slots may be considered to be a lifting wing (see Fig. 11.5)

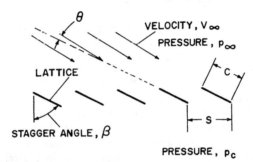

Fig. 11.5. *Sketch of lattice with small stagger angle.*

The flow inclination and lift coefficients of a lattice with a stagger angle of 90° have been determined in Ref. 4 by means of conformal mapping. For each such wing, the lift coefficient is obtained from the following equation:

$$C_L = K_L 2\pi (\sin \theta^*)(v_\infty^*/v_\infty)^2$$

with:

$$K_L = \frac{2}{\pi} \frac{1}{1-R} \cot\left(\frac{\pi}{2}R\right)$$

and the circulation around each wing from:

$$\Gamma = \pi v_\infty^* c K_L \sin \theta^*$$

where c = chord length of elementary wing. In these equations all terms with an asterisk represent the mean conditions in the plane of the lattice itself.

By applying simple laws of incompressible aerodynamics and after transformation:

Disturbance Velocity in Flow Direction in Slot Area.—

$$\frac{\Delta v_x}{v_\infty} = \sin \theta \cot\left(\frac{\pi}{2}R\right) = R \frac{v_h}{v_\infty} \cot\left(\frac{\pi}{2}R\right)$$

CROSS-FLOW CHARACTERISTICS OF WALL ELEMENTS

Pressure Drop of Cross Flow through Wall.—

$$\Delta p / q_\infty = 2 \sin \theta \cot\left(\frac{\pi}{2} R\right)$$

$$= 2R \frac{v_h}{v_\infty} \cot\left(\frac{\pi}{2} R\right) = K \frac{v_h}{v_\infty}$$

with

$$K = 2R \cot\left(\frac{\pi}{2} R\right)$$

For compressible subsonic flow, the factor K can be written[3]:

$$K = K_M / (1 - M^2)^{\frac{1}{2}} = 0$$

where K_M is the value of K at zero Mach number. These same equations are also derived in Ref. 5.

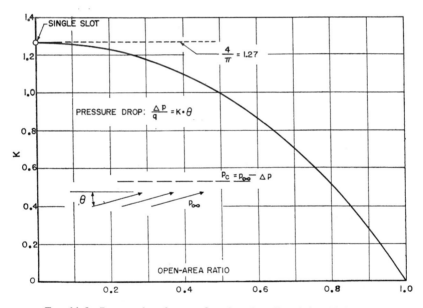

FIG. 11.6. *Pressure drop for cross flow through wall with lateral slots from lattice theory.*

The results of the above theoretical calculations are plotted in Fig. 11.6. As before, the wall cross-flow coefficient is $K = 4/\pi$ for small open-area ratios up to approximately 20 per cent; for larger open-area ratios the cross-flow coefficient rapidly decreases in magnitude. For a wall with a 50 per cent open-area ratio, for instance, the coefficient is only $K = 1.00$, that is, approximately 78 per cent of the value for small open-area ratios. The constant approaches zero as $R \to 1$, that is, in the case of the completely open wall.

d. *Single transverse slot in wall with separated flow on the plenum chamber side (subsonic flow)*

In the preceding sections it was assumed that the thin transversally slotted wall is placed in a parallel oblique flow and that potential flow without flow separation exists on both sides of the test-section wall. Another theoretical study was conducted[6] in which the more realistic assumption was made that the flow separates at the plenum chamber side and that free-boundary jet streamlines originate from each slot.

Thin Walls.—Some significant streamlines for this flow pattern are shown in Fig. 11.7a. At the upstream edge of the slot the flow separates

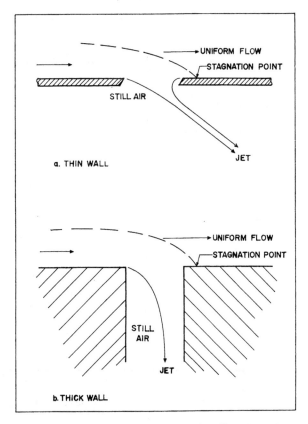

Fig. 11.7. *Streamline pattern for cross flow through wall with single transverse slot and flow separation on plenum chamber side.*[7]

and forms a free-jet streamline with a boundary pressure equal to the static pressure in the plenum chamber. Further downstream in the test section, a dividing streamline is formed with a stagnation point at the wall. Starting at this point, the flow proceeds forward against the main wind tunnel flow, separates at the edge of the slot, and proceeds into the plenum chamber to form another free-jet boundary streamline. It is believed that this flow picture corresponds more closely to the real flow conditions, and the study, therefore, serves as a check on the previous more simplified calculations. As in the previous treatment of a single slot (Section 2a of this chapter) the pressure drop was again[7]:

$$\Delta p/q_\infty = (4/\pi)(v_h/v_\infty)$$

Since the two calculations for walls with a single transverse slot arrive at the same result, it may be expected that the preceding calculations for multi-slotted walls will also give a sufficiently accurate prediction of the pressure drop through the wall.

These calculations assume potential non-viscous flow, and the results are given for small flow velocities through the wall, that is, for linearized flow only. For higher cross-flow velocities through the wall the higher-order terms of the theory will cause deviations which may be approximated by

$$\Delta p/q_\infty = (4/\pi)(v_h/v_\infty) + K_1(v_h/v_\infty)^n$$

Thick Walls.—In Fig. 11.7b, the streamline pattern is presented for a very thick wall with a single transverse slot. Again the separated flow forms a free-jet streamline at the upstream edge of the slot. On the downstream edge, however, the flow proceeds along the side wall of the slot after it has perhaps formed a small separated flow area near the downstream edge of the slot.

In either case of flow through thin or thick walls, the hodograph representation of the velocity field along the walls and along the free streamlines results in very simple curves. They are either straight lines corresponding to the flow along the surfaces of the wall and the slot, or they are arcs corresponding to the free-jet boundaries having constant velocities. Because of mathematical difficulties, no successful treatment for the thick wall has been published to date.

e. Perforated walls with circular holes

The preceding calculations have been carried out for walls with two-dimensional transverse slots. No calculations have been published for walls with more confined openings such as circles, squares or triangles. However, the main physical aspects of flow through a wall with circular holes, or holes with other shapes, should be similar to those of flow through a wall with transverse slots. Again, a "leading edge" effect will occur on the downstream edge of the holes, and a "trailing edge" effect, establishing the "Kutta" condition, will occur on the upstream edge. The wall elements between the openings may be assumed to act like wings with a small aspect ratio. Such walls will therefore develop a smaller pressure drop than walls with transverse slots which act similar to wings of an infinitely large aspect ratio.

f. Perforated walls in supersonic flow

When a perforated wall is replaced by a wall with transverse slots, the pressure drop in cross flow through such a wall can be readily determined using the method of characteristics and the oblique shock-wave theory. As shown in more detail in Chapter 8, the pressure drop through a transversally slotted wall in supersonic flow was found to be:

$$\frac{\Delta p}{q_\infty} = \frac{2}{(M^2-1)^{\frac{1}{2}}}\left(\frac{1}{R}-1\right)\theta$$

$$= \frac{2}{(M^2-1)^{\frac{1}{2}}}(1-R)\frac{(\rho v)_h}{(\rho v)_\infty}$$

$$= K(\rho v)_h/(\rho v)_\infty$$

where θ = flow angle upstream of the wall
R = open-area ratio of the wall.

Because of linearization of the flow characteristics, this equation may be expected to apply only at Mach numbers which are sufficiently higher than Mach number one. In the immediate vicinity of sonic speed, the *real flow* characteristics will cause the factor K to reach a finite plateau instead of tending toward infinity. As a consequence, in supersonic flow, a perforated wall of given geometry will not satisfy the requirements for interference-free testing through the entire Mach number range. In the immediate vicinity of Mach one a smaller open-area ratio must be employed in order to establish the high cross-flow resistance required by flow theory.

3. EXPERIMENTS TO DETERMINE THE CROSS-FLOW CHARACTERISTICS OF PERFORATED WALLS

A large number of experiments have been carried out in order to verify and supplement the theory concerning the cross-flow characteristics of perforated walls. Such experiments covered a wide range of wall geometry, in which, among other parameters, the hole arrangements, wall thickness, hole orientation, etc., were varied as were the Mach number and viscosity effects. Some typical experimental results are discussed in the following sections.

a. Simple types of perforated walls in low-speed flow ($M = 0.05$) wall with transverse slot

Since theoretical investigations (see Section 2 of this chapter) had predicted quantitatively the behavior of a single transverse slot in a thin wall, some early experiments were performed to verify the theoretical results in a low-speed ($M = 0.05$) wind tunnel.[2] Various cross-flow velocities were produced by varying the pressure in the plenum chamber which surrounded the test wall. In this manner, outflow from and flow into the test section could be established.

First, the pressure distribution along the wall upstream and downstream of the slot was determined and the data compared with the theoretical results (see Fig. 11.8a). The agreement between theoretical and experimental data is remarkable. Upstream of the slot the pressures are lower than in the parallel flow; higher pressures exist downstream of the slot. The theoretical assumptions made in the simplified theory are therefore compatible with real flow, and consequently, good agreement can also be expected in the case of the pressure drop characteristics.

The pressure change caused by the cross flow through the single slot described above is shown in Fig. 11.8b. The experiments prove definitely that the pressure drop characteristic is of the linear type. However, the slope of the curve is somewhat larger than linear theory predicts, even when quadratic terms for the pressure change are taken into consideration.

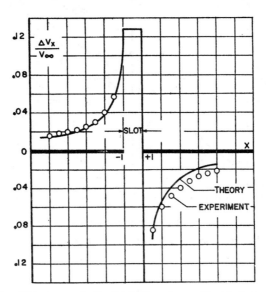

Fig. 11.8a. *Disturbance velocity distribution along wall with single transverse slot* $(M = 0.05)$.[2]

It is significant that even in the case of flow *into* the test chamber, that is, at a condition for which the theory is not expected to apply, the linear relationship still exists.

Walls with a Single Hole and Various Types of Perforations.—In the same test series, the characteristics of a wall with a single large hole and several truly perforated walls with many holes were investigated at a subsonic Mach number of approximately 0.05.

The single hole had a diameter of 2.26 in. in a test section 4 in. wide. As was the case with the single slot, the cross-flow characteristics of the single hole exhibit a linear trend (Fig. 11.9). The cross-flow pressure drop, however, is somewhat smaller than for the single slot.

The cross-flow results for the four perforated walls with open-area ratios varying between 20 and 37 per cent show the same linear trend as the other extremely idealized configurations (single slot and single hole, Fig. 11.9). The average cross-flow pressure drop for the perforated walls may be represented by:

$$\Delta p/q = 1.0(v_n/v_\infty)$$

The results for all perforated walls coincide quite well. Apparently the interference between individual holes has small influence within the range of open-area ratios tested.

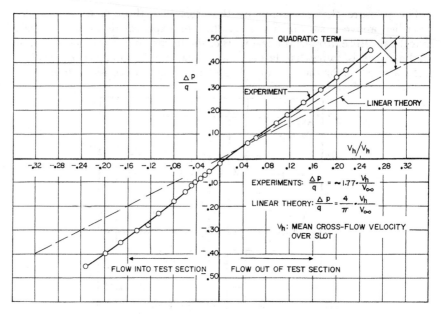

Fig. 11.8b. *Cross-flow characteristics of single transverse slot* ($M = 0.05$).[2]

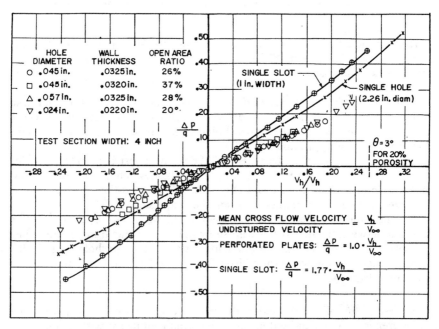

Fig. 11.9. *Cross-flow characteristics of various perforated walls* ($M = 0.05$).[2]

If the pressure-drop characteristic is related to the velocity v_y perpendicular to the perforated wall and at some distance from the wall, the following equation is obtained:

$$\Delta p/q = 1.0(1/R)(v_y/v_\infty) = 1.0(1/R)\theta$$

where R = open-area ratio of the wall
 θ = flow angle.

Again it should be noted that in the range of *inflow* to the test section the linear relationship holds approximately true; however, some pronounced differences between the individual walls are apparent. Unfortunately no measurements are reported for the boundary layer thickness along these test walls. In the light of later experiments (see Section 4 of this chapter), it may be assumed that the boundary layer thickness was not larger, but probably smaller, than the diameter of the individual holes. Also the holes tested in this series were consistently larger than the thickness of the wall, with the ratio of hole diameter to wall thickness varying between 1.1 and 1.7. As shown later, the small boundary layer thickness in conjunction with the relatively large holes was probably instrumental in establishing the linear characteristics of the perforated walls in the above test series.

b. Mach number influence of cross-flow characteristics of perforated walls (see also Section 4c of this chapter)

Subsonic Range.—Several test series have been performed to investigate the influence of Mach number on the cross-flow characteristics of perforated walls. Typical results of one such investigation are shown in Fig. 11.10

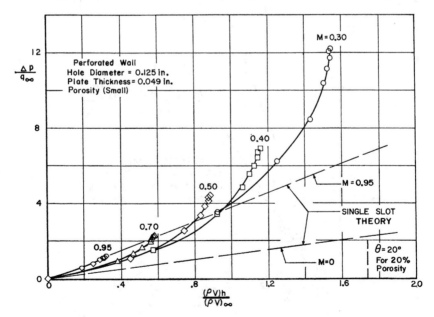

FIG. 11.10. *Cross-flow characteristics of perforated wall at different subsonic Mach numbers.*[8]

(Ref. 8). In this case the test wall had a small open-area ratio of approximately 7 per cent and holes of 0.125 in. diameter in a thin wall 0.049 in. thick. The test results show the slope of the cross-flow curves increasing with increasing subsonic Mach number at approximately the rate predicted by theory. However, the curves show a pronounced tendency to deviate from linear characteristics when large cross-flow velocities occur in the individual holes. At each Mach number the characteristic curve becomes extremely steep when large cross-flow velocities occur, indicating the proximity of choked flow for the hole. In general, however, cross-flow velocities high enough to produce choking in the holes do not occur in normal wind tunnel testing. The range of small angles of approaching flow, up to approximately 5°, are of greater interest. However, the experimental data of this test series were not obtained at small enough intervals to define wall characteristics in this range with sufficient accuracy.

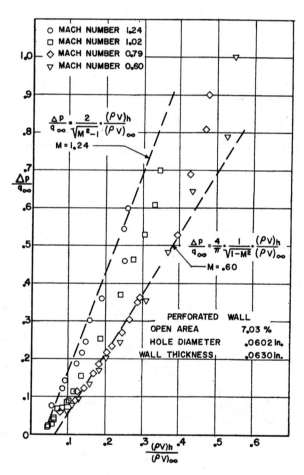

Fig. 11.11. *Cross-flow characteristics of perforated wall at subsonic, sonic and supersonic Mach numbers.*[7]

Subsonic, Sonic and Supersonic Mach Numbers.—In another test series a perforated wall with a small open-area ratio of 7 per cent, a hole diameter of 0.06 in., and a wall thickness of 0.063 in. was investigated at Mach numbers ranging from 0.60 to 1.24 (see Ref. 7). Some experimental results for this wall are shown in Fig. 11.11. Again it is apparent that with increasing Mach number the slope of the characteristic cross-flow curves becomes steeper. When the range of small cross-flow mass flows is disregarded (below approximately $(\rho v)_h/(\rho v)_\infty = 0.10$), the slope of the curve for Mach number 0.6 is in reasonable agreement with the curve predicted by theory for individual slots, that is,

$$\frac{\Delta p}{q_\infty} = \frac{4}{\pi(1-M^2)^{\frac{1}{2}}} \frac{(\rho v)_h}{(\rho v)_\infty}$$

In the case of supersonic velocity, the test curve is also in satisfactory agreement with the theoretical slope:

$$\frac{\Delta p}{q} = \frac{2}{(M^2-1)^{\frac{1}{2}}} \frac{(\rho v)_h}{(\rho v)_\infty}$$

It should be remembered that in the experiments reported in Fig. 11.11, the test-section flow direction was not oblique but was parallel to the test wall. Consequently the cross-flow pressure drop equation must be modified slightly to account for the difference from the theoretical treatment described in Chapter 8 for *oblique* supersonic flow approaching a perforated wall. Because of the small open-area ratio of the test wall in this case, however, the difference between oblique and parallel flow approaching a wall is very small.

It is interesting that in the reported experiments with perforated walls (Fig. 11.11) the characteristic curves show increased curvature in the range of small cross-flow velocity. It is believed that the curvature is caused by viscosity effects and especially by the flow of the boundary layer along the perforated wall. This consideration is in agreement with the results reported later in this section on the influence of boundary layer thickness. In the tests under discussion here the boundary layer should be rather large in comparison with the relatively small hole diameter of 0.06 in. When, at larger cross-flow velocities, the boundary layer is bled off through the perforated walls, the influence of the boundary layer is reduced, and better correlation with theory occurs. This trend is also exhibited by the test data shown in Fig. 11.12.

Supersonic Mach Numbers.—A large number of different perforated walls were investigated to study cross-flow characteristics at Mach number 1.45 (see Ref. 7). The walls had hole diameters ranging from 0.06 to 0.50 in., wall thickness from 0.02 to 0.299 in., and greatly varying open-area ratios. The experimental results presented in Fig. 11.12 indicate that in most cases the slope of the linear characteristic curve is somewhat larger than was predicted by theory. Linear theory resulted in characteristic curves according to:

$$\frac{\Delta p}{q} = \frac{2}{(M^2-1)^{\frac{1}{2}}} \frac{(\rho v)_h}{(\rho v)_\infty}$$

The slope of all curves presented is approximately 10 per cent greater than was predicted.

Some remarkable features are exhibited by certain unorthodox wall configurations. The majority of the walls had ratios of hole diameter to wall thickness considerably larger than one. However, one wall had a hole diameter only approximately 40 per cent of the wall thickness (wall A in Fig. 11.12). In this case the characteristic curve clearly has a smaller slope

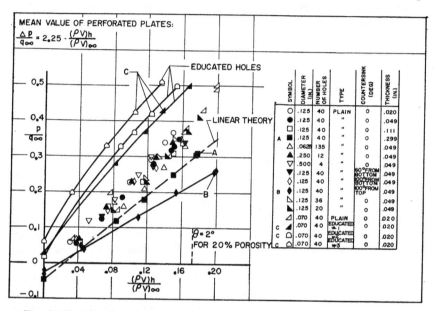

Fig. 11.12. *Cross-flow characteristics of various perforated walls at $M = 1.45$.*[7]

and also shifts in a direction which requires negative pressures in the plenum chamber to provide zero outflow from the test section. This result is in quantitative agreement with the expected results from the diffuser effect and the discussion of the physical aspects of flow through thick perforated walls (see Section 1 of this chapter).

Another wall B had holes which were large compared with the wall thickness and which were countersunk on the test-section side (Fig. 11.12). This particular shape exhibited significantly reduced resistance to cross flow, which is in agreement with qualitative thinking, since the loading effect of the holes on the test-section side is obviously disturbed. When the same wall is turned around so that the countersunk shape is located on the plenum chamber side, the walls once more exhibit the normal characteristics of thin perforated walls. The streamline pattern for a wall with conventional holes is shown schematically in Fig. 11.13.

Another interesting wall investigated had so-called "educated holes". The guiding idea for this modification of the hole shape was the development of a wall which would not produce outflow even though the pressure on the flow side of the wall was greater than in the surrounding plenum chamber.

Such a characteristic is particularly desirable in some diffuser applications when no outflow of the supersonic flow is desired, though in subsonic flow the holes should act like regular holes. The flow pattern for walls with holes of this type is also shown in Fig. 11.13. When the pressure in the plenum chamber is just slightly smaller than the static pressure in the supersonic flow, the supersonic expansion waves force the flow to turn an amount which is not sufficient to clear the downstream edge of the hole. This concept of the flow pattern is borne out by the Schlieren picture presented in Fig. 11.14.

(a) CONVENTIONAL HOLE

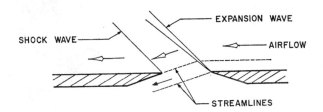

(b) "EDUCATED" HOLE

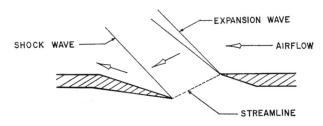

Fig. 11.13. *Flow pattern through wall with conventional holes and "educated" holes (ref. UAC Corp. Research Dept.).*

The cross-flow characteristics of the walls with "educated holes" (wall C in Fig. 11.12) indicates that, as desired, no outflow occurs at positive pressure differences, that is, at pressures which are greater in the test section than in the plenum chamber. The pressure differential may thus be as high as 6 per cent of the dynamic pressure of the supersonic flow without producing outflow from the test section. If the pressure difference is increased, outflow gradually begins to increase until finally, at a large enough outflow, the cross-flow characteristic curve is approximately parallel to the majority of perforated test walls.

c. *Summarizing remarks*

In view of the reported results of experiments to determine the cross-flow characteristics of various perforated walls, it may be concluded that as a general rule the walls exhibit the characteristics predicted by theory. It must be remembered, however, that the linearized potential flow theory can predict the wall characteristics only when the boundary layer is thin and the wall thickness is small compared with the hole diameter. Also the mean cross-flow velocities through the holes must be sufficiently lower than velocities which lead to choking of the flow through the holes. In the following sections, some parameters such as wall thickness, hole inclination, and boundary layer thickness will be discussed in more detail.

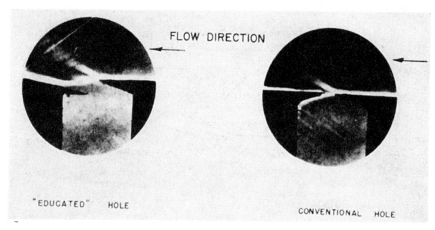

Fig. 11.14. *Schlieren pictures of flow through conventional hole and through "educated" holes at $M = 1.79$ (ref. UAC Corp. Research Dept.)*

4. EXPERIMENTS TO DETERMINE THE INFLUENCE OF DETAILED WALL GEOMETRY AND FLOW PARAMETERS

Several systematic test series have been conducted in the transonic model tunnel of the AEDC to explore the influence of wall geometry on the cross-flow characteristics of perforated walls. In the following sections some of the principal findings of these tests are discussed briefly.

a. Influence of wall thickness at constant diameter hole diameter[9]

A number of perforated test walls which had an open-area ratio of 22.5 per cent and $\frac{1}{4}$ in. diameter holes were installed in the test section of the transonic model tunnel. The thickness of the test wall was varied from $\frac{1}{16}$ to $\frac{1}{2}$ in. so that typical cases of a thin wall (wall thickness to hole diameter $= \frac{1}{4}$) and a thick wall (wall thickness to hole diameter $= 4$) were investigated. Some results are shown in Figs. 11.15a–11.15d for various Mach numbers between 0.75 and 1.175.

At subsonic Mach numbers particularly, the relatively thin test walls had consistently larger cross-flow resistance coefficients than did the relatively thick walls, and the curves were considerably straighter at the relatively thin walls. This feature is in agreement with the theoretical prediction since only a thin wall can be expected to act as a series of individual wings to produce

CROSS-FLOW CHARACTERISTICS OF WALL ELEMENTS

the desired pressure difference between the upper and lower surfaces of the plate. At supersonic Mach numbers the influence of the wall thickness is not as pronounced as at subsonic Mach numbers.

At supersonic Mach numbers, as at subsonic Mach numbers, the characteristic curves show a strong curvature at low cross-flow ratios, $(\rho v)_h/(\rho v)_\infty < 0.04$. At cross-flow ratios above 0.04, the curves tend to become more and more linear. This characteristic can be explained by consideration of the

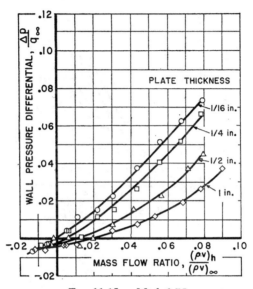

FIG. 11.15a. *Mach 0.75.*

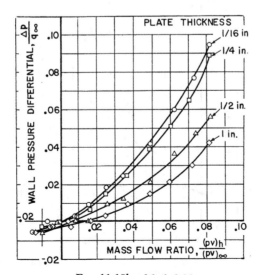

FIG. 11.15b. *Mach 0.90.*

influence of the boundary layer thickness. At small cross-flow velocities, but little boundary layer was removed through the walls. At a larger outflow velocity the boundary layer was discharged through the upstream portion of the test wall, and the major portion of the wall was exposed to oblique

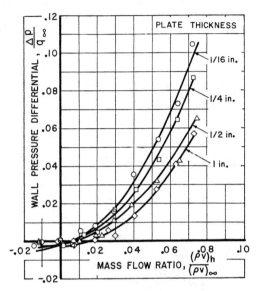

FIG. 11.15c. *Mach 1.0.*

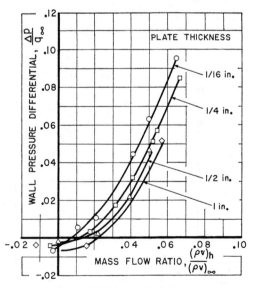

FIG. 11.15d. *Mach 1.175.*

FIG. 11.15. *Influence of wall thickness on the cross-flow characteristics of a 22.5 per cent open perforated wall with $\frac{1}{4}$ in. diam. holes.*[9]

flow having a boundary layer which had been considerably thinned. This result was caused by the fact that the test wall extended approximately 35 in. in the direction of the flow, that is, its length was three times the width of the test section. Therefore the cross-flow characteristics presented in Fig. 11.15 are the mean values obtained over the entire length of the test wall.

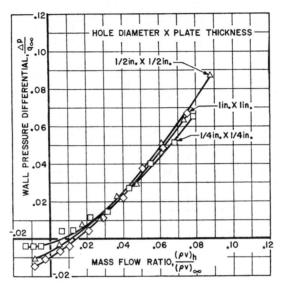

Fig. 11.16a. *Mach 0.75.*

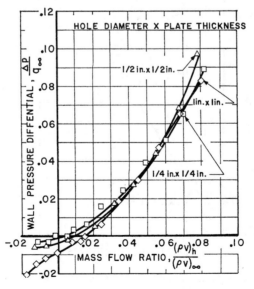

Fig. 11.16b. *Mach 0.90.*

On the basis of these experiments it may be concluded that the wall thickness should not be larger than the hole diameter, and that, if possible, it should be even smaller in order to establish consistent cross-flow characteristics.

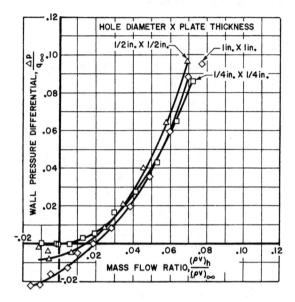

Fig. 11.16c. *Mach 1.0.*

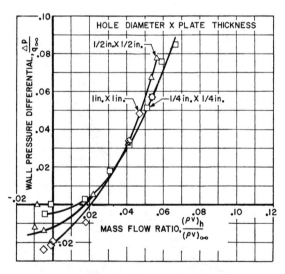

Fig. 11.16d. *Mach 1.175.*

Fig. 11.16. *Influence of hole size on the cross-flow characteristics of 22.5 per cent open wall with ratio of hole diameter to plate thickness of unity.*[9]

CROSS-FLOW CHARACTERISTICS OF WALL ELEMENTS

b. Influence of hole size at a constant ratio of hole diameter to wall thickness

In order to demonstrate that wall cross-flow characteristics are similar when the ratio of hole diameter to plate thickness is constant, a series of test walls having different hole diameters and a constant ratio of hole diameter to wall thickness were tested in the test rig discussed in the preceding section.[9] Experiments were conducted with a hole size of $\frac{1}{16}$ in. in a wall of the same thickness, and others were conducted with hole diameters up to 1 in., also in walls of the same thickness. Results at some typical Mach numbers, presented in Figs. 11.16a–11.16d, show that the curves at mass flow ratios larger than 0.02 are indeed very much alike. This is true for both the subsonic and the supersonic range. However, in the low mass flow range, that is, when $(\rho v)_h/(\rho v)_\infty < 0.02$, the curves deviate considerably. As discussed in the preceding section, this discrepancy can be traced back to the influence of the boundary layer, which, in proportion, is exceedingly thick at the small hole diameter wall ($\frac{1}{4}$ and $\frac{1}{16}$ in.) and must therefore be bled off at the higher mass flow ratios before the characteristics can approximate non-viscous characteristics closely enough. Consequently, in the low mass flow range the characteristics of the $\frac{1}{4}$ and $\frac{1}{16}$ in. walls are extremely irregular (see Figs. 11.15 and 11.16).

c. Influence of Mach number

The results obtained from tests with several perforated walls are plotted together in Fig. 11.17 for Mach numbers between 0.75 and 1.75 to show more clearly the influence of Mach number. For the wall with $\frac{1}{16}$ in. hole diameter and $\frac{1}{16}$ in. thickness the wall characteristics at all Mach numbers are irregular for outflow coefficients smaller than 0.03, that is, for a relatively thick boundary layer compared with the hole size (Fig. 11.17a). At the

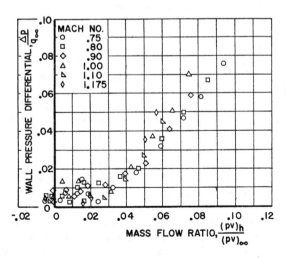

FIG. 11.17a. *Influence of Mach number on the cross-flow characteristics of 22.5 per cent open wall with $\frac{1}{16}$ in. hole diameter and $\frac{1}{16}$ in. wall thickness.*[9]

larger outflows, the wall characteristics approach linear curves having slopes, which, over a range of Mach numbers from 0.75 to 1.175, agree in trend with theory.

In Fig. 11.17b the characteristics of a wall with relatively large holes compared with the wall thickness ($\frac{1}{2}$ in. diameter hole, wall thickness $\frac{1}{16}$ in.) are shown. At large outflow ratios, that is, when $(\rho v)_h/(\rho v)_\infty$ was larger than 0.02, the slope of the characteristic curves was somewhat steeper for the large hole wall than for the small hole wall. The main

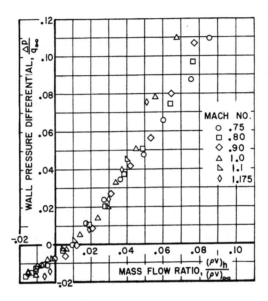

Fig. 11.17b. *Influence of Mach number on the cross-flow characteristics of 22.5 per cent open wall with $\frac{1}{2}$ in. hole diameter and $\frac{1}{16}$ in. wall thickness.*[9]

difference occurred at the small outflow ratios, that is, when $(\rho v)_h/(\rho v)_\infty$ was smaller than 0.03, and at negative values. While the $\frac{1}{16}$ in. wall exhibits irregular characteristics in this cross-flow range (see Fig. 11.17a), the wall with $\frac{1}{2}$ in. holes exhibits consistent characteristics in the same range, having a definite but reduced slope. Again the difference in the behavior of the two walls can be explained by the influence of the boundary layer thickness compared with the hole diameter.

When the effectiveness of perforated walls in the reduction of wall interference effects in wind tunnel testing is considered, it is evident that walls with large holes compared with the boundary layer and the wall thickness most nearly produced the desired characteristics. However, even with such walls the slope of the characteristic curves is so much reduced in the range of small or negative mass flow ratios that special efforts are necessary to develop walls which will also minimize this undesirable feature. Some of the experimental work in this field is presented in the following section.

d. Walls with inclined holes (differential-resistance walls)

When the holes in a perforated plate are inclined in the direction of the flow, flow out of the test section becomes easier and flow into the test section more difficult (Fig. 11.18). The latter result is due to the fact that the inflowing air must overcome part of the dynamic head of the test-section flow. On the other hand, since the important downstream edge of the holes is made sharper by inclining the holes, such a wall acts more like a thin wall and consequently exhibits more linear characteristic curves over the entire cross-flow range. These walls also produce consistently greater resistance to inflow to the test section.

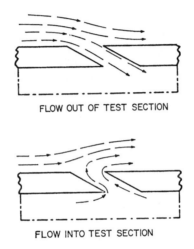

FIG. 11.18. *Streamline pattern for inflow and outflow through a wall with inclined holes.*

The feature of considerably *reduced* outflow resistance was documented, for instance, at outflow ratios $(\rho v)_y/(\rho v)_\infty$ larger than 0.005.[10] In another instance[11] the overall characteristics of walls with inclined holes were systematically investigated.

Influence of Hole Inclination Angle.—Some typical data obtained in the transonic model tunnel of the AEDC are presented in Figs. 11.19a and b for a wall having 12 per cent open-area ratio, $\frac{1}{4}$ in. hole diameter, and $\frac{1}{4}$ in. plate thickness. The holes were arranged in the walls at angles of 0° (straight holes), 30°, 40°, 45° and 60°, respectively.[11] The experimental data show that the outflow resistance is drastically reduced when the angle of the holes is inclined in the direction of the flow. More significantly, however, the characteristic curves exhibit a pronounced steeper slope at small and negative flow ratios, particularly at the inclination angle of 60°. This trend of the curves was established over the entire Mach number range of the test, as demonstrated at the two selected Mach numbers of 0.90 and 1.10.

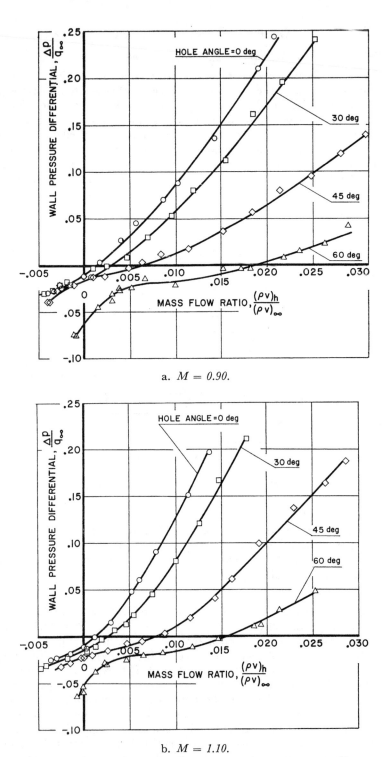

a. $M = 0.90$.

b. $M = 1.10$.

FIG. 11.19. *Influence of hole inclination on the cross-flow characteristics of 12 per cent open perforated wall.*[11]

The sharp increase in the slope of the curves in the low and negative flow region in the case of the inclined-hole walls indicates that not only can the irregular character of the curves produced by straight holes be avoided, but that, on the other hand, the often extremely desirable characteristic of *increased* resistance to flow into the test section can be realized by means of hole inclination.

The superiority of an inclined-hole wall over a straight-hole wall with respect to its nearly linear characteristics and the increased slope in the inflow region is demonstrated by the typical results presented in Figs. 11.20a and b. The open-area ratio of the wall with inclined holes was reduced to 6 per cent in order to match the outflow resistance of a wall with straight holes and a $22\frac{1}{2}$ per cent open-area ratio.

It should be noted here that the open-area ratio is always calculated on the basis of the total hole area measured perpendicular to the axis of the inclined holes.

Influence of Hole Diameter vs. Plate Thickness of Walls with Inclined Holes.—In the graphs showing the characteristics of perforated walls with inclined holes, only the test results obtained using walls with equal hole diameter and thickness are presented. Since the effectiveness of an inclined-hole wall depends upon the ability of the wall to guide the inflow against the test-section flow, the length of the guiding inclined holes required to produce this counter-flow effect of the inflow is of interest. Consequently, a special test series was initiated in the AEDC transonic model tunnel to investigate this influence.[12]

The test results for 6 per cent open-area ratio walls with 60° inclined holes indicate consistent nearly-linear characteristics when the hole diameters are equal to or twice as large as the wall thickness (Fig. 11.21). When, however, the wall thickness is increased to produce a more perfect guiding of the flow, the characteristic wall curves change in a very unfavorable manner.

If, for instance, the wall is twice as thick as the hole diameter, the pressure difference between plenum chamber and test section remains essentially constant between inflow values $(\rho v)_h/(\rho v)_\infty$ of approximately 0.002 and outflow values up to 0.014. This result can be understood when the aerodynamic action of thick walls is considered (discussed in detail in Section 1 of this chapter with Fig. 11.1). Thick walls, particularly thick walls with inclined holes, tend to turn the flow in a manner similar to the turning of the channel flow in a compressor lattice. Consequently, a diffuser effect with a pressure rise is produced which, in the ideal case of a thick wall with perpendicular holes without inflow losses, amounts to

$$\frac{\Delta p}{q_\infty} = \frac{1}{R^2}(\sin^2\theta - 1)$$

In order to produce a diffuser effect, the individual channels in the thick wall must be long enough to turn the flow efficiently. It is of interest to note that the thin wall with 60° inclined holes produced the most desirable crossflow characteristics (see Fig. 11.21), but *no complete channel* for guidance of the flow was formed (Fig. 11.22). With thick walls, corresponding to

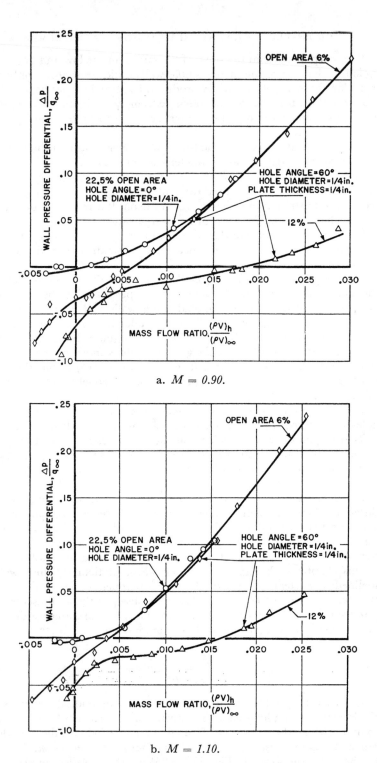

a. $M = 0.90$.

b. $M = 1.10$.

Fig. 11.20. *Comparison between cross-flow characteristics of perforated wall with straight holes and perforated wall with inclined holes.*[11]

ratios of hole diameter to wall thickness of 0.5, the guidance was considerably more complete, and, as expected, produced a sizable and extremely undesirable diffuser effect (see Fig. 11.21). When the cross-flow angle was larger than the inclination of the holes, the diffuser effect naturally vanished.

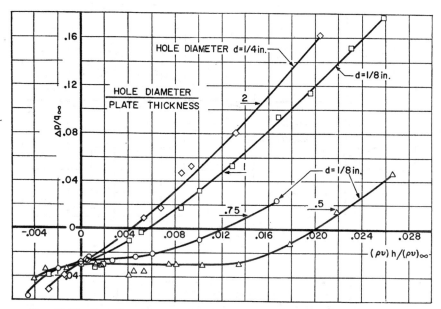

FIG. 11.21. *Cross-flow characteristics of perforated walls with 60° inclined holes for various ratios of hole diameter to wall thickness, open-area ratio 6 per cent at* $M = 0.90$.[12]

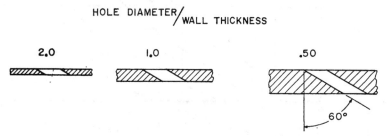

FIG. 11.22. *Flow pattern through walls with inclined holes at different ratios of hole diameter to wall thickness.*

e. Influence of boundary layer thickness

In the previous discussion of test results, it was repeatedly mentioned that the boundary layer along the perforated walls tended to interfere greatly with the desired linear characteristics of the perforated walls, particularly in the region of small and negative cross-flow values. In order to study the boundary layer effect more specifically, a special test rig was developed for the AEDC transonic model tunnel which made it possible to vary the thickness of the boundary layer approaching the perforated test wall.

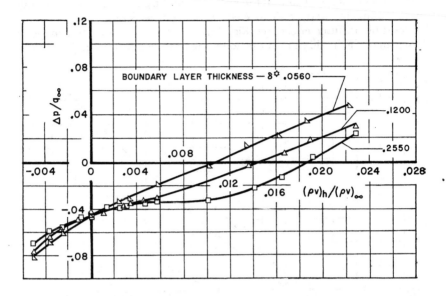

a. $M = 0.90$.

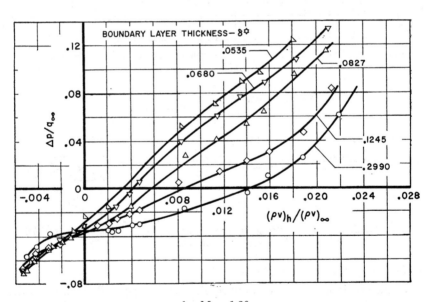

b. $M = 1.20$.

Fig. 11.23. *Influence of boundary layer displacement thickness on cross-flow characteristics of perforated wall with $60°$ inclined holes, hole diameter $\frac{1}{8}$ in., wall thickness $\frac{1}{8}$ in., open-area ratio 6 per cent.*[12]

The test wall had only a small extension in the flow direction so that, for purposes of analysis, a mean constant thickness of the boundary layer across the test plate could be assumed (see Ref. 12). Typical data obtained for walls with holes of 60° inclination, a hole diameter and wall thickness of $\frac{1}{8}$ in., and 6 per cent open-area ratio indicate that serious irregularities of the curves occur when the boundary layer displacement thickness is larger than the diameter of the holes (Figs. 11.23a and b). This fact became particularly obvious in the experiment at Mach number 1.20. Apparently the displacement thickness of the boundary layer should not exceed approximately 75 per cent of the hole diameter.

The observed change of the wall characteristics with boundary layer thickness has a significant effect on the growth of the boundary layer along the test-section walls. In well-adjusted transonic test sections the plenum chamber and test-section pressures are constant over the length of the test section. When a given pressure differential is established by such means as control of the plenum chamber suction, as is customary in transonic wind tunnel operation, the wall regions having thick boundary layer will produce

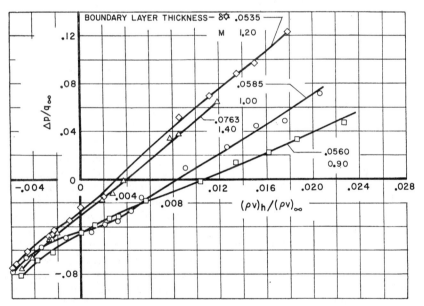

Fig. 11.24. *Influence of Mach number on cross-flow characteristics of perforated wall with 60° inclined holes; hole diameter $\frac{1}{8}$ in., wall thickness $\frac{1}{8}$ in., open-area ratio 6 per cent, for relatively thin boundary layer thickness.*[12]

a large outflow and, consequently, a reduction of the boundary layer thickness. As a result, walls with inclined holes have a tendency to establish a more uniform boundary layer thickness along the length of the test section.

With proper boundary layer control by such means as auxiliary plenum chamber suction pumps, it is possible to establish wall characteristics having nearly linear cross-flow curves over the entire Mach number range. In the experiments presented in Fig. 11.24, the boundary layer displacement thickness was controlled so that it did not exceed values of 50 per cent of the hole diameter. Over the entire range of the test, from Mach number 0.9 to 1.4, the curves exhibit nearly linear characteristics. The walls with inclined holes and proper plenum chamber suction control therefore represent a significant means of establishing wall characteristics which match the values necessary to obtain interference-free test results in transonic test sections.

5. EXPERIMENTS TO DETERMINE CROSS-FLOW CHARACTERISTICS OF WALLS WITH LONGITUDINAL SLOTS

a. Slotted walls in oblique flow without flow curvature

In order to explore the characteristics of walls with longitudinal slots, some brief experiments were conducted at several laboratories. According to non-viscous flow theory, walls with longitudinal slots should produce a pressure drop proportional, not to the cross-flow velocity itself, but to its square. As shown in Chapter 8, the cross-flow pressure drop for slotted walls in non-viscous flow should be:

$$\Delta p/q_\infty = (v_h/v_\infty)^2$$

In real viscous flow, however, an additional linear term appears which is caused either by friction on the side walls of the slot or by flow separation along the edges of the slot. The total pressure drop in parallel oblique flow can then be written in the general form:

$$\Delta p/q_\infty = K_1(\rho v)_h/(\rho v)_\infty + K_2(\rho v)_h^2/(\rho v)_\infty^2$$

Wall with Single Slot.—In the AEDC transonic model tunnel (12 in.) some exploratory tests were conducted using a test wall which had a single longitudinal slot. Purposely, the slot width was selected to be 1.3 in., that is, 11 per cent of the test section width, in order to make the boundary layer effects small. This sharp-edged slot was placed in a relatively thin wall ($\frac{1}{8}$ in. thickness—see Fig. 11.25a). The most remarkable result is that the test points for all Mach numbers between Mach number 0.75 and 1.20 form a single curve, which, though showing slight curvature, exhibits a definite linear character in the range of small cross-flow velocities. When the quadratic term is eliminated, the cross-flow pressure drop can be represented approximately by:

$$\Delta p/q = 0.4(\rho v)_h/(\rho v)_\infty$$

This value is considerably smaller than the values obtained for most perforated walls, but in view of the consistent constant characteristics through the Mach number range, the longitudinally slotted wall was subjected to more detailed exploratory investigations. Several tests (results unpublished) were conducted in the same model tunnel using a wall thickness increased considerably beyond the $\frac{1}{8}$ in. of the wall presented in Fig. 11.25a. Also, slots with the edges beveled to increase their sharpness and with rounded edges were studied. In all cases, basically similar characteristics were obtained, that is, a remarkable independence of Mach number existed as well as a predominantly linear characteristic of the cross-flow pressure drop.

The velocity patterns just inside the test section, in the slot itself, and on the plenum chamber side of the wall obtained with the single slot configuration shown in Fig. 11.25a were determined and plotted in Fig. 11.25b. Obviously flow separation occurs along the sharp edges of the slot, as expected.

Walls with Numerous Small Slots.—Several walls with short longitudinal slots equivalent to perforated walls with elongated holes were investigated in a test series conducted by United Aircraft Corporation.[10] These walls might also be considered combination slotted–perforated walls. As expected, the walls showed characteristics similar to those of conventional perforated walls, that is, a predominant linear term occurs (see Fig. 11.26). They also show only a relatively small influence of Mach number.

Such a combination of longitudinally slotted–perforated walls deserves further consideration for transonic wind tunnel testing because of its favorable linear and consistent cross-flow characteristics. The test data (Fig. 11.25) indicate consistent characteristics at small cross-flow velocities, even

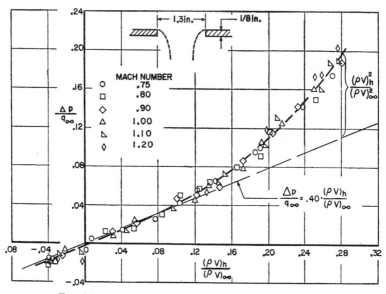

Fig. 11.25a. *Cross-flow characteristics at various Mach numbers.*

with flow into the test section, when the boundary layer is kept sufficiently thin, as in this case. Also, such combination perforated–slotted walls eliminate the disadvantage of secondary flow which was discussed in Chapter 9 and was found to be a major source of the difficulties with pure slotted walls in supersonic test sections.[1]

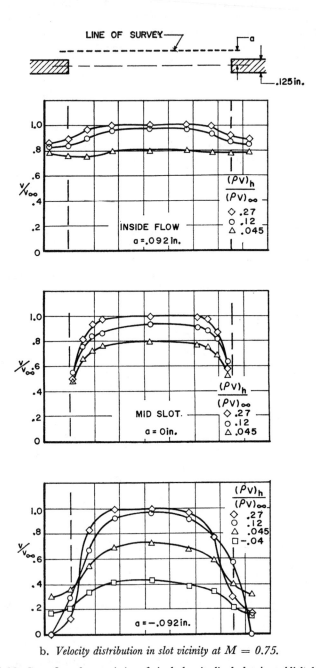

b. *Velocity distribution in slot vicinity at $M = 0.75$.*

Fig. 11.25. *Cross-flow characteristics of single longitudinal slot (unpublished data of AEDC transonic model tunnel—Gardenier and Chew).*

b. Flow curvature effect for longitudinally slotted walls

According to theory, longitudinally slotted walls should produce a pressure change in cross flow which is proportional to the curvature of the flow approaching the wall. As shown in Chapter 8, this pressure is due to the magnification of the centrifugal forces in this flow.

In order to check the predicted effect, experiments were conducted in which a two-dimensional thick airfoil was used to produce a flow curvature

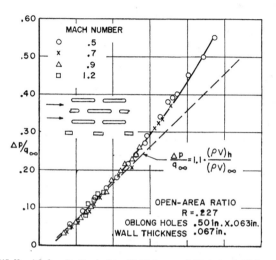

a. Wall with longitudinal slots (0.50 in. × 0.63 in.), 0.67 in. wall thickness, and 22.7 per cent open-area ratio.

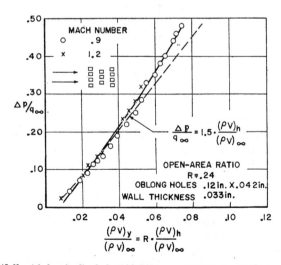

b. Wall with longitudinal slots (0.20 in. × 0.042 in.), 0.033 in. wall thickness, 22.4 per cent open-area ratio.

FIG. 11.26. Cross-flow characteristics of two walls with small-grain longitudinal slots.[10]

near the slotted test wall.[6] For these tests the wind tunnel was equipped with slotted top and bottom walls having a 14 per cent open-area ratio. The mean disturbance velocities at a small distance from the test wall were measured, both in the flow direction and perpendicular to it (see Fig. 11.27). According to theory, the disturbance pressures should be connected

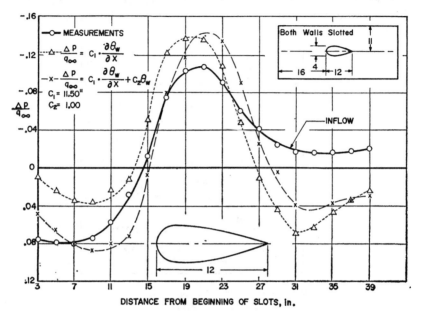

FIG. 11.27. *Two-dimensional airfoil in wind tunnel with longitudinal slots in incompressible flow*[7] (*pressure disturbances along slotted walls according to measurements and theory*).

with the flow curvature according to the following equation:

$$\Delta p/q + C(\partial \theta / \partial X) = 0$$

The constant C in the present case equals 5.75 in.

The data presented in Fig. 11.27 shows that the experimental curve and the curve calculated from the measured flow curvature coincide in their general trends. The agreement between the two can be improved somewhat when a linear pressure term is added according to

$$\Delta p/q = 0.16\theta$$

The two curves, one measured directly, the other derived from the flow inclination and curvature, agree satisfactorily upstream of the maximum thickness of the wing. In the region of flow into the test section, however, considerable discrepancy occurs. Downstream of station 21 the flow along the wall changes from outflow to inflow, and the observed discrepancy between the two disturbance curves begins in the same region.

It may be concluded, therefore, that in principle the theoretical equation for the pressure drop due to flow curvature has been verified by the experiments. This is particularly true in the region of outflow from the test section. It might be possible to reduce the inflow discrepancies by converging the walls or by using deeper slots in order to preserve the axial momentum of the flow entering the slots and to minimize viscosity effects.

6. FLOW DISTURBANCES OF WALL OPENINGS AND THEIR DECAY

Sensitive Schlieren pictures of supersonic flow in perforated test sections frequently indicate that the wall openings produce a number of small waves which propagate in the main flow over considerable distances. Naturally in supersonic flow each opening produces a system of compression and expansion waves which, at a large enough distance from the wall, eliminate each other. As pointed out in Chapter 8, the distance over which this decay occurs depends upon the size of the individual holes; consequently, a maximum hole size is defined for supersonic flow in a transonic test section since the wave disturbances of the wall will decay to very small values in the vicinity of the model and, ideally, will disappear completely.

Two typical Schlieren pictures are presented in Fig. 11.28. For this test the holes were purposely made oversize in order to produce a more pronounced disturbance effect. It is obvious that in both cases the model was exposed to flow which was disturbed by two sets of waves originating at the wall. Even if such disturbances are not large enough to produce a noticeable effect on the non-viscous flow characteristics of the model, the boundary layer might be disturbed when it flows through the region of the oscillating static pressure produced by the wave system. A particular danger is that premature transition from laminar to turbulent flow might occur.

In order to study the magnitude of the pressure disturbances produced by holes in perforated walls, several test walls with oversize holes were studied in the transonic model tunnel of the AEDC.[13] First, a conventional straight-hole pattern with an open-area ratio of 22 per cent was used. The walls were tested at parallel setting, that is, with relatively thick boundary layer, and at 30' converged setting, that is, with relatively thin boundary layer. Some typical results at Mach number 1.00 are shown in Fig. 11.29a, and for Mach number 1.20 in Fig. 11.29b. The pressure survey for Mach number 1.00 indicates that even at very small distances from the wall no pressure disturbances could be determined within the accuracy of the measurements because, in subsonic flow, localized disturbances propagate in all directions and consequently decay very rapidly.

In the supersonic case of Mach 1.2, however, appreciable disturbances of static pressure and, consequently, of the local Mach number occur near the wall. They gradually lose intensity as the distance from the wall increases. At a distance of approximately 7 in., that is, at a distance of fourteen hole diameters, the pressure distribution is reasonably uniform.

A cross plot for the same wall showing the disturbances as a function of the wall distance is presented in Fig. 11.30. Initially the disturbances decay very rapidly, and, at a distance of approximately twenty-four hole diameters, they reach a low value of $\Delta M = 0.002$. This value appears to indicate the level of flow non-uniformity in the basic test section.

Fig. 11.28. Schlieren pictures of perforated test section with large holes; hole diameter 0.062 in., test-section height 5.5 in.[14]

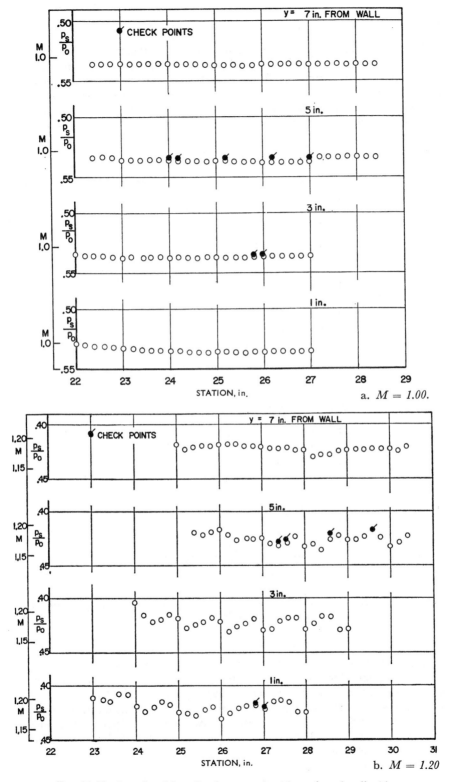

FIG. 11.29. *Intensity of flow disturbances produced by perforated wall with $\frac{1}{2}$ in. diameter holes.*[13]

Similar tests were conducted using walls with 60° inclined holes and an open-area ratio of 6 per cent. The data shown in Fig. 11.31 indicate that, at a distance of approximately thirty mean hole diameters, the disturbances decay to values of $\Delta M = 0.002$. Previous findings were also verified since, at Mach number one and at subsonic Mach numbers, the disturbances of the wall holes decayed so rapidly that they had practically no significance for transonic testing.

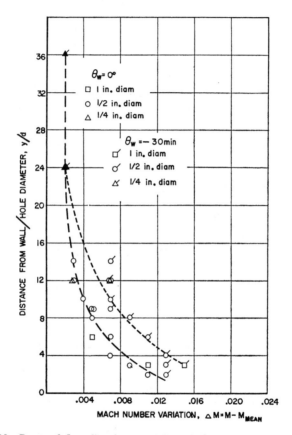

FIG. 11.30. *Decay of flow disturbances produced by various perforated walls with straight holes at $M = 1.20$.*[13]

These tests appear to indicate that the model should not be placed closer to the wall than thirty times the mean of the wall openings. To be safe, particularly in the case of full-scale wind tunnels in which the relative boundary is generally thinner than in model wind tunnels, the hole diameters should not exceed $\frac{1}{80}$ to $\frac{1}{100}$ of the test-section height. In a 12 in. model tunnel, for example, the holes should not be larger than $\frac{1}{8}$ in. In a 10 ft wind tunnel, the hole diameter may be as large as 1.2 in.

7. CONCLUDING REMARKS

The preceding excerpts from numerous investigations of the characteristics of transonic test-section walls show that both theoretical and extensive experimental investigations can be utilized as guides to determine the proper wall geometry for a transonic test section. The linear cross-flow characteristics of the walls desired for transonic testing can be closely approximated using partially open walls with openings which may be

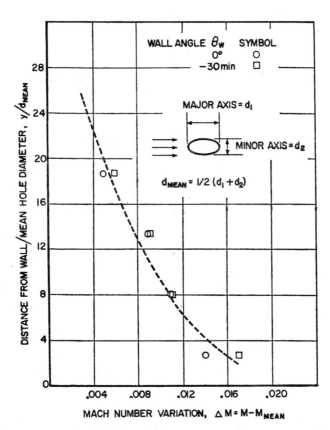

FIG. 11.31. *Decay of flow disturbances produced by perforated wall with $60°$ inclined holes at $M = 1.20$.*[13]

either circular, elliptical, or of a greatly elongated shape. The hole size must be large enough to produce consistent characteristics in spite of the boundary layer along the walls. The viscosity effect can be kept sufficiently small when the displacement thickness of the boundary layer is not thicker than approximately one-half the hole diameter. On the other hand, to eliminate wave disturbances in supersonic flow, the holes must also be small enough so that their disturbance wave system decays sufficiently before it reaches the model. This condition requires that the holes be no larger than approximately $\frac{1}{100}$ of the test-section height.

REFERENCES

[1] ALLEN, H. J. and SPIEGEL, J. M. "Transonic Wind Tunnel Development at the NACA." S.M.F. Fund Paper No. FF–12. Institute of Aeronautical Sciences publication, 1954.
[2] MAEDER, P. F. "Investigation of the Boundary Condition at a Perforated Wall." Brown University, TR WT 9, May, 1953.
[3] GOETHERT, B. H. "Plane and Three-dimensional Flow at High Subsonic Speeds." NACA TM 1105, October, 1946.
[4] WEINIG, F. *Die Stroemung um die Schaufeln von Turbomaschinen, Beitrag zur Theorie, Axial Durchstroemter Turbomaschinen.* Johann Ambrosius Barth, Leipzig 1935 (or J. W. Edwards, Ann Arbor, Michigan, 1948).
[5] MAEDER, P. F. and WOOD, A. D. "Transonic Wind Tunnel Test Sections." *Z. Angew. Math. Phys.* **7**, 1956.
[6] CHEN, C. F. and MEARS, J. W. "Experimental and Theoretical Study of Mean Boundary Conditions at Perforated and Longitudinally Slotted Wind Tunnel Walls." AEDC–TR–57–20, December, 1957.
[7] HILL, J. A. F. "An Analytical Study of Flow Through a Perforated Wall." UAC Report R–95630–9, November 13, 1953.
[8] MCLAFFERTY, G. H., JR. "A Study of Perforation Configurations for Supersonic Diffuser." UAC Report R–53372–7, December, 1950.
[9] CHEW, WILLIAM L. "Cross-Flow Calibration at Transonic Speeds of Fourteen Perforated Plates with Round Holes and Airflow Parallel to the Plates." AEDC–TR–54–65, July, 1955.
[10] RICE, JANET B. "Flow Calibration Studies of Seventeen Perforated Plates with Airflow Parallel to the Plates." UAC Report M–95630–16, May 11, 1954.
[11] CHEW, WILLIAM L. "Experimental and Theoretical Studies on Three-dimensional Wave Reflection in Transonic Test Sections—Part III: Characteristics of Perforated Test-Section Walls with Differential Resistance to Cross-Flow." AEDC–TN–55–44, March, 1956.
[12] CHEW, WILLIAM L. "Characteristics of Perforated Plates with Conventional and Differential Resistance to Cross-Flow and Airflow Parallel to the Plates." Arnold Engineering Development Center, July, 1956.
[13] GARDENIER, H. E. "The Extent and Decay of Pressure Disturbances Created by the Holes in Perforated Walls at Transonic Speeds." AEDC–TN–56–1, April, 1956.
[14] GIARDINA, WILLIAM A. "Effect of Hole Size on Characteristics of Porous Walls in Transonic Tunnel." UAC Report M–15494–1, May 15, 1952.

BIBLIOGRAPHY

CHEW, WILLIAM L. "Total Head Profiles Near the Plenum and Test Section Surfaces of Perforated Walls." AEDC–TN–55–59, December, 1955.
GOETHERT, B. H. "Flow Establishment and Wall Interference in Transonic Wind Tunnels." AEDC–TR–54–44, June, 1954.
GOLDBAUM, G. C. "Research Work on Supersonic Wind Tunnels to Investigate the Use of Perforated Walls." WADC–TR–55–185, March, 1956.
MAEDER, P. F. and STAPELTON, J. F. "Investigation of the Flow through a Perforated Wall." Brown University, TR–WT–10, May, 1953.
MAEDER, P. F. "Some Aspects of the Behavior of Perforated Transonic Wind Tunnel Walls." Brown University, TR–WT–15, OSR–TN–55–116, September, 1954.
MAEDER, P. F. and HALL, J. F. "Investigation of Flow Over Partially Open Wind Tunnel Walls, Final Report." AEDC–TR–55–67, December, 1955.
STOKES, G. M., DAVIS, D. D., JR. and SELLERS, T. B. "An Experimental Study of Porosity Characteristics of Perforated Materials in Normal and Parallel Flow." NACA RM L53H07, November, 1953.

CHAPTER 12

FLOW ESTABLISHMENT IN TRANSONIC WIND TUNNELS

1. GENERAL DISCUSSION OF THE PROBLEM AREAS

The problems associated with the establishment of uniform flow in partially open transonic test sections can best be described by discussing some typical experiments. The first accounts of experiments concerning the establishment of transonic flow in two test sections having combined open-area ratios of $12\frac{1}{2}$ per cent, one with eight longitudinal slots and one with ten, were published by the NACA.[1] The slots had a constant width through the test section, extended into the diffuser, and terminated in an effuser bell. Typical experimental Mach number distributions for these NACA test sections are presented in Fig. 12.1. In the subsonic speed range, the Mach number distributions were very uniform through the main portion of the test section. This same uniformity was maintained when the Mach number approached sonic speed, as indicated by the curve for Mach number 0.99. Even the abrupt change from the solid walls of the inlet nozzle to the slotted portion of the test section produced only a minor waviness of the Mach number distribution. However, at the downstream end of the test section a considerable dip in Mach number occurred which increased in intensity as the Mach number approached unity.

The observed uniformity of the Mach number distribution through the test section up to and including Mach number one is in decided contrast to the distributions observed in closed wind tunnels and indicates that the venting of the test section through the slots into the surrounding plenum chamber is highly effective (Fig. 12.1). Even at the slot open-area ratio of only $12\frac{1}{2}$ per cent, the constant plenum chamber pressure was impressed upon the test-section flow. It was left to later investigators to determine to what extent the open-area ratio of the test section might be reduced without interfering with the venting into the plenum chamber and, consequently, with the constant Mach number distribution through the test section.

In the supersonic Mach number range, significant non-uniformities of the Mach number distributions occurred which grew in intensity as the Mach number in the test section was increased (Fig. 12.1). For example, at an average Mach number of 1.2, the local Mach number at the centerline fluctuated between 1.27 and 1.15. Hence, such test sections are unsuitable for wind tunnel testing unless additional measures are taken to make the distribution more uniform. This erratic behavior of the Mach number distributions in the supersonic speed range is not surprising. Although at subsonic speeds the contours of the inlet nozzle preceding the test section do not greatly affect uniform test-section flow, the contours of the supersonic inlet nozzle must be shaped very carefully and in accordance with supersonic wave reflection and attenuation theory.

It is significant that supersonic Mach numbers were established in the test section simply by increasing the power input to the wind tunnel compressor, that is, by increasing the pressure ratio between the stagnation section ahead of the test section and the downstream end of the diffuser. This type of operation differs from that of conventional closed-wall supersonic test sections. In the conventional closed test sections, Mach number increases in the supersonic range can be obtained only by changing the contour of

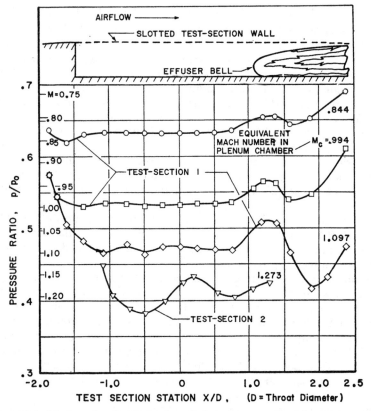

FIG. 12.1. *Mach number distribution for typical slotted test sections (slots with constant width extended into diffuser).*[1]

the supersonic inlet nozzle since, for each contour, only one specific Mach number can be obtained in the test section. The very desirable simplification of operations which is thus possible in a partially open test section over the supersonic speed range and the penalties attendant upon the simplification will be discussed later.

With reference to experiments just cited, the problem areas associated with flow establishment in transonic test sections can be characterized as follows:

(1) Minimum open-area ratio to establish uniform flow in the subsonic speed range up to Mach number one.

(2) Disturbances at the downstream end of a partially open test section leading either to strong Mach number reductions or to strong Mach number increases at the point of transition from test section to diffuser.

(3) Establishment of uniform supersonic flow without the erratic distributions obtained using simple rectangular slots.

These three problem areas will be discussed in more detail in the following paragraphs.

2. ESTABLISHMENT OF SUBSONIC FLOW IN PERFORATED TEST SECTIONS

a. Different types of test-section venting

In the experiments just discussed, the test sections had longitudinal slots which extended into the diffuser area behind the test section. The fact that the slots did extend into the diffuser section is of significance since, in this

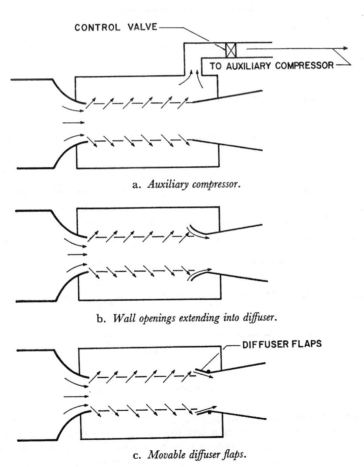

a. *Auxiliary compressor.*

b. *Wall openings extending into diffuser.*

c. *Movable diffuser flaps.*

FIG. 12.2. *Different schemes for mass flow removal from the plenum chamber.*

way, a portion of the test-section flow can enter the plenum chamber and can be discharged from there into the diffuser (see Fig. 12.2b). By changing the pressure distribution in the diffuser—for instance, by increasing the power input into the wind tunnel or, more generally, by reducing the pressure downstream of the diffuser—the amount of air removed from the plenum chamber through the diffuser end of the slots may be varied and controlled.

Control of the air removed from the plenum chamber can also be easily accomplished with an auxiliary compressor by means of throttling valves or bypass valves which are independent of the power input to the main wind tunnel compressor. It is even possible to introduce air into the plenum chamber and from there through the slots into the test section itself.

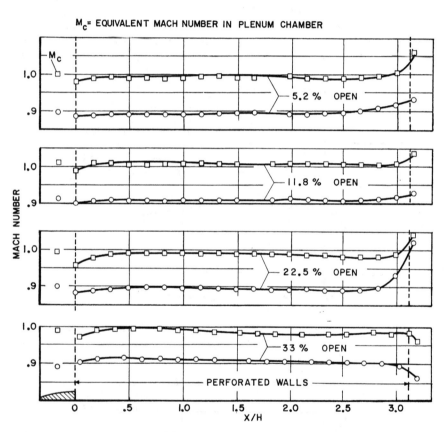

FIG. 12.3. Subsonic Mach number distribution in perforated test section with parallel walls and open-area ratios between 5.2 and 33 per cent.[2]

Diffuser suction through the slots extending into the diffuser and suction by an auxiliary compressor in a test section having openings only in the test-section area itself are simply two different methods of obtaining mass flow removal from the plenum chamber. The method, that is, the type of mass flow removal used, should not affect the Mach number distribution in the test section as long as the total mass removed from the test section remains the same. Only minor local differences could occur in the immediate vicinity of the suction device utilized, for instance, in the vicinity of the flap opening (Fig. 12.2c) or near the wall openings which extend into the diffuser (Fig. 12.2b). The following discussion will first consider the mass flow removal from the plenum chamber as a freely controllable parameter and, later, the advantages and disadvantages of the various means of establishing the desired amount of mass flow removal.

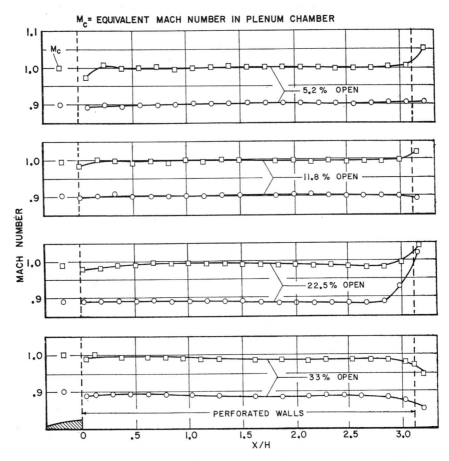

FIG. 12.4. *Subsonic Mach number distribution in perforated test section with top and bottom walls diverged 30' and open-area ratios between 5.2 and 33 per cent.*[2]

b. Experimental Mach number distribution in various perforated test sections.

In a systematic series of tests in the transonic model tunnel of the AEDC several test sections were constructed with perforated walls having an open-area ratio ranging from 5.2 to 33 per cent.[2] Some typical subsonic Mach number distributions obtained with these walls in the high subsonic Mach number range are shown in Fig. 12.3.

Parallel Wall Setting.—The walls of the test sections were set parallel for the selected series, and no allowance was made for boundary layer development along the walls. It is apparent from Fig. 12.3 that in all cases the subsonic Mach number distribution was very regular, as previously, at the upstream end and, particularly, at the downstream end of the test section.

Even at geometric open-area ratios of 5.2 per cent, the venting of the test section to the plenum chamber was so effective that the constant pressure in the plenum chamber controls the pressure in the test section itself. However, at the largest open-area ratio of 33 per cent, long wave-length disturbances in the Mach number distribution through the test section occurred which were similar to the disturbances frequently observed in completely open test sections. This observation seems to indicate that test sections with parallel walls having geometric open-area ratios of 33 per cent or more act much like completely open test sections.

Diverged and Converged Wall Settings.—The same test sections with the same walls were modified so that the top and bottom walls were each diverged 30'. Again (see Fig. 12.4) the subsonic distributions were very smooth in the extreme downstream region of the test section. When the top and bottom walls were converged 30', the same observations were made (Fig. 12.5).

It can be stated in conclusion that the Mach number distribution in perforated test sections is extremely insensitive in the subsonic speed range. Smooth distributions with constant Mach number are obtained regardless of whether the wall open-area ratio is varied between 5 and 33 per cent or the test-section walls are diverged or converged by 30'. The flow disturbances which occur at the downstream end of the test sections sometimes indicate a reduction and sometimes an increase in the local Mach number (see Figs. 12.3 and 12.4). As explained in more detail in the next paragraph, these flow disturbances can usually be attributed to incorrect setting of the auxiliary plenum chamber suction. When too much suction is employed, the Mach number at the downstream region of the test section is reduced; when not enough suction is applied, the Mach number at the downstream end is increased and might even reach choking conditions.

c. Theoretical studies of subsonic Mach number distribution in perforated test sections and comparison with experiments

In order to obtain a better understanding of the experimental results discussed in the preceding paragraphs, a theoretical investigation was conducted to determine the Mach number distribution in an idealized perforated test section as a function of plenum chamber suction and wall divergence or convergence.[3] The following simplifying assumptions were made:
(1) The flow in the test section was considered to be isentropic; this means particularly that the wall boundary layer was neglected.
(2) One-dimensional flow equations were utilized, that is, the flow parameters were assumed to be constant over any given station of the test section.

(3) The compressibility effects were linearized.
(4) The perforated test-section walls were assumed to exhibit a linear relationship between pressure drop and cross flow through the wall.

The following relationship, defined in the terms used in Chapter 11 of this AGARDograph, exists:

$$\frac{p_x - p_c}{\frac{1}{2}\rho_\infty v_\infty^2} = K\frac{(\rho v)_y}{(\rho v)_\infty} = K\frac{m_x}{(\rho v)_\infty}$$

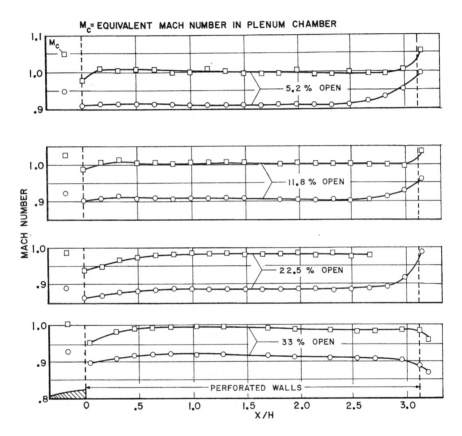

Fig. 12.5. *Subsonic Mach number distribution in perforated test section with top and bottom walls converged 30' and open-area ratios between 5.2 and 33 per cent.*[2]

Here $(\rho v)_y = m_x$ is the cross flow through the wall per unit area at station x, and K is a constant to be determined from experiments. Within the range of validity of linearized compressible flow, the product $K(1 - M_\infty^2)$ may be considered a constant (see Chapter 11).

Theory for Parallel Walls.—If a square test section with parallel walls is assumed, the continuity law provides the basic equation for determining the Mach number distribution over the axial length of the test section (see Fig. 12.6).

$$\frac{\rho_x v_x}{(\rho v)_\infty} = 1 - 4 \int_0^{x/H} \frac{m_x}{(\rho v)_\infty} d(x/H)$$

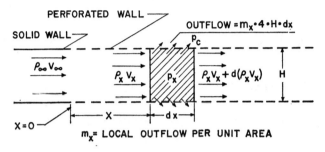

FIG. 12.6. *Perforated test section with outflow into plenum chamber at subsonic speed.*[3]

After the wall characteristics and the boundary conditions are introduced into the above equation, the following result is obtained for the local outflow of the test section.

$$\frac{m_x}{(\rho v)_\infty} = \frac{1 + K(1 - M_\infty^2) m_{x0}/(\rho v)_\infty}{K(1 - M_\infty^2)} \times$$

$$\times \left\{ 1 - \tanh^2 \left[\frac{-4[1 + K(1 - M_\infty^2) m_{x0}/(\rho v)_\infty]^{\frac{1}{2}}}{K(1 - M_\infty^2)} \frac{x}{H} + \right. \right.$$

$$\left. \left. + \tanh^{-1} \left(\frac{1}{1 + K(1 - M_\infty^2) m_{x0}/(\rho v)_\infty} \right)^{\frac{1}{2}} \right] \right\}$$

In the above equation the level outflow m_{x0} at station $x = 0$ is connected with the pressure, p_c, in the plenum chamber and the static pressure, p_{x0}, at the upstream end of the perforated test section ($x = 0$) in the following manner:

$$K \frac{m_{x0}}{(\rho v)_\infty} = \frac{p_{x0} - p_c}{\frac{1}{2} \rho_\infty v_\infty^2}$$

When the plenum chamber pressure is controlled in such a manner that it is exactly equal to the pressure at the upstream end of the test section the above equation becomes:

$$\frac{m_{x0}}{(\rho v)_\infty} = 0$$

and $m_x = 0$

That is, through the entire test section no outflow is obtained and, consequently, constant Mach number is maintained over the entire test section, a result which is plausible from simply physical reasoning. If, however, the plenum chamber pressure is controlled in such a manner that it differs even slightly from the static pressure in the test section at $x = 0$, finite values for the local outflow are established which follow a quite complicated mathematical relationship.

The pressure distribution over the length of the test section is shown in Fig. 12.7 for several quantities of mass flow removal for a test section with

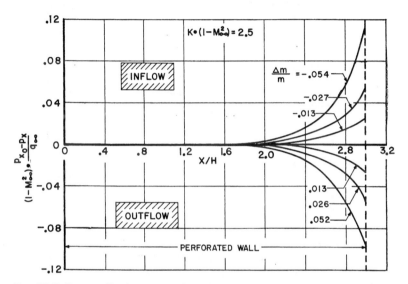

Fig. 12.7. *Pressure distribution in perforated test section with parallel walls for various plenum chamber suction quantities in subsonic flow (p_{x0} = static pressure in test section at $x = 0$).*[3]

an assumed outflow resistance characteristic of $K(1-M_\infty^2) = 2.5$. This value corresponds approximately to the characteristic parameter of a perforated wall with 3 per cent open-area ratio at Mach number 0.90. The indicated mass flow removal ratios $\Delta m/m$ are calculated assuming that the test section has a length three times its width. The figure shows that inflow or outflow of the test section influences the pressure distribution only at the downstream end. Even with 5 per cent inflow or outflow, only the last one-third of the test-section length is noticeably disturbed, and the upstream two-thirds is practically unaffected. Furthermore, at 5.2 per cent outflow the difference between plenum chamber pressure and the pressure at the upstream end of the test section at $M_\infty = 0$ is only $\Delta p/q = 7 \times 10^{-6}$, that is, for all engineering purposes the pressure difference between plenum chamber and test section is non-existent at $x = 0$.

The axial extent of the disturbance region in the case of artificially-maintained inflow or outflow of the test section naturally depends upon the open-area ratio of the wall. In the limiting case, for instance, when a finite number of infinitely small holes are provided in the perforated walls, the pressure drop through the wall reaches infinitely large values, and the outflow of the test section must obviously be uniform over its entire length. Such a case is shown in Fig. 12.8 which represents the pressure distribution through a perforated test section for 1.3 per cent inflow and outflow and for various values of the wall characteristic constant K. The smaller the cross-flow characteristic constant, for instance $K = 1$, the more concentrated the

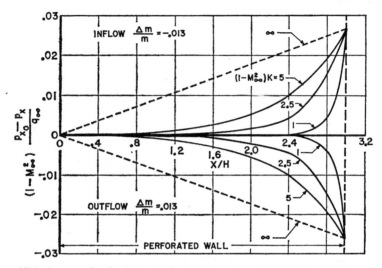

Fig. 12.8. *Pressure distribution in perforated test section with parallel walls for various wall parameters K in subsonic flow (p_{x0} = static pressure in test section at $x = 0$).*[3]

outflow in the downstream region of the test section. The larger the cross-flow characteristic constant, for instance, K approaching infinity, the farther does the disturbance region from the downstream end extend toward the upstream region of the test section. For normal wind tunnel application it may be generally assumed that outflow or inflow of the plenum chamber will affect the pressure distribution in the downstream region of the test section only. In this case it is assumed that the wind tunnel main compressor is properly adjusted to keep the mean test section Mach number constant.

Comparison with Experiments for Parallel Walls.—The Mach number distributions over a test section with typical perforated walls are shown in Fig. 12.9 for various values of plenum chamber suction.[3] In order to maintain a constant Mach number in the main portion of the test section, the wind tunnel compressor was adjusted in such a manner that, in spite of the removal of different amounts of mass flow, constant Mach number was maintained. If such an adjustment had not been made, a higher Mach number would have been produced when the plenum chamber suction was increased. This behavior can be attributed to the fact that the losses in the wind tunnel circuit, particularly in the diffuser, are considerably reduced if the boundary layer approaching the diffuser is thinned by means of auxiliary suction.

As predicted by theory, the Mach number distributions for the experiments shown in Fig. 12.9 remained practically unchanged over much of the upstream portion of the test section, even when more and more air was discharged from the plenum chamber. At small suction values the Mach number started to rise appreciably at the test-section end and approached choking even when mass flow removal was 0.8 per cent. This phenomenon is caused by the boundary layer which is formed along the perforated walls.

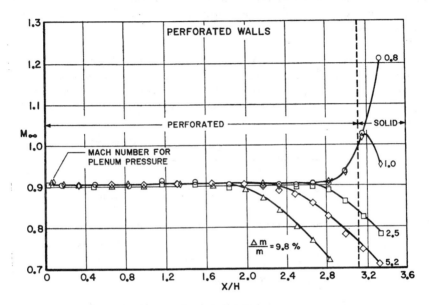

FIG. 12.9 *Mach number distributions for perforated test section with parallel walls for various plenum chamber suction quantities.*[3]

Boundary layer growth has the same effect locally as inflow to the test section: it reduces the cross-sectional area available for the main test-section flow. Consequently, the boundary layer effect changes the effective amount of suction compared with the case of non-viscous flow. From the sample experiments presented in Fig. 12.9 it may be estimated that approximately 2 per cent mass flow removal is required in order to establish parallel flow.

To make comparison possible, the plenum chamber pressures for the experiments with different amounts of suction have been converted into fictitious Mach numbers and are also shown in Fig. 12.9. It is apparent that in the case of the test section with parallel wall setting, the plenum chamber pressure was only slightly lower than the test-section pressure, even at the extremely large mass flow removal of 9.8 per cent.

The concentration of the mass flow removal in the downstream end of the test section significantly affects boundary layer removal along the walls of the test section. If shock-wave cancellation with perforated walls is to be effective (Chapters 9 and 10), it is necessary to keep the wall boundary layer thin; in most cases, it must even be thinned artificially by means o plenum chamber suction. The results of the above theoretical study, as verified by the experiments, indicate that simple mass flow removal from the plenum chamber does not produce the desired uniform thinning of the boundary layer along the test-section length. While much of the boundary layer is removed at the downstream end of the test section, practically no removal is effected in the main portion of the test section. Consequently, other means, which will be discussed subsequently, must be employed in conjunction with plenum chamber suction to accomplish the desired thinning of the boundary layer in the critical upstream and middle region of the test-section length.

Theory for Test Sections with Non-Parallel Walls.—For the following calculation it is assumed that all four walls of the experimental test section are set at the same diverging or converging angle. As is apparent from consideration of Fig. 12.10, not all test-section flow which crosses the reference

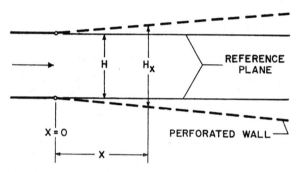

FIG. 12.10. *Sketch for definition of reference planes for diverged test section.*[3]

planes must also cross the perforated walls. A certain amount will flow between the reference planes and the perforated walls. The same calculations previously derived for parallel walls may be applied to the non-parallel walls, but only cross flow through the walls must be considered as producing the pressure drop through the walls. The unit mass flow, m_{x_t}, accounting for the flow between the reference plane and the perforated walls, can be obtained from the following equation for flow continuity:

$$m_{x_t}(4H_x\,dx) = \frac{d(H_x^2)}{dx}dx(\rho_x v_x)$$

or in a simplified linearized manner:

$$m_{x_t} = \frac{1}{2}\frac{dH_x}{dx}(\rho_x v_x) \simeq \frac{1}{2}\frac{dH_x}{dx}(\rho v)_\infty$$

With the above approximate version of the mass flow between the reference plane and the perforated wall, the results of the parallel wall calculations can be adapted by simply correcting the local outflow of the test section, m_x, by the value, m_{x_t}, shown above.[3] In a similar manner the pressure level can be easily corrected by use of the linearized pressure equations.

Some orienting calculations for converged and diverged wall settings were carried out according to the relations described above. The results are shown in Fig. 12.11 for the case of no plenum chamber mass flow removal, that is, for $\Delta m/m = 0$ and a typical wall porosity constant $K(1-M^2) = 2.5$. As in the case of parallel wall settings, practically constant pressure distribution is maintained in the forward and middle portions of

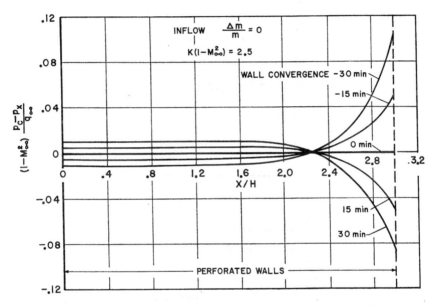

FIG. 12.11. *Pressure distribution in perforated test section for various wall settings without inflow into the test section (subsonic flow, two walls converged or diverged, p_c = plenum chamber pressure).*[3]

the test section, while large disturbances (either toward higher Mach numbers in the case of wall convergence or toward lower Mach numbers in the case of wall divergence) are established at the downstream end. It is significant that with the converged wall, nearly constant outflow is established over the main forward test-section region, as indicated by the lower pressure in the test section compared with that of the plenum chamber. The overall picture of the outflow and inflow distribution is shown schematically in Fig. 12.12b for the case of a convergent wall setting. While a uniform *outflow* occurs over the main portion of the test section, a concentrated inflow of the same total amount is established in the downstream portion of the test section (no mass flow removal).

Local wall cross-flow distribution is shown in Fig. 12.12c for the case of divergence. In this case, uniform *inflow* occurs over the main portion of the test section while a strong outflow is established at the downstream end. It is important to note that in the case of both the converged and the diverged test sections, no uniform pressure distribution through the test section is possible without auxiliary suction or bleeding.

The fact that uniform flow in the entire test section, including its downstream portion, can be obtained in a converged or diverged wall test section only by the proper amount of plenum flow removal is demonstrated in

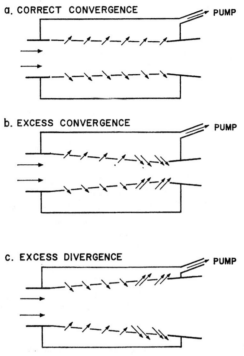

FIG. 12.12. *Flow pattern for correct and incorrect wall setting of perforated test section.*[3]

Fig. 12.13. For this sample calculation a constant inflow of $\Delta m/m = -0.026$ was established. It may be seen that with a wall setting of 15' divergence a constant pressure is established over the entire test section. If the wall setting is changed to either larger divergence or to convergence, pressure disturbances similar to those previously encountered are developed. Correct matching of test-section wall convergence and mass flow removal will result in the desired uniform distribution of local mass flow as shown in Fig. 12.12a.

Comparison of Experiments with Convergent and Divergent Walls.—Uniform flow distribution through the test section is possible only when wall convergence or divergence and plenum chamber mass flow removal are properly matched. This conclusion from the simplified theory is verified by the experimental results presented in Fig. 12.14 (Ref. 3). This figure shows the Mach number distribution in a typical perforated wall test section for no mass flow removal from the test section and various wall

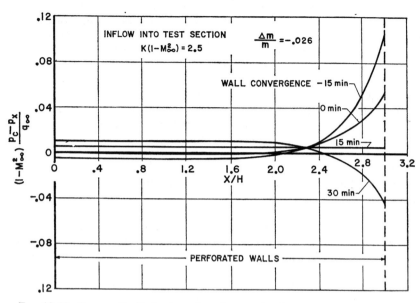

FIG. 12.13. *Pressure distribution in perforated test section for various wall settings and constant inflow into the test section (subsonic flow, two walls converged or diverged, $_{c}p$ = plenum chamber pressure).*[3]

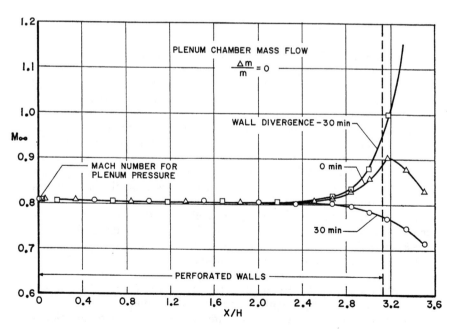

FIG. 12.14. *Mach number distribution through perforated test section without plenum chamber suction for various wall settings.*[3]

divergent or convergent settings. As predicted by theory, locally larger or smaller Mach numbers are established by the downstream end of the test section when the walls are either diverged or converged. However, even in such cases of mismatch, the Mach number distribution through the test section is practically unaffected and uniform at the extreme downstream end. When no mass is removed from the plenum chamber, uniform flow can be maintained to the end of the test section only with a wall divergence of approximately 25'. It is to be noted that in the experiments presented in Fig. 12.14 only two walls of the test section were converged or diverged; the two side walls remained at the parallel setting.

d. Summary

The preceding theoretical and experimental investigations of uniform thinning of the boundary layer in the test section may be summarized as follows:
 (1) Mass flow removal from the plenum chamber will result in uniform boundary layer thinning through the entire test-section length, only when the correct amount of wall convergence is simultaneously applied.
 (2) Merely discharging air from the plenum chamber without properly converging the walls will not produce uniform thinning of the boundary layer, as is required for shock-wave reflection consideration. At subsonic speeds, concentrated local disturbances in the downstream portion of the test section are established.

3. ESTABLISHMENT OF SUBSONIC FLOW IN SLOTTED TEST SECTIONS

Basically the same relationships presented for perforated test sections are also valid for slotted test sections, as numerous experiments with slotted test sections show. As in the case of perforated test sections, the Mach number distribution in the subsonic speed range is not very sensitive to the slot shape and the open area of the slots. Only at the downstream end of the test section do local disturbances in Mach number distribution frequently occur, and these can be effectively controlled by plenum chamber suction.

Typical Mach number distributions in the subsonic speed range for the NACA Langley Field 16 ft transonic wind tunnel are shown in Fig. 12.15 (Ref. 4). In this wind tunnel numerous tapered longitudinal slot configurations having different distributions of width were investigated. Though this width distribution of slots has a significant effect on the quality of the test-section flow in the supersonic speed range, the two drastically different slot shapes selected for the subsonic testing shown in Fig. 12.15 both show uniform Mach number distribution within the accuracy of normal testing. Only at the downstream end of the test section do disturbances occur.

The open-area ratio of the eight slots used in the NACA experiments described above was $12\frac{1}{2}$ per cent. In other test series uniform Mach number distribution was obtained, even when the open-area ratio of the slotted wind tunnels was drastically reduced. For example, in some systematic experiments using wind tunnels equipped with slotted test sections of

7.5 × 3 in. cross section, the open-area ratio of the slots was reduced from 8.0 per cent to 0.5 per cent.[5] Only the upper and lower walls of the test section were slotted; the two side walls were kept solid (see Figs. 12.16 and 12.17a). This model test section was equipped with slots of constant width which extended from the beginning of the test section some distance into the diffuser. In this way, a certain amount of plenum chamber suction was provided which depended upon the pressure distribution in the diffuser. A uniform Mach number distribution through the test section was obtained

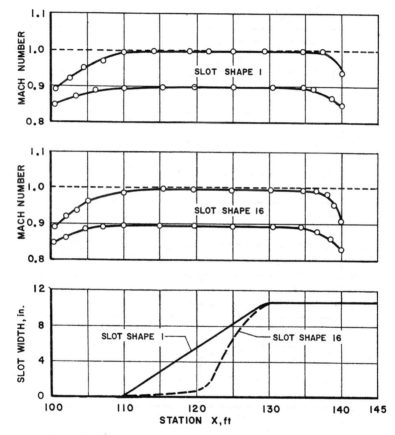

FIG. 12.15. *Mach number distribution at centerline of NACA Langley 16 ft transonic tunnel for two slot shapes in subsonic speed range.*[4]

with the open-area ratio between 8 and 2.7 per cent. Also with the open-area ratio reduced to 1 per cent, uniform Mach number distribution in the test section proper was established except for a local disturbance near $x/H = 0.7$ (Fig. 12.16). In the latter case, however, a strong disturbance did develop in the downstream region of the test section, as was to be expected.

When, in the above test series, the area ratio of the longitudinal slots was reduced to one-half per cent (Fig. 12.17a), the Mach number distribution

297

in the forward portion of the test section was still quite uniform. However, the strong disturbance downstream of the test section extended forward through the entire rear half of the test section (Ref. 5).

According to the theory described for perforated test sections, this downstream test-section disturbance could be greatly reduced in intensity if air were removed from the plenum chamber and the choking condition at the downstream end of the working section were thus alleviated. This result was actually experimentally observed (Fig. 12.17b) when additional openings were provided in the diffuser through which air was sucked from the test

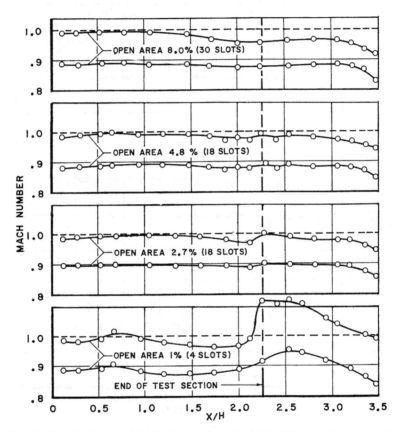

Fig. 12.16. *Mach number distribution along wall of 7.5 × 3 in. slotted wind tunnel for different slot open areas.*[5]

section into the low pressure region of the diffuser. The Mach number distribution is much smoother with the additional plenum chamber suction (see Fig. 12.17a for comparison). Since the slotted test sections were identical for both series of experiments (Figs. 12.17a and b) the only difference was the open area added in the diffuser to increase plenum chamber suction.

The fact that the flow disturbances in the downstream portion of slotted test sections depend decisively upon the amount of air which is either removed from or added to the plenum chamber was demonstrated clearly in a test series conducted in the transonic model tunnel of the AEDC. In these tests, a slotted test section was equipped with slots extending only through the test section itself.[6] Plenum chamber suction or bleed-in was provided by either an auxiliary suction source or by simple bleed-in from a high-pressure area. Some typical Mach number distributions for a test-section configuration having sixteen slots with a total open-area ratio of 11 per cent are shown in Fig. 12.18. While the Mach number was kept constant by proper control of the pressure at the downstream end of the diffuser, the mass removal from the test section (parallel walls) was varied from no removal to 2.8 and 5.0 per cent of the test-section flow. Without plenum chamber suction, the Mach number at the downstream end increased greatly as a result of the boundary layer development along the test-section

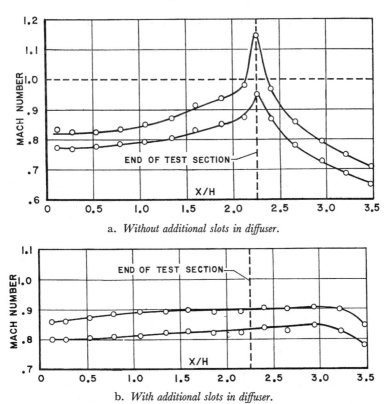

a. *Without additional slots in diffuser.*

b. *With additional slots in diffuser.*

FIG. 12.17. *Mach number distribution along wall of 7.5 × 3 in. slotted wind tunnel with and without additional slots in diffuser.*[5]

walls. The results from 5 per cent mass flow removal indicate that too much air was removed through the slots and that, consequently, the Mach number in the downstream region was drastically reduced. The proper amount of suction, that is, approximately 2.8 per cent, resulted in uniform Mach number distribution through the entire test section.

The comparison of the test results for various slotted test sections with the theory, presented in Section 2c of this chapter, indicates clearly that in slotted test sections, also, plenum chamber suction or bleed-in affects only the flow in the downstream end of the test section; the main portion of the test-section flow is practically unaffected. Furthermore, when mass flow is removed from the plenum chamber without simultaneously converging the test-section walls, the boundary layer along the wall is not thinned in the main portion of the test section but only in the extreme downstream region.

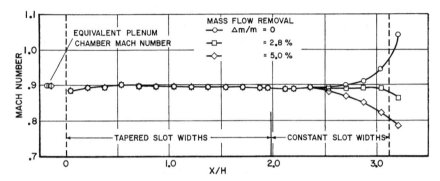

Fig. 12.18. *Influence of plenum chamber suction on Mach number distribution at centerline of slotted test section (solid diffuser walls, sixteen slots with 11 per cent open-area, parallel walls).*[6]

In order to provide an effective thinning of the boundary layer through the entire test section a proper combination of wall convergence and mass flow removal from the plenum chamber must be applied. In view of the nonlinear relationship between local mass flow removal and local pressure difference between test section and plenum chamber, however, it is probable that the plenum chamber pressure of slotted test sections differs even less from the mean test-section pressure than does the plenum chamber pressure of perforated test sections.

4. SUPERSONIC FLOW ESTABLISHMENT IN PERFORATED TEST SECTIONS

a. *Theoretical considerations*

In suitably designed partially open test sections, the Mach number in the test section can be increased beyond sonic speed by merely increasing the pressure rise of the wind tunnel compressor or the plenum chamber mass flow removal. This fact is very significant because in conventional supersonic wind tunnel operation, supersonic Mach numbers can be established only by placing ahead of the test section a supersonic nozzle which has a definite contour for each individual Mach number. Moreover, the necessary change in contour is so small near Mach number one that it is practically impossible in a closed test section to establish the very small changes in this Mach number range. Also, it is extremely difficult in such tunnels to control the degree of test-section divergence needed for boundary layer compensation in such a manner that no Mach number gradients of significance occur. In practical operation of partially open wind tunnels, therefore, the supersonic test Mach numbers near Mach number one are always established by mass flow removal from the plenum chamber. This is true regardless of whether or not a supersonic nozzle is provided.

A distinct difference exists in the effects of mass flow removal from the plenum chamber for the supersonic and the subsonic speed ranges. In subsonic flow, a change in the mass flow removal affects only the Mach number distribution in the downstream portion of the test section, but this is not true in the test section proper. (Again it should be remembered that increased mass flow removal generally leads to a reduction of the flow losses in the wind tunnel circuit and thus might indirectly cause a change in the test-section Mach number.) In supersonic flow, the relationships are fundamentally different, as can be readily demonstrated by consideration of the simplified theory.

The supersonic flow pattern for plenum chamber suction is presented schematically in Fig. 12.19. This pattern is the equivalent of the subsonic

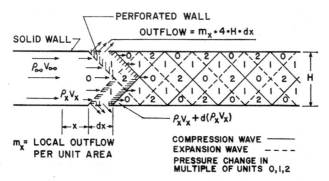

Fig. 12.19. *Perforated test section with outflow into plenum chamber at supersonic speed.*[3]

flow pattern shown in Fig. 12.6. Local outflow through the perforated wall creates locally lower pressures by means of Mach waves. When the same simplified assumptions are made for these theoretical calculations as for subsonic flow (see Section 2c of this chapter), the following equation for local outflow through a wall element as a function of the axial station is obtained:

$$\frac{m_x}{(\rho v)_\infty} = \frac{1 - K(M_\infty^2 - 1)m_{x0}/(\rho v)_\infty}{K(M_\infty^2 - 1)} \times$$

$$\times \left\{ 1 - \coth^2\left[\frac{4\{1 - K(M_\infty^2 - 1)m_{x0}/(\rho v)_\infty\}^{\frac{1}{2}}}{K(M_\infty^2 - 1)}\frac{x}{H} + \right.\right.$$

$$\left.\left. + \coth^{-1}\left\{\frac{1}{1 - K(M_\infty^2 - 1)m_{x0}/(\rho v)_\infty}\right\}^{\frac{1}{2}}\right]\right\}$$

The local pressure difference between test section and plenum chamber can then be readily obtained by means of the assumed linearized wall characteristic and linearized compressible flow equations:

$$(M_\infty^2 - 1)\frac{p_x - p_c}{\tfrac{1}{2}\rho_\infty v_\infty^2} = K(M_\infty^2 - 1) m_x/(\rho v)_\infty$$

The pressure distribution obtained from the above equations for a typical case of a perforated wall is presented in Fig. 12.20 for various amounts of

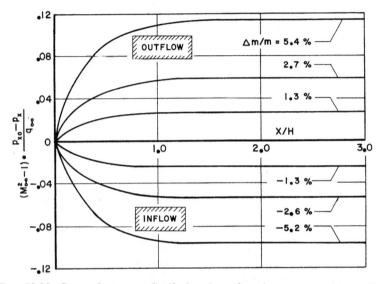

FIG. 12.20. *Supersonic pressure distributions in perforated test section with parallel walls for various plenum chamber suction quantities for* $K(M^2 - 1) = 2.5$ ($p_{x0} =$ *static pressure in test section at* $x = 0$).[3]

mass flow removal from and bleed-in to, the plenum chamber. Unlike the subsonic case (see Fig. 12.7), all non-uniformities in the supersonic flow are concentrated in the upstream portion of the test section. Hence, if the flow approaching the test section is assumed to be choked, a variation in the plenum chamber suction will result in a change of Mach number in the test section proper. The upstream portion of the test section acts as a supersonic nozzle which establishes flow in the test section. Hence control of the mass flow from the plenum chamber represents a convenient means of adjusting the test Mach number to the desired value.

One basic assumption made in the derivation of the equation for Mach number of pressure distribution through the test section (Fig. 12.20) is that the pressure in the test section is influenced only by the primary waves produced by local mass outflow or inflow. It is specifically assumed that all reflections of the primary expansion or compression waves are cancelled completely by the partially open wall. If cancellation of waves is not completely achieved, the resulting wave reflections will cause fluctuation in the pressure distribution which could make a test section unsuitable for wind tunnel testing.

From these theoretical calculations it can be concluded that only in the most upstream portion of the test section is a thinning of the boundary layer produced by mass flow removal from the plenum chamber; in the test section proper the growth of the boundary layer along the perforated walls remains unchecked. In order to provide for boundary layer removal through the entire length of the test section, the walls must be converged and a suitable matching between convergence angle and mass flow removal must be established, as in the subsonic case (see Section 2a of this chapter).

b. Test section with sonic nozzle

Theoretical Relationships.—In conventional supersonic wind tunnels supersonic flow is established by means of a contoured nozzle. The initial portion of the nozzle beyond the throat may be of any shape, within certain limits. However, the downstream portion is predetermined by the condition that all waves initiated in the upstream portion must be cancelled by proper curvature of the wall element so that parallel flow in the test section is established. As a consequence of this well-known rule, supersonic nozzles have a minimum length which depends upon the Mach number (Fig. 12.21). Detailed consideration of the flow pattern of one such nozzle proves

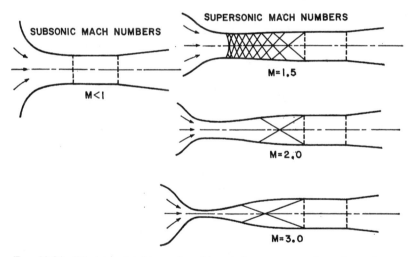

FIG. 12.21. *Wind tunnel inlet nozzles with typical wave systems for various subsonic and supersonic Mach numbers.*

that the expanding nozzle flow crosses an imaginary partially open wall, parallel to the centerline of the nozzle, at normal velocities which reach a maximum near the inflection point of the contoured nozzle (see Fig. 12.22). Downstream of the inflection point, they gradually decrease to zero. If a partially open wall were provided with a suction system which could control the local outflow through the wall in the manner indicated by the normal velocity distributions, the flow within the test section would be the same as that which occurs in a closed test section with a supersonic nozzle.

Consequently, to establish smooth supersonic flow in transonic test sections by means of a sonic nozzle, the distribution of the wall cross flow must be controlled precisely to simulate the normal cross-flow velocity distribution of a conventional supersonic nozzle.

Since the distribution of the normal velocity components along a partially open wall can be readily calculated, by the method of characteristics, for example; it is possible to determine theoretically the open-area ratio distribution in the flow-establishing section of a perforated test section when suitable assumptions are made concerning the wall characteristics. Fortunately the conditions in partially open wind tunnels are considerably less critical than in closed wall wind tunnels because a partially open wall is automatically, to a large extent, capable of eliminating waves. As a result of this fortunate characteristic of partially open walls, the greatly simplified pattern of wall opening distribution in the flow-establishing section provides the desired uniform supersonic flow. Some typical tests showing this effect will be discussed in the following sections.

The method of establishing supersonic flow in partially open test sections with sonic wind tunnel nozzles is practical only over a relatively narrow Mach number range, slightly above Mach number one, in spite of the significant advantages of the method with respect to the simplicity of the required equipment (no precision supersonic nozzle is required). The limits to practical application are set by the rapidly increasing requirement for mass flow removal when the supersonic Mach number is increased much beyond Mach number one. The amounts of air which must be bypassed are shown in the following table:

Mass Flow Removal Requirements for Partially Open Test Sections Having Sonic Nozzles

(*Isentropic Flow*)

Mach Number	Mass Flow Removal, $\Delta m/m$, per cent*
1.0	0.0
1.1	0.8
1.2	3.0
1.3	6.6
1.4	11.5
1.5	17.6

* Related to total test section flow.

To establish a Mach number of 1.1, only approximately 0.8 per cent of the test-section flow need be bypassed; this value increases to 3.0 per cent at Mach number 1.2 and to 11.5 per cent at Mach number 1.4. Since the bypassed air cannot avoid losing most of its kinetic energy because of vortex formation in the plenum chamber, the diffuser efficiency of the wind tunnel at a Mach number of 1.4 is automatically reduced by the factor of approximately $1/1.11 = 0.9$. In addition, the large volume flow of the bypass air which must be handled rapidly exceeds the limits of practical operation at higher Mach numbers. Consequently, only relatively small transonic wind tunnels which have a test-section height not exceeding approximately 1 ft

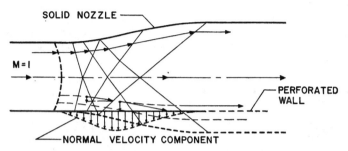

Fig. 12.22. *Supersonic wind tunnel nozzle with conventional solid walls and with partially open walls in conjunction with controlled local outflow.*

are built for operation at supersonic Mach numbers up to (and even exceeding) 1.25, if the supersonic flow is to be established by means of a sonic nozzle used in conjunction with plenum chamber suction. All larger wind tunnels either are restricted in maximum Mach number to approximately 1.2 when sonic nozzles are employed or use supersonic nozzles to establish supersonic flow in the conventional manner.

Experiments with Various Perforated Test Sections.—In the systematic studies of perforated wall configurations conducted in the AEDC transonic model tunnel, walls with approximately 22.5 per cent open-area ratio were examined at parallel wall settings.[7] At subsonic Mach numbers, uniform flow could be readily established in the test sections, but the flow at supersonic Mach numbers was extremely wavy and consequently unsuited for wind tunnel testing (see Fig. 12.23a). The initial extensive over-expansion in the forward portion of the test section which produced a strong wave pattern throughout the test section was particularly disturbing. This wave system was only gradually eliminated by the effect of the perforated walls.

In order to provide a simple control of the outflow of the test section as a first step toward achievement of the required theoretical outflow distributions, the upstream portion of all four perforated walls was partially closed by a set of solid taper strips (see Fig. 12.24). This extremely simple configuration change resulted in very smooth Mach number distributions in the test section (Fig. 12.23).

Similar results were obtained in numerous other test series with different walls. At the most common open-area ratios, from 20 to 33 per cent, the simple straight taper strips always proved effective up to Mach numbers of approximately 1.2, the limit of most of these investigations. However, with wall open-area ratios of from 5 to 10 per cent, no taper strips were required to establish the desired uniform flow.

Experiments with Tapered Porosity Sections of Different Lengths.—In the British ARA 9 × 8 ft transonic wind tunnel, an extensive investigation was carried out to determine a suitable tapered porosity distribution for the perforated test section[8] of this wind tunnel. This test section is characterized by a rather short length of only 1.44 × mean width of the tunnel working section. For the Mach number range up to 1.1 the flexible supersonic nozzle preceding the working section is set at a *sonic* contour because of design limitations. Also, in order to utilize as much as possible of the perforated test section as a working section, the region of perforated porosity used for

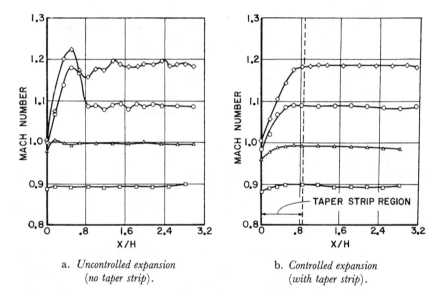

a. *Uncontrolled expansion (no taper strip).*   b. *Controlled expansion (with taper strip).*

FIG. 12.23. *Mach number distribution in perforated test section with uncontrolled and controlled supersonic flow expansion (walls: 22 per cent open, parallel setting).*[12,7]

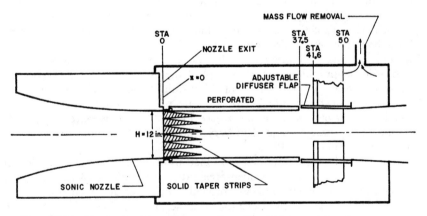

FIG. 12.24. *Sketch of perforated test section with solid taper strips for controlled supersonic expansion.*[12]

the flow build-up had a length 0.65 × the mean test-section height instead of 0.83 × the mean height as was generally the case in the previously cited AEDC tests (see Fig. 12.25). The porosity of both vertical walls started further upstream than did that of the horizontal walls (Fig. 12.25a).

Some typical Mach number distributions are shown in Fig. 12.26 for the wall open-area ratio of 22.5 per cent with both diffuser and auxiliary suction applied. Even in the subsonic Mach number range between 0.9 and 1.0, the Mach number did not reach a uniform value upstream of $x/H = 0.40$, measured from the beginning of the porous test section. At supersonic

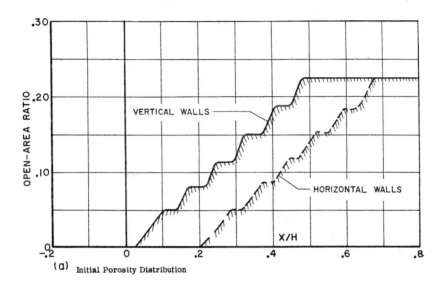

(a) Initial Porosity Distribution

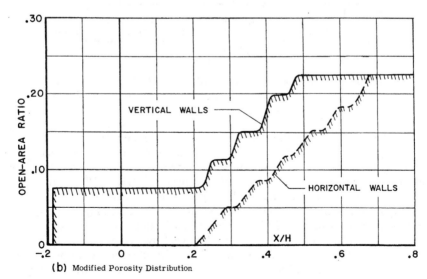

(b) Modified Porosity Distribution

Fig. 12.25. *Porosity distribution of British ARA 9 × 8 ft transonic wind tunnel.*[8]

Mach numbers, the establishment of the final test-section flow was shifted even further downstream, and the uniformity of the Mach number curves was not satisfactory. It is suspected that not only the shorter axial extent of the tapered porosity section, but also the relatively thicker boundary layer at the beginning of the working section might be responsible for the observed non-uniformities.

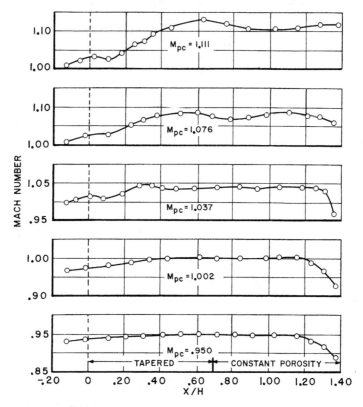

FIG. 12.26. *Mach number distribution in ARA transonic 9×8 ft wind tunnel with initial porosity distribution, parallel walls, and sonic nozzle (M_{pc} = equivalent Mach number in plenum chamber).*[8]

On the basis of the initial investigations of the Mach number distribution, a redesign of the tapered porosity section was gradually accomplished. The guiding idea was to increase the axial extent of the tapered section by extending the tapered porosity section approximately $0.18 \times H$ upstream and, more significantly, to change the approximately linear growth of porosity, as employed initially, to a more refined rate of growth (see Fig. 12.25b). The final distribution of porosity consisted of an initial rapid increase, then a region with nearly constant porosity, followed by a final region with rapid increase of approximately linear rate. The middle region with nearly constant porosity was found to be most effective in eliminating over-expansion of the test section flow. As intended, the Mach number distribution in the test section at both subsonic and supersonic Mach numbers was considerably improved (see Fig. 12.27). At a test-section Mach number of approximately 1.10, the region with satisfactorily uniform flow begins at the station located approximately $0.6 \times H$ downstream of the upstream end of the porosity section. This axial extent is in good agreement with the data cited for the AEDC transonic model tunnel and shown in Fig. 12.23b.

c. Test sections equipped with supersonic nozzles

When a conventional supersonic nozzle is used to establish supersonic flow, the problem of flow uniformity in perforated test sections is greatly simplified. In the absence of wall contour irregularities or boundary layer

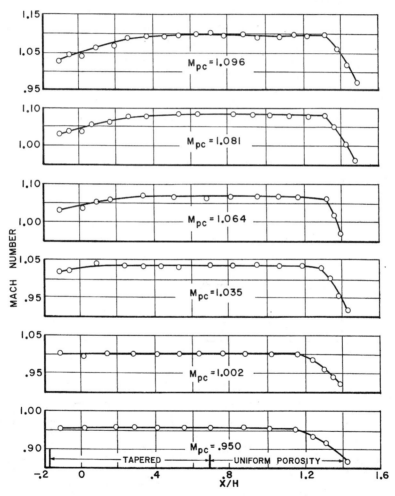

FIG. 12.27. *Mach number distribution in ARA transonic 9×8 ft wind tunnel with modified porosity distribution, parallel walls, and sonic nozzle (M_{pc} = equivalent Mach number in plenum chamber).*[8]

disturbances along the supersonic nozzle and the perforated test-section walls, there should be no flow disturbances at all. However, the transition from the solid walls of the supersonic nozzle to the partially open walls of the test section establishes different conditions for the boundary layer development. In numerous tests it was found necessary to cushion this transition by providing tapered perforations such as were used to establish supersonic flow without a preceding supersonic nozzle.

Some typical Mach number distributions determined in the ARA 9×8 ft transonic wind tunnel are presented in Fig. 12.28 for the porosity configuration developed for subsonic and low supersonic Mach numbers (Fig. 12.25b). The distribution was generally uniform, but some distinct disturbances did exist which were traced back by the staff of the ARA wind tunnel to irregularities in the internal contours of the tunnel. These contour irregularities occurred at the junction between the flexible nozzle and the perforated side walls and at some stations in the flexible nozzle itself. It appears, therefore, that the problem of establishing smooth supersonic Mach number distributions in perforated test sections having supersonic nozzles consists mainly in controlling the irregularities in the internal contours of the wind tunnel itself.

The same conclusion as to the significance of internal contour smoothness in the establishment of uniform supersonic flow in perforated test sections

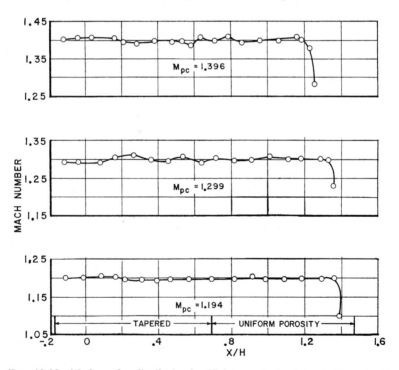

FIG. 12.28. *Mach number distribution in ARA transonic 9×8 ft wind tunnel with modified porosity distribution, parallel walls, and supersonic nozzle (M_{pc} = equivalent Mach number distribution in plenum chamber).*[8]

was also drawn from the calibration tests of the AEDC 16 ft transonic wind tunnel.[9] For this test series, the tunnel was equipped with perforated walls having inclined holes with a geometric open-area ratio of approximately

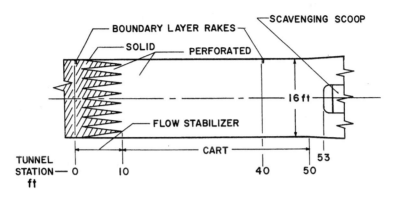

FIG. 12.29. *Sketch of test section of AEDC 16 ft transonic wind tunnel.*[9]

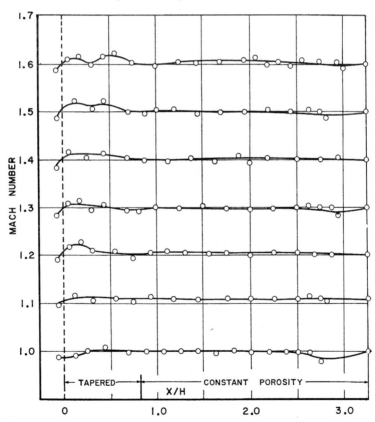

FIG. 12.30. *Mach number distribution in AEDC 16 ft transonic wind tunnel with supersonic nozzle (parallel perforated walls, $60°$ inclined holes).*[9]

6 per cent. Each wall had a tapered porosity section 0.62 × the tunnel height; each wall was equipped with eight tapered solid sections (Fig. 12.29). Typical Mach number distributions to Mach 1.6 are shown in Fig. 12.30. Some of the most pronounced Mach number disturbances can be traced back to irregularities of internal shape, for instance, to the juncture between the removable test section and the fixed flow stabilization section. The test results indicated were obtained for a parallel wall setting and a medium amount of suction. Extensive full-scale tests showed that, for each given wall setting, any suitable combination of pressure rise of the wind tunnel compressor and mass flow removal yielded an identical Mach number distribution in the test section. A change of wall setting to $\frac{1}{2}°$ converged had some noticeable but minor effects on the distribution.[9]

These tests in the AEDC transonic wind tunnel also indicated that supersonic flow of acceptable uniformity can be readily established in perforated test sections when a supersonic nozzle is provided upstream of the working section.

5. SUPERSONIC FLOW ESTABLISHMENT IN SLOTTED TEST SECTIONS

It is possible in slotted test sections, as in perforated test sections, to establish supersonic flow with either a sonic nozzle and subsequent flow removal through the slots or with a conventional supersonic nozzle which precedes the slotted test section. In the latter case, the problem of uniform flow depends upon the quality of the supersonic nozzle, upon the uniformity of the internal wall contours, and upon the smoothness of the transition from the solid to the slotted test-section wall regions. In this respect, the problem of establishing uniform flow in slotted test sections with supersonic nozzles does not differ very much from the problem in perforated test sections. Consequently, the following discussion of supersonic flow establishment in slotted test sections will be restricted to a discussion of slotted test sections having sonic nozzles.

a. Slots with straight taper

During a preliminary test series conducted in the transonic model tunnel of the AEDC, the test section shown in Fig. 12.24 was converted from a perforated to a slotted tunnel. The square cross section was equipped with a total of sixteen tapered slots, four slots on each wall, for a total open-area ratio of 11 per cent (see Ref. 6). Initial tests with this configuration indicated that simple straight tapering of the slot width in the upstream portion of the test section did not eliminate over-expansion and irregular Mach number distribution in the test section.

In the case of perforated test sections having straight tapering extending over an axial distance approximately 80 per cent of the test-section height, uniform distribution in the low supersonic Mach number range was established (see Fig. 12.23b). The slotted test section in the present test series was equipped with slots having a straight taper extending over an axial distance corresponding to 1.0 × the test-section height. The results shown in Fig. 12.31a indicate large irregularities which made the test-section configuration completely unsuitable for testing. Even when the straight slot tapering was extended over a length 2 × the test-section height, that is, to a length which comprised the main portion of the wind tunnel test section, the distribution was still irregular and barely usable for wind tunnel testing below Mach number 1.2 (see Fig. 12.31b).

The reason for the drastic difference between the behavior of perforated and slotted test sections in regard to the establishment of regular supersonic

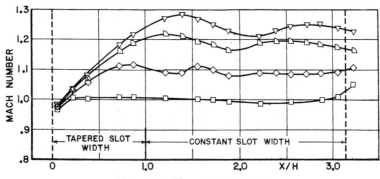

a. *Slots with x/H = 1.0 straight taper.*

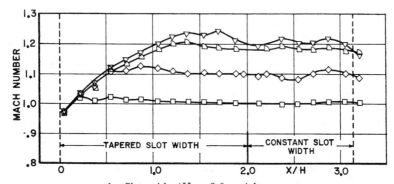

b. *Slots with x/H = 2.0 straight taper.*

FIG. 12.31. *Supersonic Mach number distribution in slotted test section with 11 per cent open-area ratio and sonic nozzle.*[6]

flow by a simple tapering of the open wall area can be attributed to the fundamentally different cross-flow characteristics of perforated and slotted walls. As shown in Chapter 11, a perforated wall has a roughly linear relationship, that is, the cross flow through the wall is roughly proportional to the pressure difference between the test section and the surrounding plenum chamber. A slotted wall, however, exhibits what is basically a quadratic relationship, that is, the pressure drop through the wall is approximately proportional to the square of the cross-flow velocity. As a result, very small differences between the pressures inside and outside the slotted test section produce sizable local outflows; hence the control of the local outflow is much more difficult in a slotted than in a perforated test section.

Also in subsonic flow in a slotted test section the relationship between cross flow and pressure difference is basically quadratic since, as is shown in Chapter 11 in more detail, a slotted test section with constant slot width behaves identically in supersonic and subsonic flow. When the slots are tapered, the basically subsonic quadratic relationship is still maintained as long as the change of slot width in the flow direction is sufficiently small and the local predominant flow parameter is the local open-area ratio.

In view of the considerably greater sensitivity of the flow in a slotted test section to the local slot width, extensive programs have become necessary to develop a slot width distribution which yields uniform supersonic flow.

b. Non-linear slot width distribution

Various attempts have been made to arrive theoretically at a slot width distribution which will establish the desired uniform supersonic flow. If the basic cross-flow characteristic of a slotted wall in non-viscous flow is assumed and three-dimensional effects caused by the finite number of slots are neglected, it is possible to shape the slot widths in such a manner that the desired outflow distribution is established within wide limits. With this method, an outflow velocity distribution such as is shown schematically in Fig. 12.22 results in a slot width distribution which is very narrow at the upstream end, reaches a maximum slightly downstream of the maximum outflow velocity, and thereafter becomes narrow again. Some test series have been conducted with slots shaped according to this method, and it is reported that encouraging results have been obtained. However, no experimental results for such theoretically-determined slot shapes appear to have been published.

In most cases the proper slot width distribution has been determined on the basis of systematic wind tunnel tests using progressively improved slot shapes. The slot width development program for the NACA 16 ft transonic wind tunnel at Langley Field may be cited as a typical example of such a development series. The octagonal test section of this wind tunnel had eight slots with a maximum open-area ratio of 12.5 per cent and a wall divergence of 5' or 10' (Ref. 4). Initially, a straight tapered slot shape (1) was utilized (Fig. 12.32a). The Mach number distribution at the low supersonic Mach number of approximately 1.05 was quite uniform in the working section area, but a pronounced over-expansion of the flow and subsequent wave disturbances in the downstream region occurred at a mean Mach number of approximately 1.10 (Fig. 12.32a).

In order to eliminate these strong over-expansions of flow in the upstream portion of the test section, the slot width was progressively reduced in the upstream portion until an extremely narrow initial slot shape (16) finally emerged. With this modified slot width, the Mach number distribution at an average Mach number of approximately 1.10 was considerably more uniform than with slot shape 1 (Fig. 12.32b), but a noticeable over-expansion at the upstream portion of the test section remained. Also, since the flow with slot shape 16 expanded much more slowly, the improvement in Mach number distribution was achieved at the expense of a very undesirable shortening of the potential working section.

In order to compensate for the observed reduction in test-section length another slot configuration (18) was developed which employed slots beginning several feet further upstream than the slots of configurations 1 and 16.

The results (Fig. 12.32c) gave the expected effect. The test-section Mach number for a wall divergence of 5′ was very uniform, an average value of $M = 1.08$. Also the usable test-section length was greatly increased, as intended, over that available with slot shape 16.

Then in order to obtain higher Mach numbers without auxiliary suction, which was not available at the time of the calibration of this tunnel, the walls

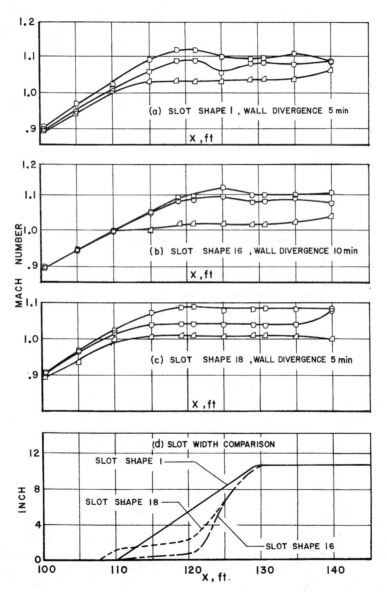

FIG. 12.32. *Influence of slot shape on supersonic Mach number distributions in Langley 16 ft transonic wind tunnel.*[4]

of the test section were diverged from 4′ to 20′. Though the expected increase of the mean Mach number was accomplished, the flow distribution deteriorated very quickly with increasing divergence (see Fig. 12.33).

Another example of the Mach number distribution in a slotted wind tunnel which does not have a supersonic nozzle preceding the slotted working section is shown in Fig. 12.34 for the NACA 8 ft transonic wind tunnel at Langley Field.[10] The slot shape, selected in this case also after an extensive slot development program, finally resulted in the Mach number distribution shown in Fig. 12.34. In the working section, which is relatively short (a length approximately 40 per cent of the test-section diameter), the Mach number distribution is satisfactorily uniform, even at the maximum Mach number of 1.13

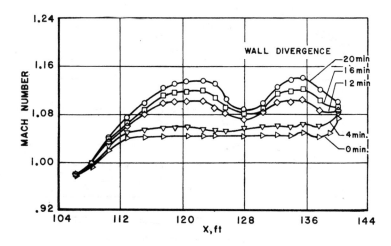

Fig. 12.33. *Influence of wall divergence on supersonic Mach number distributions in Langley 16 ft transonic wind tunnel (slot shape 18 of Fig. 12.32).*[4]

It may be concluded that a delicate and extensive experimental program is generally required to determine a slot configuration which will satisfy the various conflicting requirements for flow establishment in the supersonic speed range. This somewhat complex procedure is in contrast to the very easy development of suitable open-area distributions for perforated test sections described previously.

c. Slotted test sections with perforated cover plates

As discussed previously in detail, the difficulties associated with the great sensitivity of the slot width distribution in slotted test sections are apparently caused by the quadratic relationship between pressure drop and local outflow from slotted test sections. In a development program conducted in the Wright Field 10 ft wind tunnel these inherent difficulties were successfully overcome by incorporating the basic features of a perforated test section in the slotted test-section design. In this wind tunnel the slots were covered with perforated plates which could be readily changed to establish any desired distribution of the effective open-area ratio.[11] The sensitive quadratic characteristic of slotted walls was thus replaced by a more linear relationship which is characteristic of perforated plates. In these Wright Field tests, both in the model and in the full-scale 10 ft wind tunnel, it was very easy to establish a uniform flow which had no tendency to initial over-expansion or to waviness of the Mach number distribution in the test section. The 10 ft wind tunnel with perforated cover plates over its slots is also noted for its extremely stable operating characteristics. The frequently observed tendency of slotted wind tunnels toward Mach number instability (timewise fluctuations) was completely eliminated.

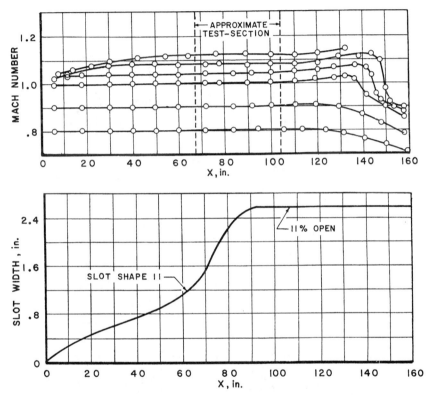

FIG. 12.34. *Supersonic Mach number distribution in Langley 8 ft transonic wind tunnel.*[10]

The Wright Field 10 ft wind tunnel had neither a supersonic nozzle preceding the test section nor sufficient auxiliary suction to establish high enough supersonic Mach numbers. Therefore the test section was equipped with movable side walls with which the divergence of the test section could be effectively changed. Typical Mach number distributions obtained in this tunnel for proper combinations of wall setting and auxiliary suction are shown in Fig. 12.35. It is apparent that in spite of the extremely limited means available for establishment of supersonic flow, the Mach number distributions at the centerline are satisfactorily uniform for normal wind tunnel testing up to the peak Mach number of 1.23.

It was pointed out in Chapter 9 that neither pure slotted nor pure perforated test sections are able to eliminate shock reflections from three-dimensional model configurations. It was shown that combined slotted–perforated test sections can be designed which much more closely approach the required reflection-free characteristics. In view of this previous finding and the experimentally-determined superiority of the supersonic flow established in the combined slotted–perforated test sections cited above, the combined perforated–slotted test section offers a potential which should be further explored.

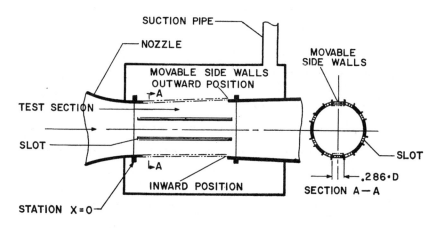

a. *Sketch of test section.*

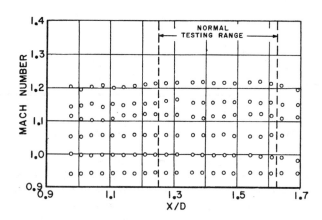

b. *Mach number distribution.*

Fig. 12.35. *Wright Field 10 ft transonic wind tunnel with slotted test section and perforated cover plates.*[7]

6. REQUIREMENTS FOR MASS FLOW REMOVAL FROM PARTIALLY OPEN TEST SECTIONS

a. *Boundary layer compensation in closed test sections*

In closed test sections with parallel walls, the effective cross-sectional area of the test section gradually decreases in the flow direction because of the continuous boundary layer growth along the test-section walls. Therefore, such a test section will choke, that is, sonic velocities will be established at the downstream end of the test section, while the Mach number in the upstream and center portion of the test section is still considerably below Mach number one. In order to obtain a Mach number close to one in an empty closed-wall test section, it is necessary to diverge the walls. A typical value for a circular test section is a divergence angle of approximately 5–7′, depending upon the smoothness of the walls and the Reynolds number and Mach number of the flow.

b. *Influence of specific wall parameters in subsonic speed range*

A similar phenomenon also occurs in partially open test sections. When no plenum chamber suction is applied by auxiliary compressors or diffuser flaps, the boundary layer inside the test section thickens along the perforated walls and is partially discharged into the plenum chamber, and from there into the downstream region of the test section. It follows from continuity conditions that the total discharge of boundary layer into the plenum chamber for a parallel wall test section with uniform Mach number distribution is:

$$\Delta m = (\rho v)_\infty \delta^* C_T$$

where δ^* = boundary layer displacement thickness at the station considered
C_T = circumference of test section.

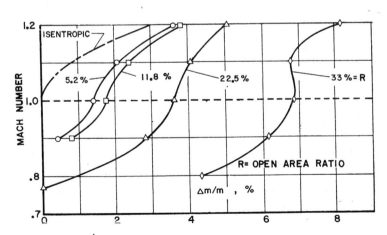

Fig. 12.36a. *Parallel walls (hole diameter and wall thickness = $\frac{1}{16}$ in.)*.

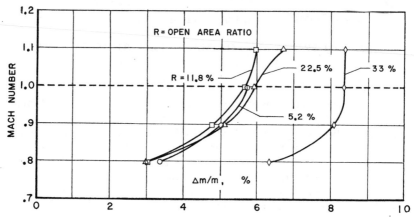

FIG. 12.36b. *Top and bottom walls converged 30′ (hole diameter and wall thickness = $\frac{1}{16}$ in.)*.

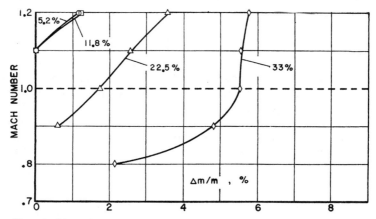

FIG. 12.36c. *Top and bottom walls diverged 30′ (hole diameter and wall thickness = $\frac{1}{16}$ in.)*.

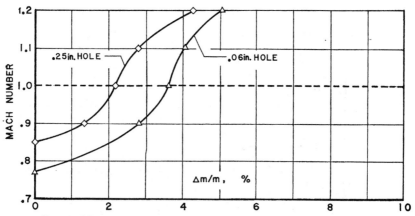

FIG. 12.36d. *Parallel walls: influence of hole size (wall thickness $\frac{1}{16}$ in., open-area ratio 22.5 per cent)*.

FIG. 12.36. *Minimum plenum chamber suction requirements for perforated test sections (length = $3.1 \cdot H$)*.[2]

As discussed in detail in Section 2b of this chapter, choking at the downstream test section restricts the Mach number in the main portion of the test section to subsonic values. This choking phenomenon can be considerably more severe with a partially open than with a closed wall test section because of the multiple disturbances produced along the wall by the individual openings. In the case of a test section having a parallel perforated wall with 22 per cent open-area ratio and no auxiliary suction (test-section length = 3 × height), choking was observed at Mach numbers as low as 0.77 (Ref. 12 and Fig. 12.36a). In order to alleviate this condition it was necessary to use approximately $\Delta m/m = 3.6$ per cent mass flow removal.

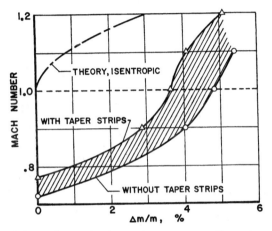

Fig. 12.37. *Minimum plenum chamber suction requirements for perforated test section with parallel walls and sonic nozzle (walls 22.5 per cent open, holes $\frac{1}{16}$ in. diameter, (test-section length = 3.1 × test-section height).*[12]

When more auxiliary suction is employed than is required to alleviate choking, the Mach number is increased beyond Mach number one in the test section proper, as pointed out previously. For instance, in order to establish a Mach number of 1.2 in the test section, the auxiliary suction must be increased from $\Delta m/m = 3.6$ per cent, required for Mach number one, to 5.1 per cent—an increase of 1.5 per cent. This value compares with an increase of approximately 3 per cent mass flow removal which would be required in isentropic flow. The difference is undoubtedly due to the fact that auxiliary suction mainly removes air from the boundary layer which, for the same displacement thickness, has a considerably smaller mass flow density than isentropic flow.

The curves for walls with 22.5 per cent open-area ratio were obtained for a test section which was equipped with tapered porosity, as shown in Fig. 12.24. When no taper strips were applied and Mach number distributions as irregular as those shown in Fig. 12.23a were established, the mass removal requirements were much greater (Fig. 12.37). This difference can be readily attributed to the local alternating inflow and outflow of the test section which depends on the varying pressure difference between plenum chamber and test section. The alternating flows naturally cause much greater losses, which, in turn, must be compensated for by increased mass flow removal from the plenum chamber, to obtain the area expansion required to establish the desired Mach number.

Influence of Open-Area Ratio.—Various perforated walls with different open-area ratios were investigated in the AEDC transonic model tunnel[12,2] to explore the influence of open-area ratio on the choking phenomenon and the mass flow removal requirement for such walls. The results (Fig. 12.36a) indicate that with walls of 5.2 per cent open-area ratio the choking Mach number is approximately 0.89 in a test section without suction and with parallel walls. To establish a Mach number of one, a mass flow removal of 1.4 per cent was required. With increasing open-area ratio the losses in the flow along the wall were naturally increased. For example, a mass flow removal of 6.9 per cent was necessary to establish sonic flow with walls having 33 per cent open-area ratio. Such values approach the limits of practical wind tunnel design and make the application of such large open-area ratios very difficult.*

Influence of Convergence and Divergence of Test-Section Walls.—When the walls of a test section are converged, as is desirable in order to keep the boundary layer thin in the test section proper, the mass flow removal requirement is increased (see Fig. 12.36b). With 30' convergence, all walls with an open-area ratio between 5.2 and 22.5 per cent require approximately the same mass flow removal to establish Mach number one, that is, approximately 5.8 per cent. Only the 33 per cent open walls require the much larger removal value of 8.3 per cent. On the other hand, if the test section is diverged 30', choking is alleviated because of the larger area available at the downstream end of the test section, and thus the mass flow requirements are also considerably reduced (see Fig. 12.36c).

In the experiments just cited for converged and diverged walls, only the upper and lower walls were converged or diverged; the two side walls were maintained at a parallel setting. In spite of this non-uniform setting of the walls, the boundary layer thickness was found to be equal along all four walls. This result was expected because the main parameters governing boundary layer growth (wall roughness and local pressure difference between the inside of the test section and the plenum chamber) are identical for all four walls, in spite of their different settings.

Influence of Hole Size.—The results previously cited for the mass flow requirements of perforated walls were obtained in the AEDC transonic model tunnel using $\frac{1}{16}$ in. thick walls equipped with holes of $\frac{1}{16}$ in. diam. As discussed in Chapter 11, the hole size and the wall thickness have a large influence on the effectiveness of a perforated wall and, consequently, a significant influence on boundary layer development and auxiliary mass flow requirements. This influence was documented during an investigation of a wall with a 22.5 per cent open-area ratio, holes enlarged to a diameter of $\frac{1}{4}$ in., and the same wall thickness of $\frac{1}{16}$ in. (see Refs. 12 and 2). No taper strips were required to eliminate over-expansion because of the larger cross-

---

* The change in the mass flow requirements due to installation of models is discussed later in Chapter 14.

flow resistance of the walls with the larger holes. As expected, the mass flow requirements were considerably reduced (Fig. 12.36d).

All values cited previously for auxiliary suction were obtained using model test sections equipped with a sonic nozzle; in other words, supersonic flow was established by removal of air through the wall perforations. These values represent the minimum values of suction with which it is possible to establish the desired Mach number in the main portion of the test section. Since with minimum suction, the main wind tunnel compressor has large power requirements, it is usually more advisable to increase auxiliary suction beyond the values cited (see Chapter 14).

c. Test sections with supersonic nozzles in supersonic speed range

When the perforated test section is preceded by a supersonic nozzle which is used in establishing supersonic flow, mass flow removal requirements are considerably reduced because only the boundary layer growth along the perforated walls must be removed by suction; mass flow removal to establish flow is no longer necessary. The minimum flow removal values for a perforated wall with 11.8 per cent open-area ratio (no supersonic nozzle) are shown in Fig. 12.38 from tests in the AEDC transonic model tunnel.[13] It

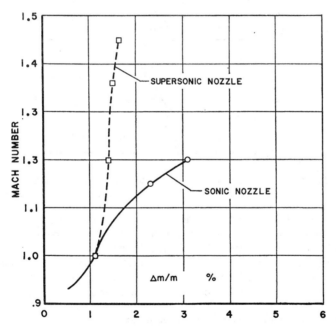

FIG. 12.38. *Minimum plenum chamber suction requirements for perforated test section with 11.8 per cent open-area ratio and perforated walls with sonic and supersonic nozzles (test-section length $3.1 \times H$).*[13]

is apparent that for this particular test-section configuration, the mass flow removal requirements differed only slightly over the test Mach number range, 1.7 per cent at Mach number 1.45 compared with 1.1 per cent at Mach number 1.0.

The values above were obtained in a relatively small transonic model tunnel of 1 ft test-section height. The trend of the mass flow removal requirements is parallel to the trend of the removal requirements determined for the full-scale British ARA 9 × 8 ft transonic wind tunnel and presented in Fig. 12.39 (see Ref. 8). The British values were obtained by setting the test-section Mach number approximately equal to the supersonic nozzle exit Mach number. This type of operation yields roughly minimum flow removal requirements. At Mach number one, it was necessary to remove

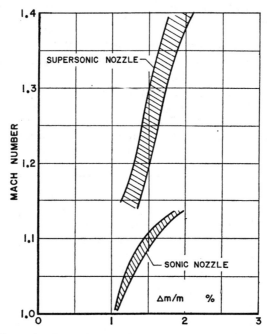

FIG. 12.39. *Plenum chamber suction requirements for perforated test section with 22.5 per cent open-area ratio and parallel walls (nozzle set at test-section Mach number).*[8]

approximately 1.1 per cent of the total test-section mass flow; at Mach number 1.4 a mass flow removal of approximately 1.9 per cent was required. However, it should be remembered that the perforated walls in the British tunnel had a total surface area of approximately 5 × the cross-sectional area of the wind tunnel; the perforated wall area of the transonic model tunnel (Figs. 12.36a and 12.38) was approximately 10 × cross-sectional area of the tunnel.

d. Slotted test sections compared with perforated test sections

The principal reasons for mass flow removal from the plenum chamber of a slotted wind tunnel are the same as those previously detailed for removal from perforated test sections. In the case of parallel walls, a sonic nozzle, and no suction, choking would occur at the downstream end of the test section due to the boundary layer growth along the solid walls, elements, and open slots. Also, in order to build up supersonic flow without using a conventional supersonic nozzle, a certain amount of air must be removed from the test section by plenum chamber suction. For example, in the AEDC transonic model tunnel a test section configuration with sixteen slots and approximately 11 per cent open-area ratio[6] was investigated, as noted in connection with Fig. 12.31. The minimum mass flow removal requirements for this slotted test section are shown in Fig. 12.40. Throughout

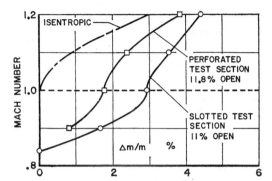

FIG. 12.40. *Minimum mass flow removal requirements for slotted and perforated test sections with parallel walls.*[6,2]

the entire Mach number range from subsonic choking, through Mach number one, to the maximum Mach number of 1.2, the plenum chamber suction requirements are noticeably larger than those for a perforated test section with the same amount of open-area ratio. This result was something of a surprise since it had been expected that part of the kinetic energy of the flow entering the plenum chamber, particularly in the supersonic Mach number range, would retain a portion of its momentum until it re-entered the test section at the downstream end. As the suction requirements show, this was obviously not accomplished. Detailed momentum surveys of the flow in the slot area indicated that the flow very quickly lost its momentum so that no advantage over perforated test sections was established (see Fig. 11.25b).

7. CALIBRATION OF PARTIALLY OPEN TEST SECTIONS
a. General relationships

Operational procedures must be set up for partially open test sections of either the perforated or the slotted wall type to determine the proper combinations of pressure boost from the main drive system, auxiliary suction, and setting of the wind tunnel nozzle. In the numerous test series conducted with partially open test sections, it was determined that in the region of usable test-section flow, the pressure in the plenum chamber is very nearly equal to the mean pressure in the test section. Consequently, it has become customary to base the calibration of a partially open wind tunnel on the static pressure in the plenum chamber and to apply suitable corrections to this pressure in order to determine the correct test-section pressure. The small magnitude of the difference between the two pressures is readily apparent from the typical Mach number distributions presented previously in Figs. 12.3–12.5.

b. Subsonic speed range

In the subsonic speed range the difference between plenum chamber and test-section pressure is practically independent of mass flow removal and pressure boost of the main wind tunnel drive. When the Mach number is kept constant in a given wind tunnel, the difference is merely a function of the setting (parallel, converged or diverged) of the walls. The small deviations between test-section and plenum chamber pressures can be determined from calibrations of the empty test section. This was done for a typical test section of the full-scale transonic wind tunnel of the AEDC.[9] The pressure differences determined for this tunnel are shown in Fig. 12.41 converted

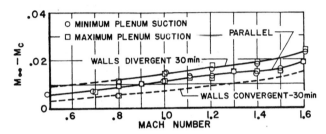

FIG. 12.41. *Empty tunnel calibration for AEDC 16 ft wind tunnel with inclined-hole-perforated walls* (M = test-section Mach number, M_{pc} = equivalent plenum chamber Mach number).[9]

into equivalent Mach number differences. In this test section with perforated walls having inclined holes of 60° inclination angle and 6 per cent nominal open-area ratio, the maximum Mach number deviation was on the order of $\Delta m = 0.015$ at Mach number one.

When a model is installed in a test section, it becomes necessary to determine the proper Mach number correction for the blockage of the model. If the area ratio of a well-designed, partially open wind tunnel is selected properly, the Mach number or pressure gradient correction is very small or may even disappear completely. Hence, in actual wind tunnel operation, when an open-area ratio is selected which is nearly equal to the theoretical open-area ratio for zero correction, the empty wind tunnel calibration may also be utilized for tests with a model installed.

If the boundary layer effects can be neglected, the theoretical open-area ratio for zero correction will produce correct values. This, however, may not be entirely true, particularly if the wall boundary layer is very thick; for instance, in test sections with divergent walls. A partial reduction of the errors involved can be accomplished by selecting the open-area ratio of the walls on the basis of the correct cross-flow resistance coefficient, instead of on the basis of the correct geometrical open-area ratio. However, no theoretical or experimental investigations are known to have been made to determine the magnitude of the boundary layer effect. From the results of comparative tests with different models in different wind tunnels, it may be assumed that the boundary layer has only minor influence on the accuracy of transonic wind tunnel test results (see Chapter 10).

c. Supersonic speed range

In the supersonic, as in the subsonic speed ranges, numerous combinations of plenum chamber suction and pressure boost of the main wind tunnel compressor produce the same Mach number distribution in the test section and the same difference between plenum chamber and mean test-section pressure, as long as the setting of the wind tunnel nozzle and the test-section walls is not changed. In the low supersonic Mach number range, it is frequently customary to use a sonic nozzle and to determine the Mach number in the test section from the difference between the static pressure in the plenum chamber and the mean static pressure in the test section. Such a calibration depends only upon the wall setting. that is, upon a parallel, convergent, or divergent setting.

When a supersonic nozzle precedes the test section, it is customary to select the proper setting of the nozzle for the desired Mach number and to adjust plenum chamber suction and main wind tunnel pressure boost in such a manner that uniform flow is obtained in the test section. Empty tunnel calibrations were conducted in the AEDC transonic wind tunnel with the supersonic nozzle adjusted so as to maintain the nozzle exit Mach number equal to the mean Mach number in the test section.[9] At constant Mach numbers the resultant differences between plenum chamber and test-section pressure converted into Mach numbers were found to depend solely upon the convergence angle of the walls (Fig. 12.41).

When a model is installed in a partially open test section, the supersonic flow approaching the model will not be changed as long as the empty tunnel calibration defined by the wind tunnel nozzle setting and the plenum chamber pressure is employed. Compensation for the displacement effect of the model is then automatically made by the larger mass removal from the plenum chamber required to establish the correct plenum chamber pressure. In the region downstream of model-produced waves which might be reflected from the tunnel walls, the above method of setting the Mach number remains correct as long as a wall is used which absorbs the impinging waves perfectly. If, however, the wave reflections are not completely cancelled by the wall, deviations from correct flow will occur. These deviations cannot be reduced by a change in the overall parameters of the test-section flow. It is necessary in this case to correct the boundary conditions of the walls locally to prevent the wave reflections.

d. Perforated walls with conventional and inclined holes

The results discussed and presented previously for the AEDC transonic wind tunnel were obtained with special walls, that is, with walls having inclined holes. Such holes have the tendency to establish a pressure difference between the plenum chamber and the test-section sides of the walls, even when the flow is essentially parallel to the walls. Consequently, the relatively large difference between plenum chamber and test-section pressure for parallel wall setting may be attributed to this special type of walls.

In the case of conventional perforated walls, that is, in the case of walls with straight holes, the differences in pressure between the plenum chamber and the test section are usually small when the walls are set divergent or parallel to the mean flow direction (Fig. 12.42). The results shown in this figure were obtained using a test section with 22.5 per cent open conventional straight hole walls in the AEDC transonic wind tunnel.[14] When the wall setting was changed to 30′ convergent, corrections as great as $\Delta m = 0.027$ were required at Mach number 1.4.

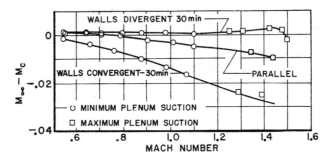

FIG. 12.42. Empty tunnel calibration for AEDC 16 ft wind runnel with straight hole perforated walls (M = test-section Mach number, M_{pc} = equivalent Mach number in plenum chamber).[14]

e. *Control of mass flow removal and wind tunnel compressor for tests with models installed*

It was stated previously that the Mach number distribution in a perforated or slotted test section is practically equal for many combinations of plenum chamber suction and main wind tunnel pressure boost. However, this statement is true only for the main region of the test section.

Supersonic Speed Range.—Flow non-uniformities occur in the upstream portion of test sections, particularly when supersonic speeds are established with sonic nozzles. They are caused by the transition from the solid wall to the perforated wall portion and by the supersonic flow establishment itself. It is therefore necessary to determine the range of these non-uniformities and to select model installations which do not extend into this range.

Deviations from the uniform mean Mach number frequently occur at the downstream end of the test section also, and can be attributed directly to incorrect setting of the plenum chamber mass flow removal (see Figs. 12.20, 12.23 and 12.27). The operating principle is established that enough auxiliary suction must be applied so that no Mach number reduction occurs in the downstream portion of the test section, that is, so that no oblique or normal shocks are generated in this portion of the test section. This condition can be checked by pressure orifices located at the upstream end of the diffuser. As long as these orifices indicate that the local pressure is smaller than the mean test-section pressure, the condition of uniform flow is satisfied almost as far as the end of the test section. When it is not possible to produce the necessary pressure rise in the main wind tunnel compressor to draw the terminating shock system far enough into the diffuser, it may be necessary to have the drop-off in Mach number occur in the test section itself, with subsequent increases in the amount of auxiliary suction. If such a condition is unavoidable, the downstream portion of the test section is not usable for testing, and the extent of this region must be determined by detailed measurements.

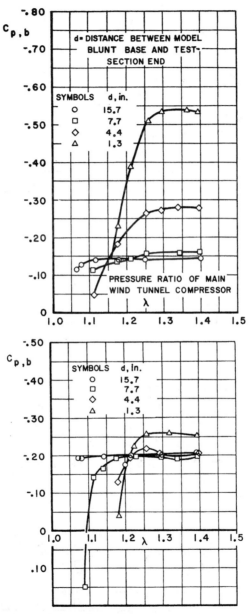

Fig. 12.43. *Base pressure of blunt-base cylindrical model in 1 ft perforated test section at different locations and different combinations of tunnel pressure ratio and auxiliary plenum chamber suction.*[15]

Subsonic Speed Range.—In the subsonic speed range, the rule for control of auxiliary suction is simply that neither too much nor too little mass flow removal be applied; otherwise Mach number reductions or Mach number increases occur in the downstream portion of the test section (see Figs. 12.7 and 12.9). In empty test sections it is easy to adjust plenum chamber suction to the proper amount by observing pressure orifices located at the downstream portion of the test-section walls. However, when a model is installed, this method of determining correct auxiliary suction may not always be

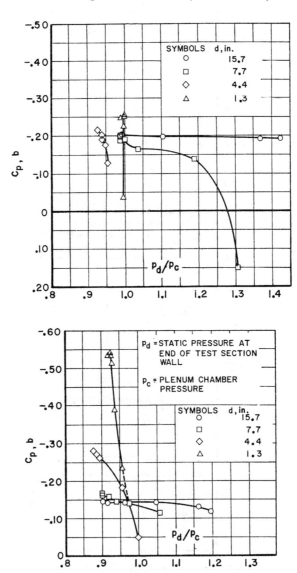

Fig. 12.44. Base pressure of a blunt-base cylindrical model in *1 ft perforated test section* at different locations as a function of static pressure at end of test section.[15]

reliable since the static pressure along the wall in the downstream portion of the test section is not only influenced by inflow and outflow conditions of the test section but also by the disturbances of the model.

Non-uniform Mach number distribution in the downstream portion of a perforated test section is particularly noticeable in experiments to determine the base pressure of models. In a series of systematic experiments conducted at the AEDC transonic model tunnel[15] cylindrical models were placed in a perforated wind tunnel with their blunt bases at different locations. The base pressure was then determined for the various locations as a function of combinations of plenum chamber suction and pressure boost of the main wind tunnel compressor. It was apparent that the base pressure was largely dependent upon the proper combination of main compressor boost and auxiliary suction (Fig. 12.43). In order to obtain an indication of the correct setting for these bases, several orifices were placed in the downstream portion of the test-section wall, and the base pressures were again determined, this time as a function of the downstream test-section static pressure. At all subsonic Mach numbers, the pressure ratio $p_d/p_c = 0.975$ gave a correct indication of the proper combination of mass flow removal and wind tunnel compressor boost (see Fig. 12.44). Consequently, in the subsonic range, models can be extended into the rear portion of the test section when this operating principle is employed. In the supersonic range, however, this principle breaks down (Fig. 12.44b). It is suspected that the base pressure p_B is also affected by strong diffuser shock waves which can influence the base pressure through the subsonic wake. Further experiments proved that the base of blunt bodies should not approach the downstream portion of the test section by more than $2\frac{1}{2} \times$ the diameter of the blunt base.[15] If this rule is observed and the end pressure p_d in the test section is smaller than the plenum chamber pressure p_c, interference-free data can be expected.

In summary it may be stated that the local disturbance which develops in the downstream portion of a partially open test section can be avoided quite effectively in the subsonic range by simply using the static pressure in the downstream test-section region as a reference. In the supersonic Mach number range, not only must the static pressure in the upstream diffuser section be observed, but special care must also be taken to avoid placing the model base too close to the end of the test section. As a practical rule, it appears that the base of blunt models should not approach the test-section end by more than $2\frac{1}{2} \times$ the diameter of the model base in order to avoid disturbances which might be produced by strong shock or expansion waves impinging upon the wake.

REFERENCES

[1] WRIGHT, RAY H. and WARD, V. G. "NACA Transonic Wind Tunnel Sections." NACA RM L8J06, October 25, 1948.

[2] CHEW, WILLIAM L. "Wind Tunnel Investigations of Transonic Test Sections—Phase II: Comparison of Results of Tests on Five Perforated-Wall Test Sections in Conjunction with a Sonic Nozzle." AEDC–TR–54–52, March, 1955.

[3] GOETHERT, B. H. "Influence of Plenum Chamber Suction and Wall Convergence on the Mach Number Distribution in Partially Open Test Sections of Wind Tunnels at Subsonic and Supersonic Speeds." AEDC–TR–54–42, March, 1955.

[4] WARD, V. G., WHITCOMB, C. F. and PEARSON, M. D. "Air Flow and Power Characteristics of the Langley 16 ft Transonic Tunnel with Slotted Test Section." NACA RM L52E01, July, 1952.

[5] HOLDER, D. W., NORTH, R. D. and CHINNECK, A. "Experiments with Slotted and Perforated Walls in a Two-dimensional High-Speed Tunnel." T.P. 352 F.M. 1631, November 14, 1951.

[6] ALLEN, EDWIN C. "Wind Tunnel Investigations of Transonic Test Sections—Phase III: Tests of a Slotted-Wall Test Section in Conjunction with a Sonic Nozzle." AEDC–TR–54–53, February, 1955.

[7] GOETHERT, B. H. "Flow Establishment and Wall Interference in Transonic Tunnels." AEDC–TR–54–44, June, 1954.

[8] HAINES, A. B. and JONES, J. C. M. "The Centre Line Mach Number Distributions and Auxiliary Suction Requirement for the A.R.A. 9 ft × 8 ft Transonic Wind Tunnel." Aircraft Research Association Ltd. (G.B.), ARA–R–2, April, 1958.

[9] DICK, R. S. "Calibration of the PWT 16 ft Transonic Circuit with a Full Test Cart Having Inclined-Hole Perforated Walls." AEDC–TN–58–97, January, 1959.

[10] RITCHIE, V. S. and PEARSON, A. O. "National Advisory Committee for Aeronautics Calibration of the Slotted Test Section of the Langley 8 ft Transonic Tunnel and Preliminary Experimental Investigation of Boundary Reflected Disturbances." NACA RM L51K14, July 7, 1952.

[11] GOETHERT, B. H. "Development of the New Test Section With Movable Side Walls of the Wright Field 10 ft Wind Tunnel (Phase A—Operation With Slots Closed)." WADC TR–52–296, November, 1952.

[12] CHEW, WILLIAM L. "Wind Tunnel Investigations of Transonic Test Sections—Phase I: Tests of a 22.5 per cent Open-Area Perforated Wall Test Section in Conjunction with a Sonic Nozzle." AEDC–TR–53–10, October, 1953.

[13] GARDENIER, H. E. "Flow Establishment and Power Requirements in Slotted and Perforated Wind Tunnels." Arnold Engineering Development Center, July, 1956.

[14] DICK, R. S. "Calibration of the PWT 16 ft Transonic Circuit With an Aerodynamic Test Cart Having 20 per cent Open Perforated Walls and Without Plenum Auxiliary Suction." AEDC–TN–58–24, June, 1958.

[15] RITTENHOUSE, L. "Base Pressure Effects Resulting from Changes in Tunnel Pressure Ratio in a Transonic Wind Tunnel." AEDC–TN–58–88, January, 1959.

BIBLIOGRAPHY

ALLEN, H. J. and SPIEGEL, J. M. "Transonic Wind Tunnel Development at the NACA." S.M.F. Fund Paper No. FF–12, Institute of Aeronautical Sciences publication, 1954.

BATES, GEORGE P. "Preliminary Investigations of 3 in. Slotted Transonic Wind-Tunnel Test Sections." NACA RM–L9D18, September, 1949.

BAUER, R. C. and RIDDLE, C. D. "Transonic Dynamic Stability Tests of a Small-Scale Cone Model in Combination with Three Afterbodies." Arnold Engineering Development Center, July, 1958.

CHEW, WILLIAM L. "Experimental and Theoretical Studies on Three-dimensional Wave Reflection in Transonic Test Sections—Part III: Characteristics of Perforated Test-Section Walls with Differential Resistance to Cross-Flow." AEDC–TN–55–44, March, 1956.

CHEW, WILLIAM L. "Total Head Profiles Near the Plenum and Test Section Surfaces of Perforated Walls." AEDC–TN–55–59, December, 1955.

CUSHMAN, H. T. "Small-Scale Transonic Interference Studies with Perforated Wall Test Sections." UAC R-95538-16, June, 1953.

DICK, R. S. "Calibration of the 16 ft Transonic Circuit of the Propulsion Wind Tunnel With an Aerodynamic Test Cart Having 6 per cent Open-Inclined-Hole Walls." AEDC-TN-58-90, November, 1958.

GOODMAN, T. R. "The Porous Wall Wind Tunnel—Part I: One-dimensional Supersonic Flow Analysis." CAL Report No. AD-594-A-2, October, 1950.

HARRIS, W. G. and LESKO, JAMES S. "Development of a New Test Section with Movable Sidewalls, Wright Field 10 ft Wind Tunnel." WADC-TR-296-Part II, May, 1954.

MAEDER, P. F. and STAPLETON, J. F. "Investigation of the Flow Through a Perforated Wall." Brown University, TR-WT-10, May, 1953.

MAEDER, P. F. and HALL, J. F. "Investigation of Flow Over Partially Open Wind Tunnel Walls, Final Report." AEDC-TR-55-67, December, 1955.

NELSON, WILLIAM J. and KLEVATT, P. L. "Preliminary Investigation of Constant-Geometry Variable Mach Number Supersonic Tunnel with Porous Walls." NACA RM L50B01, May, 1950.

NICHOLS, JAMES H. "An Investigation of the PWT Transonic Circuit Full Test Cart with Inclined-Hole Perforated Walls and without Plenum Auxiliary Suction." AEDC-TN-58-86, October, 1958.

RITTENHOUSE, L. "Effects on Base Pressures Resulting from Changes in Tunnel Pressure Ratio and Scavenging Scoop Flow Rate for the PWT Test Section Configuration." Arnold Engineering Development Center, March, 1958.

STOKES, G. M., DAVID, D. D., JR. and SELLERS, T. B. "An Experimental Study of Porosity Characteristics of Perforated Materials in Normal and Parallel Flow." NACA RM L53H07, November, 1953.

TAYLOR, H. D. "Status Report of Transonic Wind Tunnel Development at United Aircraft Corp." UAC R-95295-16, September 12, 1950.

TAYLOR, H. D. "Small-Scale Studies of a Porous Wall Transonic Test Section in Combination with a Laval Nozzle, Concluding Report." UAC R-95434-19, April 4, 1952.

WARD, V. G. and WHITCOMB, C. F. "An NACA Transonic Test Section with Tapered Slots Tested at Mach Numbers to 1.26." NACA RM L9J06, October 25, 1948.

WRIGHT, RAY H. and RITCHIE, V. S. "Characteristics of a Transonic Test Section with Various Slot Shapes in the Langley 8 ft High-Speed Tunnel." NACA RM L51H10, October 18, 1951.

CHAPTER 13

BOUNDARY LAYER GROWTH ALONG PARTIALLY OPEN WALLS

1. SIMPLIFIED THEORETICAL CONSIDERATIONS

a. Physical parameters

According to the numerous experiments discussed in the preceding section, the boundary layer growth along partially open test section walls is very much larger than the growth along solid walls. At the open-area ratios of major interest in wind tunnel testing with perforated test sections, this rapid boundary layer growth produces conditions very detrimental to wind tunnel operation. The power required to operate such wind tunnels is greatly increased by the thick wall boundary layers. Besides the direct influence of the momentum loss in the boundary layer itself, the effect on the efficiency of the wind tunnel diffuser is of even greater importance. Also, because thick boundary layers must be avoided if effective shock-wave cancellation at the perforated walls is to be accomplished, it is usually necessary to reduce the boundary layer thickness along the walls by auxiliary suction. In view of these significant ramifications, the parameters governing the growth of the boundary layer along partially open walls have been studied in detail by several groups of investigators.

Detailed consideration of boundary layer development along a perforated wall reveals that thin elementary boundary layers are produced by the normal friction along the solid portions of the wall. In addition, a free-jet mixing is established along the open portions of the walls that is similar to the mixing which occurs in the boundary region of a free jet exhausting into still air. The friction coefficient along the solid portions of the perforated wall is generally quite high due to the turbulent boundary layer produced by the many small disturbances along the edges of each individual wall opening. Furthermore, because a thin elementary boundary layer for each solid wall element starts anew at each opening, the local Reynolds number of these sub-boundary layers are small, and, thus, according to normal boundary layer theory, high local friction coefficients result.

b. Theoretical calculations of wall boundary layer growth along perforated walls[1]

Since the boundary layer growth along open areas of a wall is much faster than the boundary layer growth along solid portions, it can be expected that the boundary layer growth along perforated walls is predominantly controlled by the mixing phenomenon along the open wall portions. Consequently, an attempt was made to calculate theoretically the boundary profile and the resulting discharge of air through the perforated wall into the plenum chamber by means of the methods used to calculate turbulent mixing along free-jet boundaries.[2]

The theory for turbulent free-jet mixing introduces a mixing length, l, which is proportional to the distance x in the flow direction, that is:

$$l = cx$$

with c an experimental constant.

It was intuitively assumed that in the case of perforated walls the mixing length does not grow in direct proportion to x but is proportional to Rx, that is, proportional only to the accumulative length of the openings in the wall:

$$l = cRx$$

with R the open-area ratio of the perforated wall.

According to this equation, the mixing length assumes a value $l = 0$ when R approaches zero. In other words, for a wall with an infinitely small open-area ratio the free-jet mixing effect has no significance. On the other hand, when R approaches l, that is, with conditions similar to the case of a completely open wall, the mixing length l tends towards the value which occurs in the case of free jets.

The friction along the solid wall elements was not introduced into this simplified theoretical investigation since its influence was assumed to be small for open-area ratios having practical importance.

The theoretical problem, therefore, was reduced to the solution of the mixing layer equations for an assumed free jet with a mixing length reduced according to $l = cRx$. It was also assumed that the wall itself did not disturb the mixing layer growth and that the mixing profile extended internally into the test-section flow and externally into the plenum chamber air. The boundary conditions for the mixing layer are assumed to be the same as those for a free-jet mixing layer without a perforated wall, as follows:

Mixing Layer Boundary at the Test-Section Side ($y = y_1$).—The local velocity is assumed to be equal to the velocity of the free stream and has no component perpendicular to the free stream, that is

$$v_{x_1} = v_\infty \text{ and } v_{y_1} = 0$$

The change of the velocity at this station in the y-direction is assumed to be zero, that is:

$$(\partial v_x/\partial y)_1 = 0$$

Mixing Layer Boundary at the Plenum Chamber Side ($y = y_2$).—The local velocity and the change in the y-direction must be zero, that is:

$$v_{x_2} = 0 \text{ and } (\partial v_x/\partial y)_2 = 0$$

According to mixing layer theory, the velocity profiles along lines $y/x = $ constant are the same. Consequently, a new variable is introduced:

$$\eta = y/x$$

The stream function is then:

$$\psi = v_\infty x F(\eta)$$

The equation of motion is based on Euler's equation:

$$v_x \frac{\partial v_x}{\partial x} + v_y \frac{\partial v_x}{\partial y} = \frac{1}{\rho} \frac{\partial \tau}{\partial y}$$

with τ the apparent shear stress.

After introduction of the stream function and several transformations:

$$FF'' + (cR)^2 F'' F''' = 0$$

with c the proportionality constant in the mixing length equation. This differential equation has been solved and the following solution obtained:[2,1]

with
$$F = C_1 e^{-\alpha\eta} + C_2 e^{\alpha\eta/2} \cos(\alpha\eta\sqrt{3}/2) + C_3 e^{\alpha\eta/2} \sin(\alpha\eta\sqrt{3}/2)$$

$$\alpha = (cR)^{-2/3} = \alpha_0 R^{-2/3}$$

The coefficient α_0 is a constant which refers to the conditions along a completely open wall ($R = 1$). Its value has been determined experimentally[2] to be:

$$\alpha_0 = 11.8$$

The constants C_1, C_2 and C_3 can be determined from consideration of the boundary conditions for the above solution as:

$$\alpha C_1 = -0.0165$$
$$\alpha C_2 = 0.137$$
$$\alpha C_3 = 0.6918$$

Also the boundaries of the mixing layer, that is, the coordinates $\eta_1 = y_1/x$, and $\eta_2 = y_2/x$, have been determined to be:

$$\alpha\eta_1 = 0.981 \qquad (v_x = v_\infty)$$
$$\alpha\eta_2 = -2.04 \qquad (v_x = 0)$$

These equations describe in detail the velocity profile in a mixing zone for completely open walls ($R = 1$) and for perforated walls with $R < 1$. With the velocity profile thus known and the usual assumption in boundary layer theory that the static pressure is constant at any given station, $x =$ constant, it is possible to calculate the displacement thickness or any other parameters of the boundary layer, as desired.

It is particularly important to know how much air will pass through the perforated wall as a result of the turbulent mixing along the open areas when the static pressure in the test section is kept constant. Then, by applying the continuity equation, the mass flow passing through a perforated wall having a parallel wall setting is:

$$\Delta m = xw \int_0^{\eta_1} (\rho v_\infty - \rho v_x)\, d\eta$$

with w the width of the perforated plate.

When this equation was evaluated[1] the following numerical results were determined:

$$m' = \frac{\Delta m}{m}\frac{A_T}{A_{wp}} = \frac{0.1209}{\alpha} = \frac{0.1209}{\alpha_0}R^{2/3}$$

where: m = mass flow through test section
A_T = cross-sectional area of test section
A_{wp} = area of perforated wall.

Finally, with $\alpha = 11.8$, the unit mass flow m' is:

$$m' = (\Delta m/m)(A_T/A_{wp}) = 0.01023\,R^{2/3}$$

The mass flow entering the plenum chamber can be determined in a similar manner if the perforated wall is placed at a diverged or converged setting corresponding to $\eta_w = y/x = $ constant.

The basic equation is then:

$$\Delta m = xw\left[\int_0^{\eta_1} \rho v_\infty\,d\eta - \int_{\eta_w}^{\eta_1} \rho v_x\,d\eta\right]$$

This equation can be solved in much the same way as the equation for walls having a parallel setting.

c. Experimental determination of suction requirements and comparison with theory

Experiments were conducted in the low-speed 4×8 in. wind tunnel of the Brown University[1] using six perforated walls with open-area ratios ranging between 0.196 and 0.370. The characteristics of the individual walls tested are shown in the following table:

Wall Characteristics

Wall	Hole diameter, in.	Holes/in.[2]	R	Hole shape
A	0.057	108	0.276	hexagonal
B	0.057	144	0.367	square
C	0.045	165	0.260	hexagonal
D	0.045	225	0.370	square
E	0.023	420	0.196	hexagonal
F	0.023	576	0.250	square

Boundary layer velocity profiles along these walls were measured, and the mass flow deficiencies, that is, the flow entering the plenum chamber, were determined from the experimental data.

In order to check the experimental configurations and the measuring equipment, the boundary layer mixing zone along a free jet was determined after removing the perforated wall from the test section. The velocity profiles measured at the various stations along the free jet are presented in Fig. 13.1. For this figure, a value of $\alpha_0 = 11.0$ instead of 11.8 was used to obtain a matching between the experimental data and the theory. When this correction was made, the comparison between theory and experiment was reasonably good, indicating that reliable data can be obtained with such a test setup.

The unit mass flow m' of the six perforated walls is plotted in Fig. 13.2. At the upstream end of all walls, the experimental mass flow removal values are considerably larger than the theoretical values. Since, in this region, all six walls exhibit values of almost the same magnitude, it may be assumed that the deviation between theory and experiment is caused by the initial boundary layer formed along the solid portions of the tunnel which precede

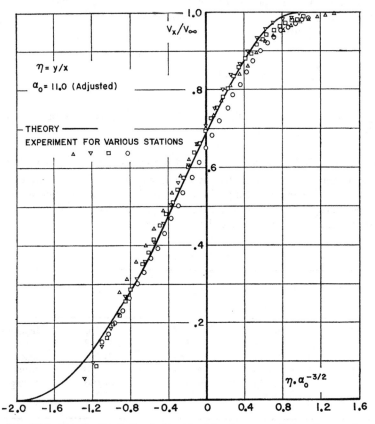

FIG. 13.1. *Velocity distribution in free-jet mixing layer according to theory and experiments (Brown University 4 × 8 in. wind tunnel*[1]*).*

the perforated section. At the downstream end of the perforated walls where the influence of the initial boundary layer thickness is considerably reduced, the walls with open-area ratios ranging between 0.196 and 0.276 show unit mass flows of the same order of magnitude as predicted by theory. However, in this region the walls with large open-area ratios of approximately 0.37 also had much greater mass flow into the plenum chamber, indicating that a considerably more intense mixing occurs than theory assumes.

In summary it may be stated that the extremely simplified theory of Ref. 1 succeeds in predicting the order of magnitude of mass flow removal through perforated walls having open-area ratios ranging between 0.20 and 0.28. This result seems to prove that as in the case of free jets the boundary layer growth along perforated walls is controlled predominantly by the mixing phenomenon.

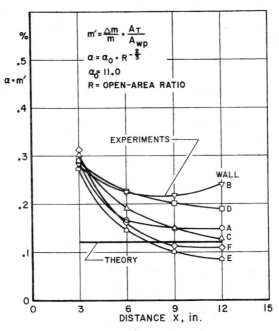

Fig. 13.2. *Mass flow through several perforated walls according to theory and experiments (parallel walls) (Brown Univeristy 4 × 8 in. wind tunnel[1]).*

2. REFINED THEORY FOR BOUNDARY LAYER GROWTH ALONG PERFORATED WALLS

a. Basis of refined theory

In an attempt to improve the theoretical calculations presented above, a more refined theory was developed which again is based on the turbulent mixing which occurs along the boundaries of free jets.[3] Turbulent mixing with a modified mixing length was assumed as before. However, instead of the intuitive assumption that the mixing length is reduced in proportion to the open-area ratio R, the reduction was more accurately calculated on the basis of the mean shear stress along the perforated wall. This stress, which consists of the combined mean shear stress along both the solid and open portions of the wall, was calculated according to well-known theories. The mixing flow adjacent to the wall was then fitted to the shear stress by a suitable modification of the mixing length. In other words, it was assumed that the entire mixing profile can be divided into two separate regions.

One region is composed of the main large-scale mixing layer which is not influenced by either the individual wall openings or the solid wall elements. It corresponds to a homogeneous fictitious wall with a mean shear force of known quantity. The second region includes the small boundary layers along the solid wall elements and the small mixing layers along each individual opening. These small-grain layers are assumed to be submerged in a uniform flow which has less velocity than the free-stream velocity. The magnitude of the effective velocity for each small-grain layer is determined from calculations for the large-scale mixing layer of the perforated wall.

A discussion of the more refined theory will be presented in the following paragraphs.

b. Relationships derived from refined theory

The width of the mixing layer along an open boundary or a perforated wall is represented by $\eta_1 = y_1/x$, that is, at station y_1 the local velocity v_x is equal to the velocity v_∞ of the undisturbed flow. The assumption was made that the mixing length l is proportional to the thickness y_1 of the mixing layer, or:

$$l = C\eta_1 x$$

In the experiments with free jets and wakes behind bodies it was found that the constant C in the above equation is approximately:

$$C = 0.290$$

This value corresponds to the constant $\alpha_0 = 11.8$ used in the simplified theory, Section 1 of this chapter.

When the stream function is again introduced,

$$\psi = v_\infty x F(\eta)$$

and a new variable is established:

$$\xi = \eta/(C\eta_1)^{2/3}$$

the following equation for function F is obtained, which describes the stream function as well as the velocity distribution in the mixing layer:[3]

$$\frac{3}{C^2 \xi_1^2} F(\xi) = (\xi_1 - 1) e^{\xi_1 - \xi} + e^{-(\xi_1 - \xi)/2} \times$$
$$\times \{(1 + 2\xi_1) \cos[(\xi_1 - \xi) 3^{\frac{1}{2}}/2] - \sqrt{3} \sin[(\xi_1 - \xi) 3^{\frac{1}{2}}/2]\}$$

Open Jet.—When the above equation is evaluated for mass flow through a reference line ($\eta_w = (y/x)_w = $ constant) and for the effective shear stress τ, by using a series development, the following solutions are obtained:

Mass Flow
$$m' = \frac{\Delta m}{m} \frac{A_T}{A_{wp}} = C^2[0.1265 + 0.661\xi_w + 0.282\xi_w^2]$$

Shear Stress
$$C(\tau/q)^{\frac{1}{2}} = C^2(0.318 - 0.1426\xi_w - 0.346\xi_w^2)^{\frac{1}{2}}$$

with:

$$\xi_w = \frac{1}{(C\eta_1)^{2/3}}\eta_w$$

and

$$q = \tfrac{1}{2}\rho v_\infty^2$$

These two equations yield a relationship between the relative thickness of the mixing layer, η_1, the mean shear stress, and the mass flow through the line along which the shear stress occurs.

Perforated Walls.—Other approximate relationships have been derived in Ref. 3 for application to perforated walls. With the general equation used previously for the stream function F, the following simplified relations between velocity profile, cross flow, and shear stress along a fictitious perforated wall with a mixing layer thickness η_1 were determined:

Velocity Distribution

$$1 - \frac{v_{xw}}{v_\infty} = \frac{\eta_1}{6C^2}\left(1 - \frac{\eta_w}{\eta_1}\right)^2\left(2 + \frac{\eta_w}{\eta_1}\right)$$

Unit Mass Flow through Wall

$$m' = \frac{\Delta m}{m}\frac{A_T}{A_{wp}} = \eta_w + \frac{\eta_1^2}{24C^2}\left(1 - \frac{\eta_w}{\eta_1}\right)^3\left(3 + \frac{\eta_w}{\eta_1}\right)$$

Shear Stress

$$C(\tau_w/q)^{\frac{1}{2}} = \frac{\eta_1}{2}\left[1 - \left(\frac{\eta_w}{\eta_1}\right)^2\right]$$

With these equations it is possible to determine the thickness of the mixing layer η_1, the velocity profile, and the mass flow through the line $\eta_w = $ constant, when the effective shear stress at the wall τ_w is known.

Determination of Effective Shear Stress for Perforated Walls.—The effective mean shear stress along a perforated wall consists of the stress along the solid portions and the apparent shear stress along the wall openings. The combined effective mean shear stress can be approximated in the following form:

$$\tau_w = R\tau_{w\text{ open}} + (1-R)\tau_{w\text{ solid}}$$

The effective shear stress along the wall openings as a function of the mass flow removal can be determined using the previously derived equations for an open jet, that is:

$$C\frac{\tau_{w\text{ open}}}{q} = f(m'_{w\text{ open}})$$

with:

$$m'_{w\text{ open}} = m'\frac{v_\infty}{v_{x_w}}\frac{1}{R}$$

The turbulent friction stress for the solid portions of the wall can be determined from the theory for friction along flat plates:

$$\tau_{w\,\text{solid}} = C_D \frac{\rho}{2} v_{xw}^2$$

The friction coefficient C_D for both turbulent and laminar boundary layers is known from experiments with flat plates to be a function of the Reynolds number.

These equations were evaluated for various open-area ratios and for various friction coefficients at the solid portions of the wall.[3] Some results are represented in Fig. 13.3 for parallel walls, in Fig. 13.4 for 30′ diverged walls and in Fig. 13.5 for 30′ converged walls. The graphs indicate that a considerable increase in suction requirements occurs when the open-area

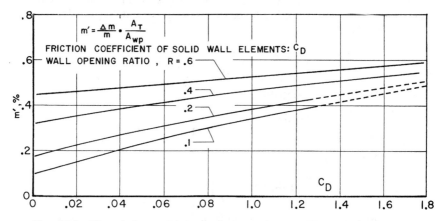

FIG. 13.3. *Theoretical mass flow removal requirements for perforated walls of various open-area ratios as function of solid wall friction coefficient; parallel wall setting.*[3]

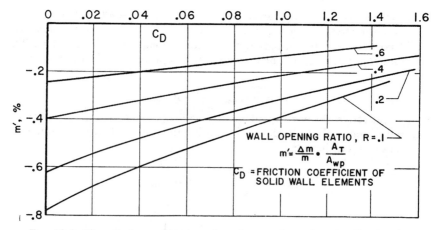

FIG. 13.4. *Theoretical mass flow removal requirements for perforated walls of various open-area ratios as function of solid wall friction coefficient; wall diverged 30′.*[3]

ratio of the wall is increased from the lowest value of the calculation $R = 0.1$ to $R = 0.6$. It is also apparent that, even with the parallel wall setting, the friction along the solid portions of the wall is noticeable and cannot be neglected. Only at large open-area ratios ($R = 0.4$ to 0.6) does the influence of wall friction along the solid portions become minor when the walls are set parallel or converged.

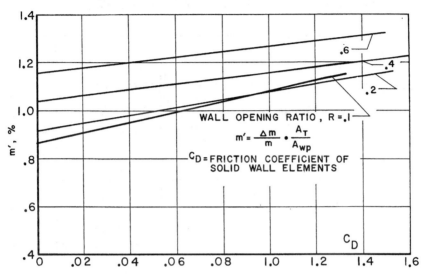

FIG. 13.5. *Theoretical mass flow removal requirements for perforated walls of various open-area ratios as function of solid wall friction coefficient; walls converged 30'.*[3]

c. Comparison between theory and experiments

On the basis of the preceding calculations the suction requirements for a typical perforated test section with a geometry corresponding to that of the AEDC transonic model tunnel were calculated for various open-area ratios of the walls.[4] The results are shown in Fig. 13.6. It was assumed that constant

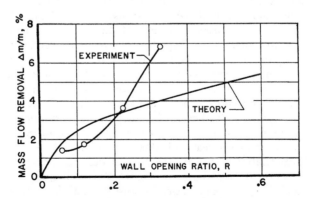

FIG. 13.6. *Comparison between theory and experiments for mass flow removal requirements of perforated walls with different open-area ratios (parallel walls).*[4,5]

pressure throughout the test section was maintained by a proper combination of mass flow removal through the walls and convergence or divergence of the walls. When the wall opening ratio was changed from $R = 0.1$ to $R = 0.6$, the suction requirements, according to theory, are more than doubled. This is mainly due to the increased mixing losses along the open areas of the wall and, to a smaller extent, to the higher friction coefficients along the solid wall elements caused by the smaller Reynolds number.

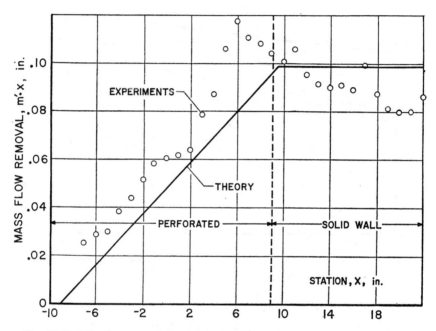

Fig. 13.7. *Mass flow removal for perforated wall according to theory and experiments, parallel walls, 40 per cent open walls, 0.31 in. holes.*[4]

When the results of the calculations are compared with earlier experimental results obtained in the AEDC transonic model tunnel,[5] the agreement is satisfactory at an open-area ratio of 22.5 per cent, where experiment indicates a requirement for 3.6 per cent mass flow removal and theory indicates a requirement for 3.3 per cent. At the lower open-area ratios, the suction requirements were slightly overestimated by theory; at the larger open-area ratio of 33 per cent a large discrepancy occurred. While the increase of suction requirements with an increase of open-area ratio is properly presented by the theory, the magnitude of the increase is underestimated. Instead of the measured 6.8 per cent mass flow removal required for the 33 per cent open-area wall, theory predicts only 4.1 per cent. This difference is probably caused by the lack of uniform pressure distribution in the test section during the experiments. Small fluctuations in pressure result in much larger mixing losses because of the alternating inflow and outflow through the wall. The theoretical values which are valid for smooth flow in the test section do not account for these increased losses.

BOUNDARY LAYER GROWTH ALONG PARTIALLY OPEN WALLS

In another series of experiments, a perforated wall with holes of 0.31 in. diam and an open-area ratio of $R = 0.40$ was investigated at low Mach numbers.[4] The suction requirements for uniform pressure in the test section at a parallel wall setting and the velocity profiles in the vicinity of the perforated wall were determined. From the velocity profile the local suction requirements and the local shear stress were calculated using the refined theory (Section 2 of this chapter). For this perforated wall the simplified theory (Section 1 of this chapter) indicates a unit mass flow removal value

$$m' = (\Delta m/m)(A_T/A_{wp}) = 5.5 \times 10^{-3}$$

The refined theory, which considers the solid wall friction, resulted in

$$m' = 5.62 \times 10^{-3}$$

indicating only a small difference at this open-area ratio and Reynolds number. Some experimental results of this test series are compared with calculated values of local suction requirements and local shear force in Figs. 13.7 and 13.8. The local suction requirements are in reasonably good agreement with theory; however, the local shear stress is approximately 40 to 50 per cent greater than theory predicts.

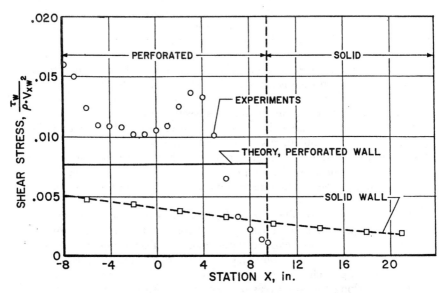

FIG. 13.8. *Mean shear stress along perforated wall according to theory and experiments, parallel walls with 40 per cent open-area ratio, 0.31 in. holes.*[4]

In spite of the observed differences between theory and experiment it is believed that the theoretical treatment, particularly in the refined version, can serve as a useful tool in predicting the suction requirements and the thickness of the mixing layer near perforated walls. It is particularly significant that the suction requirements for full-scale tunnels can be expected to be smaller than the experimentally-determined requirements for model tunnels. This favorable effect can be deduced from theory and is due to the increase of the Reynolds number in the larger wind tunnels and the consequent reduction of the friction coefficient along the solid wall elements.

3. EXPERIMENTAL DETERMINATION OF VELOCITY PROFILES AND DISPLACEMENT THICKNESS OF BOUNDARY LAYERS ADJACENT TO PERFORATED WALLS

a. Velocity profiles

The velocity distribution in the boundary layer adjacent to perforated walls has been measured in numerous experimental series. Some typical distributions for perforated walls with open-area ratios varying between 5.2 and 33 per cent obtained in the AEDC transonic model tunnel (Ref. 5) are presented in Figs. 13.9a, b, and c. At the parallel wall setting (Fig. 13.9a), the velocity profiles for all perforated walls tested were identical at the upstream end of the perforated walls (station $x = 2.3$ in. from the upstream end of the test section). Downstream, toward the end of the test section, the velocity profiles indicate the anticipated boundary layer growth at stations $x = 16.8$ in. and 30.6 in. Though the velocity profiles and boundary layer displacement thickness of the walls with open-area ratios between 5 and 22.5 per cent do not differ greatly, the 33 per cent open wall has not only a larger width of boundary layer but also a different velocity profile shape in the region near the wall. This difference is particularly pronounced at the most rearward position, $x = 30.6$ in. However, the tendency toward a different velocity profile is already apparent at the intermediate position $x = 16.8$ in. Thus it is not likely that the difference is caused by an end-effect in the transition region between test section and diffuser.

The same general tendencies in shape are exhibited by the boundary layer profiles at the 30' diverged and 30' converged wall settings (Figs. 13.9b and c, respectively). Again, the walls with open-area ratios of 22 per cent or less have quite similar shapes, and the wall with 33 per cent open-area ratio shows the characteristic reversal of curvature in the velocity profile.

When the velocity profiles for perforated walls are compared with the profiles measured along solid walls, it is apparent that in the immediate vicinity of the wall a noticeable difference occurs. Because of the "partially open-jet" character of the boundary layer near perforated walls, the velocities do not approach zero at the wall itself, as is the case with solid walls. The velocity along the *open* areas of a perforated wall are finite because of the mixing effect (see Section 2b of this chapter). Individual "solid wall" boundary layers start to build up at each wall element, but they are so thin that in practice a finite velocity is always measured near the wall. This phenomenon is particularly evident in the case of the most open wall (33 per cent open-area ratio) where the mixing layer character of the boundary layer is very pronounced, and the velocities at the wall have values in excess of $v_x/v_\infty = 0.7$, whether the wall is set parallel, diverged, or converged (see Fig. 13.9).

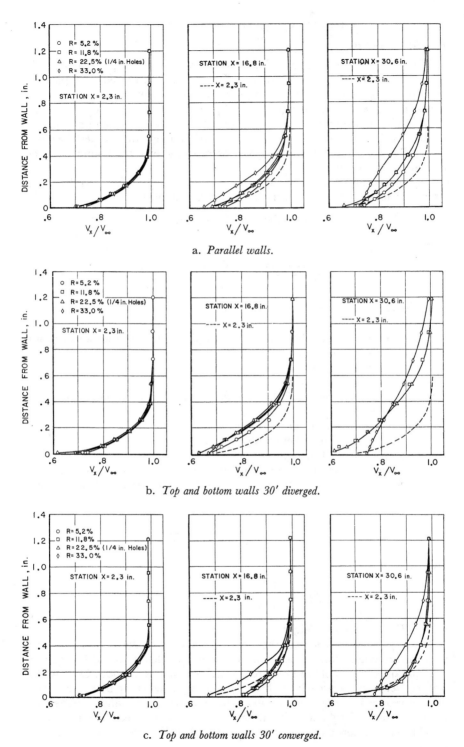

FIG. 13.9. *Boundary layer profiles along perforated walls with different open-area ratios, R, at various stations; M = 1.0 (AEDC 1 ft transonic model tunnel[5]).*

b. Displacement thickness

The displacement thickness of the boundary layer along perforated walls grows in the flow direction unless the walls are converged and plenum chamber mass flow removal is applied. When a constant static pressure exists in test sections with parallel walls, the boundary layer displacement thickness becomes greater as the mass flow removal from the plenum chamber is increased to maintain constant pressure throughout the test section. Naturally, a diverged wall setting also helps to increase the boundary layer displacement thickness.

The preceding statements are based on simple continuity considerations from which the following relationships can be established:

$$\Delta m = (\rho v)_\infty \Delta \delta^* (C_T - \eta_w) L C_{\text{Div}}$$

and

$$m' = \frac{\Delta m}{m} \frac{A_T}{A_{wp}} = \frac{\Delta \delta^* C_T}{A_{wp}} - \eta_w \frac{L C_{\text{Div}}}{A_{wp}}$$

where m = mass flow through test section
  Δm = mass flow removal from test section
  $\Delta \delta^*$ = change of the boundary layer displacement thickness throughout the test section
  A_T = area of test section
  A_{wp} = area of perforated walls
  C_T = circumference of the test section at upstream end of the test section
  C_{Div} = sum of widths of divergent walls
  L = length of divergent walls
  $\eta_w = y/x$ = wall divergence.

In the special case of all four walls of a square test section perforated and only two walls set divergent or convergent, the above equation simplifies to:

$$m' = \frac{\Delta m}{m} \frac{H}{4L} = \frac{\Delta \delta^*}{L} - \frac{1}{2}\eta_w$$

and:

$$\Delta \delta^* = \frac{H}{4} \frac{\Delta m}{m} + \frac{L}{2}\eta_w$$

where H = test-section height.

This equation proves that the greater the rate of the boundary layer growth throughout the test section, the more mass flow removal is required to maintain constant static pressure throughout the test section. Only when wall convergence (a negative value of η_w) is employed is the boundary layer growth reduced so that the boundary layer remains constant or lessens in thickness, depending upon the amount of suction applied. These general conclusions are borne out by the experimental results cited in the following paragraph.

BOUNDARY LAYER GROWTH ALONG PARTIALLY OPEN WALLS

In a test series at the AEDC transonic model tunnel,[5] the boundary layer displacement thickness at various stations in the test section was measured as shown in Fig. 13.10. The larger the open-area ratio of the wall, the larger the boundary layer displacement thickness. This graph supports the conclusion drawn from the preceding equation that in a test section with parallel walls the boundary layer thickness cannot remain constant throughout the test section even though large amounts of plenum chamber suction are

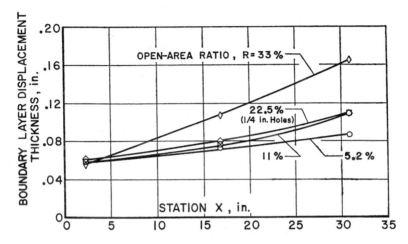

FIG. 13.10. *Influence of open-area ratio of perforated walls on boundary layer growth at $M = 1.0$ (AEDC 1 ft transonic model tunnel[5]).*

applied. For example, a mass flow removal of 1.45 per cent of the test-section flow was required to maintain constant static pressure through a test section with a wall having a 5.2 per cent open-area ratio. According to the above equations, this value of mass flow removal should result in a growth of boundary layer displacement thickness of $\Delta\delta^* = 0.035$ in. The value compares with the experimental boundary layer growth of 0.028 in. measured between stations $x = 2.3$ and 30.8 in. A comparison between calculated and measured boundary layer displacement thickness is shown for the other perforated walls in the following table.

Experimental and Theoretical Boundary Layer Displacement Thickness, $\Delta\delta^*$, for Various Perforated Walls

Wall open-area ratio, per cent	5.2	11.8	22.5	33.0
Theory, $\Delta\delta^*$, in.	0.035	0.042	0.053	0.165
Experiment, $\Delta\delta^*$, in.	0.028	0.048	0.050	0.108

($\Delta\delta^*$ theory calculated from measurements of suction requirements for parallel walls[5].)

The comparison between the theoretical and experimental values for boundary layer growth indicate that the equation predicts the growth of the displacement thickness through a test section within the accuracy of boundary layer displacement thickness measurements. The discrepancies between theory and experiments are undoubtedly caused by non-uniformities of both the boundary layer and the main flow outside the boundary layer. Very careful and detailed measurements of the velocity profile in the

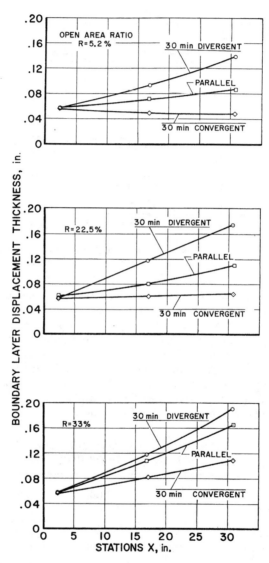

Fig. 13.11. *Comparison between calculated and measured mass flow removal from test sections with perforated walls (17 per cent open, $M = 0.98$) (NACA 3×3 in. transonic model tunnel[6]).*

boundary layer should eliminate these discrepancies so that a much better correlation could be obtained. This supposition is supported by an analysis of the boundary layer profiles and the mass flow removal values for some experiments conducted by the NACA.[6] The comparison between measured and calculated outflow* (see Fig. 13.11) establishes good correlation between theory and experiment.

During the experiments in the AEDC transonic model tunnel, the wall setting was varied from parallel to 30' diverged and 30' converged, and the boundary layer growth through the test section along the perforated walls was determined. The results, presented in Fig. 13.12, show that for

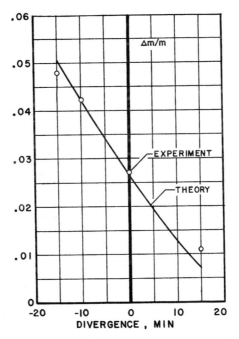

Fig. 13.12. *Influence of wall setting on boundary layer growth for different perforated walls at $M = 1.0$ (AEDC 1 ft transonic model tunnel[5])*.

walls with open-area ratios of 22.5 per cent a wall convergence of 30' merely maintains the boundary layer displacement thickness constant throughout the test section; at 5.2 per cent open-area ratio the same wall setting even shows a slight reduction of the displacement thickness. Through extrapolation of the results for the wall with 33 per cent open-area ratio it can be estimated that more than 1° of wall convergence would be required to maintain the boundary layer of this wall at a constant displacement thickness.

It should be remembered that the relationship between boundary layer thickness and wall setting depends upon the Reynolds number of the test,

---

* Comparison between measured and calculated outflow is equivalent to comparison between measured and calculated boundary layer thickness since it is: $(\Delta m/m) = 4(\Delta \delta^*/H) 2(L/H) \eta_w$.

as explained in detail in Section 2 of this chapter. Also, the shear stress along perforated walls will be smaller in full-scale wind tunnels than in model tunnels. Consequently, in full-scale wind tunnels smaller convergence angles and lower mass flow removals can be expected than in the corresponding model tunnels.

4. EXPERIMENTAL DETERMINATION OF BOUNDARY LAYER DEVELOPMENT IN SLOTTED TEST SECTIONS

a. Velocity profiles

Experiments in perforated test sections indicate that the flow which passes the perforated walls loses its momentum very rapidly after it enters the plenum chamber (see Figs. 14.5 and 14.6). In slotted test sections, on the plenum chamber side of the slots, there is a confined region of flow which only gradually loses its momentum. This external flow in the slot region not only may affect the suction requirements of slotted test sections but also causes a change in the boundary conditions since the pressure on both sides of the tunnel wall is no longer equal to the plenum chamber pressure.

Some typical full-scale measurements of the mixing in the slot region are shown in Fig. 13.13 for a slotted test section in the NACA Langley 16 ft transonic tunnel.[7] It is apparent that even at a station approximately 30 ft downstream of the leading edge of the slots, flow with considerable dynamic pressure exists in the region near the wall outside the test section.

A systematic investigation of the velocity profile in the slot region of a slotted test section at different wall settings was conducted in the AEDC transonic model tunnel (see Ref. 8, and Figs. 13.14a, b and c). The gradual penetration of flow through the slots into the plenum chamber was again

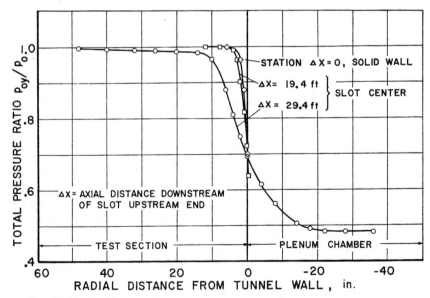

FIG. 13.13. *Total pressure distribution in slot region of NACA Langley 16 ft transonic wind tunnel, $M = 1.07$, open-area ratio 12.5 per cent, eight slots.*[7]

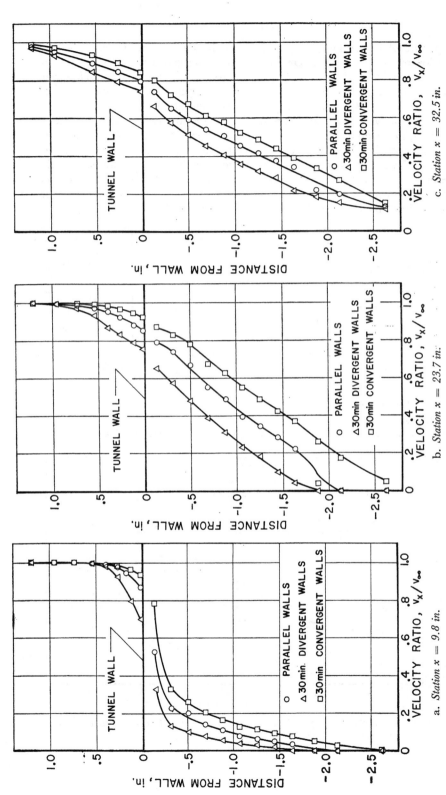

Fig. 13.14. Velocity distributions in slot region of slotted test sections at $M = 1.0$, sixteen slots, 11 per cent open-area ratio (AEDC 1 ft transonic model tunnel[8]).

a. Station $x = 9.8$ in.

b. Station $x = 23.7$ in.

c. Station $x = 32.5$ in.

apparent when the measuring station was shifted from the front to the rear portion of the test section. Divergence or convergence of the walls did not change the basic character of the velocity distribution curves. However, convergence did induce more of the flow to penetrate further into the plenum chamber, which is logical from physical considerations.

Local inflow or outflow affects the flow in the immediate slot region most severely, but even at the center of the solid portions of the wall its influence is still apparent (Fig. 13.15). Velocity profiles measured at the center of the solid wall region of a 1 ft slotted test section[9] at 2.9 in. behind the beginning of the slats are practically uninfluenced by the setting of the wall. However, at the downstream end of the test section $x = 31.9$ in., the profiles vary considerably, as shown by Fig. 13.15c.

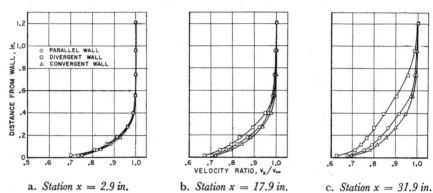

a. Station $x = 2.9$ in. b. Station $x = 17.9$ in. c. Station $x = 31.9$ in.

FIG. 13.15. *Velocity distributions along slat centerline of slotted test section at $M = 1.0$, sixteen slots, 11 per cent open-area ratio (AEDC 1 ft transonic model tunnel[8]).*

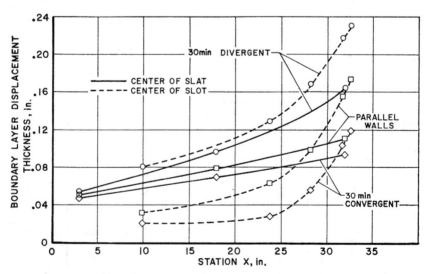

FIG. 13.16. *Boundary layer growth at slot and slat centerlines of slotted test section with sixteen slots, 11 per cent open-area ratio, $M = 1.0$ (AECD 1 ft transonic model tunnel[8]).*

b. Displacement thickness

The boundary layer displacement thickness at the centers of the slots and slats of the same slotted test section was measured at various wall settings (Fig. 13.16). At the centers of the solid portions of the wall the boundary layer displacement thickness grew steadily when the measuring station was shifted downstream. The displacement thickness in the region of the slots also shows steady growth, but in this case the magnitude depends largely upon the wall setting. In the more rearward portion of the slotted test section, a very rapid growth of displacement thickness was observed at all wall settings. This rapid growth may be caused in part by the difference in slot width (tapered as far as station $x = 2.0H$ and constant at its maximum value further downstream). However, the major reason for the rapid growth is probably local flow into the test section which may be caused by non-uniformities of the test-section flow or by insufficient mass flow removal from the test section.

REFERENCES

[1] MAEDER, P. F. and STAPLETON, J. F. "Investigation of the Flow Through a Perforated Wall." Brown University, TR–WT–10, May, 1953.

[2] DURAND, WILLIAM F. ed. *Aerodynamic Theory*. Division G: "The Mechanics of Viscous Fluids," pp. 170, by L. PRANDTL. Reprinted by California Institute of Technology, January, 1943.

[3] MAEDER, P. F. "Turbulent Mixing Along Perforated Walls." AEDC Sub-contract, Maeder, July 31, 1954.

[4] MAEDER, P. F. and HALL, J. F. "Investigation of Flow Over Partially Open Wind Tunnel Walls, Final Report." AEDC–TR–55–67, December, 1955.

[5] CHEW, WILLIAM L. "Wind Tunnel Investigations of Transonic Test Sections—Phase II: Comparison of Results of Tests on Five Perforated-Wall Test Sections in Conjunction with a Sonic Nozzle." AEDC–TR–54–52, March, 1955.

[6] DAVIS, D. D., JR., SELLARS, T. B. and STOKES, G. M. "An Experimental Investigation of the Transonic-Flow-Generation and Shock-Wave Reflection Characteristics of a Two-dimensional Wind Tunnel with 17 per cent Open Perforated Walls." NACA RM L54B15a, April, 1954.

[7] WARD, V. G., WHITCOMB, C. F. and PEARSON, M. D. "Air Flow and Power Characteristics of the Langley 16 ft Transonic Tunnel with Slotted Test Section." NACA RM L52E01, July, 1952.

[8] ALLEN, EDWIN C. "Wind Tunnel Investigations of Transonic Test Sections—Phase III: Tests of a Slotted-Wall Test Section in Conjunction With a Sonic Nozzle." AEDC–TR–54–53, February, 1955.

[9] OSBORNE, JAMES I. and ZECKY, HOWARD. "Progress Report on Development of a Transonic Test Section for the Boeing Wind Tunnel (BWT 188), Vol. II—June 1951 to September 1951." Boeing Airplane Company, D–11955, October 14, 1952.

BIBLIOGRAPHY

CHEW, WILLIAM L. "Total Head Profiles Near the Plenum and Test Section Surfaces of Perforated Walls." AEDC–TN–55–59, December, 1955.

DICK, R. S. "Calibration of the PWT 16 ft Transonic Circuit with an Aerodynamic Test Cart Having Inclined-Hole Walls." AEDC–TN–58–97, January, 1959.

OSBORNE, JAMES I. "Distribution of Losses in Model Transonic Wind Tunnel at a Mach Number of 1.25." Boeing Airplane Company, D–11924, June 19, 1952.

CHAPTER 14

POWER REQUIREMENTS OF TRANSONIC WIND TUNNELS

1. INFLUENCE OF AUXILIARY PLENUM CHAMBER SUCTION ON POWER REQUIREMENTS

As discussed in some detail in Chapter 12, the same Mach number can be established in either the perforated or slotted type of partially open test sections by using various amounts of auxiliary plenum chamber suction. Results from typical tests made to determine the Mach number distribution in a perforated test section are shown again in Fig. 14.1. As previously explained, a change in the amount of suction affects the Mach number distribution only in the extreme downstream region of the test section, and a constant Mach number can be maintained through the entire test section by proper adjustment of the pressure rise ratio of the wind tunnel compressor. When a minimum amount of suction is applied, that is, when the boundary layer at the rear portion of the test section and at the beginning of the diffuser reaches its maximum thickness, the pressure ratio of the wind tunnel compressor is a maximum. When large amounts of auxiliary plenum chamber suction are applied, and consequently the boundary layer at the test-section end is thinned considerably, the pressure rise required of the wind tunnel compressor is considerably reduced (see Fig. 14.2).

These results are explained by the well-known fact that most wind tunnel losses are closely related to the pressure recovery in the diffuser. Thus, since diffuser efficiency depends decisively upon the boundary layer thickness at the upstream end of the diffuser, the pressure ratio required of the wind tunnel drive system is gradually reduced as the amount of auxiliary suction is increased. Figure 14.2 also indicates that it is useless to apply more than the minimum compressor pressure rise at minimum mass flow removal (in this case 2.2 per cent), since it is not possible to influence the Mach number distribution and boundary layer growth caused by choking at the downstream end of the test section. Instead, a supersonic field of increasing extent is established in the diffuser when more pressure boost is provided than is required at the minimum suction point.

Because of this noticeable interrelationship between main compressor pressure rise and auxiliary suction, it is possible to increase the operating range of a wind tunnel either by increasing the power input to the main drive system or by increasing the auxiliary plenum chamber suction. Unless disturbances are produced by imperfect matching of wind tunnel pressure ratio and auxiliary suction, a partially open wind tunnel can be operated using many combinations of main drive pressure ratio and auxiliary suction. This is particularly convenient whenever it is difficult to increase the power input to the main drive system but easy to increase the mass flow removal

POWER REQUIREMENTS OF TRANSONIC WIND TUNNELS

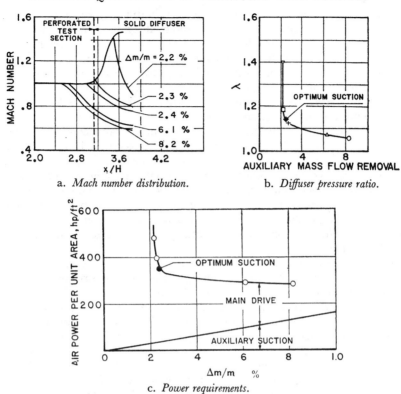

a. Mach number distribution.

b. Diffuser pressure ratio.

c. Power requirements.

FIG. 14.1. *Power requirements of perforated test section (22.5 per cent open, $\frac{1}{4}$ in. holes) for various plenum chamber suctions at $M = 1.0$, $p_0 = 1$ atm, and $T_0 = 520°R$ (AEDC 1 ft transonic model tunnel[12]).*

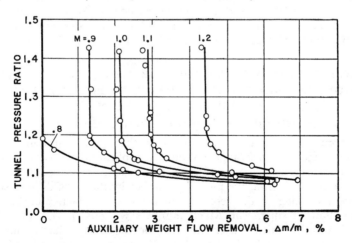

FIG. 14.2. *Tunnel pressure ratio as function of plenum chamber suction for various Mach numbers (perforated test section with sonic nozzles, 6 per cent open, inclined holes, parallel walls) (AEDC 1 ft transonic model tunnel[1]).*

capacity of the auxiliary suction equipment. Naturally, for only one combination of pressure ratio and mass flow removal is a uniform Mach number distribution established which extends to the extreme downstream end of the test section. This condition, which might be characterized as the point of optimum suction operation, requires only slightly more auxiliary suction than the minimum amount needed to establish the required Mach number flow.

The "air-power" requirements for both main drive and auxiliary suction are presented in Fig. 14.1c, based on the pressure ratio requirements of the auxiliary suction and main drive compressors for the same conditions shown in Figs. 14.1a and b. When the same efficiency is assured for both main drive and auxiliary suction compressors, the values of "air-power" are directly indicative of the shaft-power requirements. The results indicate that a trade-off between auxiliary suction power and main wind tunnel drive power is possible over a wide range of mass flow removal. It is even possible for the auxiliary suction to equal the power input into the main drive system without increasing the total power required of the auxiliary suction and main drive compressors.

Obviously, near the optimum operating point, large savings in total power can be accomplished by increasing the amount of auxiliary suction.

The relationships presented in Fig. 14.1 are typical of both perforated and slotted wind tunnels. They also hold true, not only for Mach number one, as discussed above, but also for the entire subsonic and supersonic Mach number range, regardless of whether a sonic or a contoured supersonic nozzle is employed to establish the supersonic flow. This fact is demonstrated by data obtained from a test series conducted in the AEDC transonic model tunnel (Fig. 14.2). In this case the perforated test section was equipped with inclined holes having a nominal open-area ratio of 6 per cent.[1] The variation of tunnel pressure with auxiliary weight flow removal is similar to the variation obtained at Mach number one (Fig. 14.1). The large increase in suction required to establish the supersonic flow was caused by the use of a sonic instead of a contoured nozzle. Again, operation with optimum suction, also indicated in Fig. 14.2, required somewhat more than the minimum suction needed to establish the particular Mach number.

2. POWER REQUIREMENTS OF PERFORATED TEST SECTIONS FOR VARIOUS COMBINATIONS OF AUXILIARY PLENUM CHAMBER AND DIFFUSER SUCTION

a. Power requirements without auxiliary plenum chamber suction

A perforated test section can be operated through the subsonic and low supersonic Mach number range without the application of auxiliary plenum chamber suction (see Chapter 12). In such a case, however, it is necessary to remove some air from the plenum chamber by means of diffuser suction to prevent choking of the test section at subsonic Mach numbers or to establish flow at supersonic Mach numbers without the use of a Laval nozzle. Naturally, in the latter case, the suction depends largely upon the type of openings provided in the diffuser. These openings must be of a type which will retain as much as possible of the kinetic energy of the suction flow and disturb the diffuser pressure recovery as little as possible.

One variable diffuser suction device consists of remotely controllable diffuser flaps, suggested initially by the staff of the Boeing Airplane Co., Seattle.[2] This device is shown schematically in Fig. 14.3. When the flaps are closed, a normal diffuser contour is established. When the flaps are gradually opened, various amounts of air are sucked into the diffuser by the ejector effect of the main wind tunnel flow assisted by the low pressure area in this section of the diffuser.

With diffuser flaps of this type, test-section Mach numbers can be established through the range given in Fig. 14.4. When the flaps are closed, the empty test section is choked at a Mach number of 0.89. As the flaps are gradually opened to a maximum of 8.3 per cent of the tunnel height, a test-section Mach number of 1.25 is established. Further opening of the

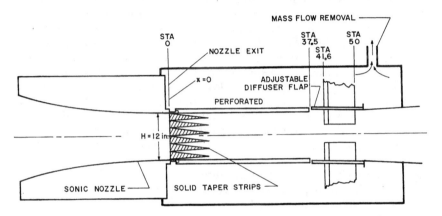

Fig. 14.3. *Sketch of perforated test section with solid taper strips and diffuser flaps (AEDC 1 ft transonic model tunnel[13]).*

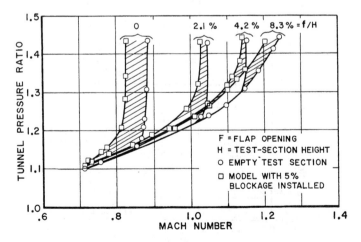

Fig. 14.4. *Tunnel pressure ratio for perforated test section with diffuser flap sections (no auxiliary suction) with and without model installed (parallel walls, 11.8 per cent open, all four flaps open, sonic nozzle) (AEDC 1 ft transonic model tunnel[1]).*

flaps does not increase the test-section Mach number noticeably but does raise the tunnel pressure ratio greatly. At the 8.3 per cent flap opening and a Mach number of 1.25, the required tunnel pressure ratio is increased to 1.45.

The Mach numbers established for an empty test section shift to somewhat lower values when a model (i.e. a body of rotational symmetry with a fineness ratio of 10 and a blockage ratio of 5 per cent) is installed (see Fig. 14.4). The difference in power requirements with and without the model is particularly noticeable in the low Mach number range.

The large tunnel pressure ratio required to establish flow in a partially open test section using diffuser suction without auxiliary suction can be understood when it is realized that an appreciable amount of air must be removed in order to prevent choking at subsonic Mach numbers or to establish supersonic flow without contoured nozzles (see Chapter 12). Therefore, for operation without auxiliary suction some air must flow into the diffuser, which, because of its low total pressure, increases the effective boundary layer thickness at the diffuser entrance and, consequently, the large tunnel pressure ratio. Experiments in the AEDC transonic model tunnel[3] show that on the plenum chamber side of perforated walls most of the kinetic energy of the inflow is lost due to eddy formation; the total pressure at the plenum chamber side of the perforated walls is practically equal to the static pressure in the plenum chamber. Some typical results of experiments on this effect for a wall with 22 per cent open-area ratio are presented in Fig. 14.5 at station $x = 1.4H$ downstream of the test-section inlet. The usual boundary layer profile was established on the test-section side of the walls, but the total pressure at the plenum chamber side was practically identical to the static pressure in the plenum chamber at the test Mach numbers of 0.9 and 1.2.

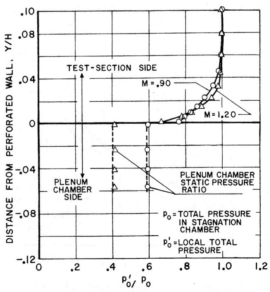

Fig. 14.5. *Total pressure distribution near perforated wall at Station $x/H = 1.4$ (22.5 per cent open, $\frac{1}{4}$ in. holes, parallel walls) (AEDC 1 ft transonic model tunnel[3]).*

POWER REQUIREMENTS OF TRANSONIC WIND TUNNELS

The total pressure distribution on both sides of a perforated wall with inclined holes shows the same basic characteristics, that is, at station $1.4H$, the total pressure at the plenum chamber side was practically equal to the static pressure in the chamber (Fig. 14.6b). This is also true at the most

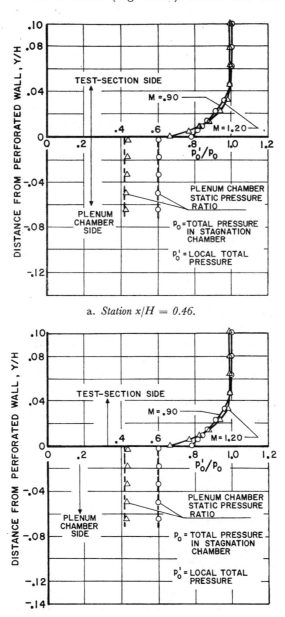

a. Station $x/H = 0.46$.

b. Station $x/H = 1.4$.

FIG. 14.6. Total pressure distribution near perforated wall with 60° inclined holes, 6 per cent open, parallel walls (AEDC 1 ft transonic model tunnel[3]).

forward station, $x = 0.46H$, Mach number 0.9 (see Fig. 14.6a). At this latter station and a Mach number of 1.2, however, the total pressure *immediately adjacent* to the wall at the plenum chamber side was slightly larger than the plenum chamber static pressure. At a lateral distance of $y = 2$ per cent of the test-section height, the kinetic energy was almost completely dissipated so that once again the local pressure approached the static pressure. This phenomenon is easily understood. To establish a Mach number of 1.2, approximately 3 per cent of the test-section flow must be removed by suction in the forward portion of the test section; thus in this region, the removed air still maintains part of its kinetic energy until it is dissipated further downstream.

On the basis of the cited test results and other unpublished data, it may be concluded that plenum chamber suction requirements are not reduced by any kinetic energy remaining in the suction flow. Thus, when diffuser flaps are used, the thickness of the boundary layer entering the diffuser is greatly increased.

b. Power requirements for various combinations of auxiliary plenum chamber and diffuser suction

In order to compare the relative merits of diffuser and auxiliary plenum chamber suction, various perforated walls were studied in the AEDC transonic model tunnel, and tunnel pressure ratio requirements were determined as a function of flap opening and auxiliary plenum suction. Since all these experiments showed the same trend, only the results at the test-section Mach number of one will be discussed.

Results obtained with a perforated wall having a 22.5 per cent open-area ratio and $\frac{1}{4}$ in. holes[4] indicate that when the flaps are open 8.3 per cent of the test-section height, the test section can be operated without auxiliary plenum chamber suction at a tunnel pressure ratio of 1.24. On the other hand, when 2.5 per cent auxiliary suction is applied, the tunnel pressure ratio is reduced to 1.18, as indicated in Fig. 14.7. It is evident that operation without auxiliary plenum chamber suction is possible only when the main wind tunnel drive is capable of producing extraordinarily large pressure ratios.

Operation without auxiliary plenum chamber suction is very uneconomical in comparison with operation with a modest amount of suction (Fig. 14.7b). For a test-section Mach number of one, 580 h.p./ft^2 of test-section area are required to maintain the flow at 1-atm stagnation pressure and 520°R stagnation temperature; this value is reduced to 320 h.p./ft^2 when 3 per cent auxiliary suction is employed. As mentioned before, the most economical operation from the standpoint of power consumption is obviously obtained when very large amounts of auxiliary suction are employed, that is, when the auxiliary suction horsepower is equal to or even greater than the main drive horsepower. In actual practice, however, the amount of auxiliary suction applied is generally only slightly larger than the minimum required to prevent choking without opening the flaps.

Basically similar results were obtained when a perforated test section with 33 per cent open-area ratio was considered (Fig. 14.8). Even with a flap opening of $0.083H$ and a tunnel pressure ratio of 1.43, it was not possible to establish Mach number one flow in the test section without using auxiliary plenum chamber suction. With the flaps closed and 8 per cent suction applied, the Mach number one flow was established with a pressure ratio of only 1.16. The power requirements for this wall with its unusually large

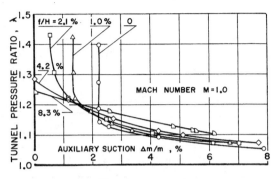

a. *Tunnel pressure ratio.*

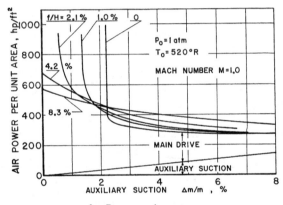

b. *Power requirements.*

Fig. 14.7 *Tunnel pressure ratio and power requirements for perforated test section with different combinations of auxiliary and diffuser suction (22.5 per cent open, $\frac{1}{4}$ in. holes, parallel walls (AEDC 1 ft transonic model tunnel[4]).*

open-area ratio are nearly twice as large as those for walls of 22 per cent open-area ratio (see Fig. 14.8b). Without auxiliary suction it should have been possible, on the basis of extrapolation of the experimental curves, to establish Mach number one flow with a power consumption of approximately 1200 h.p./ft² area. If 8 per cent auxiliary suction were employed, the requirements would be reduced to 470 h.p./ft² of test-section area. While these values are excessive for most wind tunnel operation, they do indicate the great superiority of auxiliary plenum chamber suction over diffuser flap suction.

In summary, for economical operation of partially open wind tunnels, it is considerably more advantageous to provide the necessary amount of plenum chamber mass flow removal by means of an auxiliary compressor rather than by using the diffuser as an ejection pump. This advantage is due mainly to the fact that the test-section flow passing through the walls into the plenum chamber loses its kinetic energy so quickly that any mass flow from the plenum chamber into the diffuser causes an increase in the

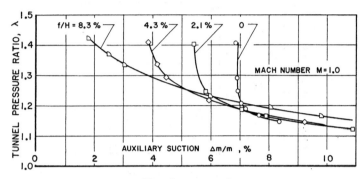

a. *Tunnel pressure ratio.*

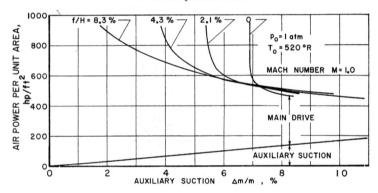

b. *Power requirements.*

Fig. 14.8. *Tunnel pressure ratio and power requirements for perforated test section with different combinations of auxiliary and diffuser suctions (33 per cent open, $\frac{1}{16}$ in. holes, parallel walls) (AEDC 1 ft transonic model tunnel[4]).*

diffuser inlet boundary layer. It is generally advisable to apply somewhat more auxiliary suction than the minimum required to avoid choking in the subsonic speed range or to establish the test-section flow in the supersonic range.

3. INFLUENCE OF WALL SETTING ON POWER REQUIREMENTS (PERFORATED TEST SECTIONS)

The results presented previously regarding the tunnel pressure ratio and power requirements of transonic wind tunnels were obtained using test sections with parallel wall settings. In order to explore the effect of converged or diverged settings, extensive experimental investigations were carried out in numerous laboratories. Some results of the experiments conducted in the

AEDC transonic model tunnel[4] are presented here to demonstrate the magnitude of the wall setting effect.

The tunnel pressure ratio, as a function of auxiliary suction required to establish Mach number 1.0 flow in a transonic test section with perforated walls having a $22\frac{1}{2}$ per cent open-area ratio ($\frac{1}{4}$ in. holes), is presented in Fig. 14.9. As in all AEDC experiments, only the upper and lower test-section

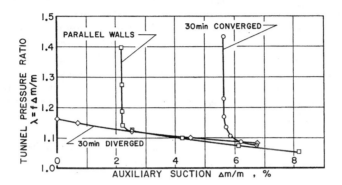

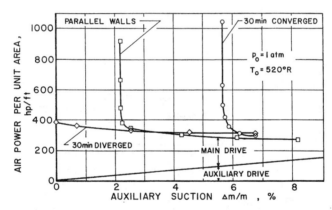

Fig. 14.9. *Influence of wall setting on tunnel pressure ratio and power requirements (diffuser flaps closed) at $M = 1.0$, 22.5 per cent open perforated walls, $\frac{1}{4}$ in. holes, top and bottom walls adjusted (AEDC 1 ft transonic model tunnel[4]).*

walls were moved; the sidewalls remained parallel. In this case, Mach number one flow was established without auxiliary plenum chamber suction (diffuser flaps closed) when the walls were diverged 30'. For parallel and 30' converged wall settings, auxiliary suction amounting to 2.3 and 5.7 per cent, respectively, was required. As expected, the tunnel pressure ratio was decreased when the auxiliary suction was increased, that is, when the wall boundary layer approaching the diffuser was thinned.

When the results obtained for tunnel pressure ratio and auxiliary suction are converted into power requirements, it can be seen that the total power required, including both main and auxiliary compressor power, is slightly decreased when wall suction is applied (see Fig. 14.9). With larger amounts of suction, the curves for all three wall settings converged into a common line.

From these results, which are typical of the conditions occurring in perforated test sections, it may be concluded that total power requirements are only slightly influenced by the wall setting. Only a small reduction in total power is experienced when the setting is changed from the diverged to the converged configuration. The main difference is the amount of auxiliary suction required to establish the desired test-section Mach number. Thus, if sufficient capacity were not available in the auxiliary compressors, it might be necessary to select a less converged or even somewhat diverged wall setting in order to establish the desired flow. However, when boundary layer thickness along the walls is considered as a function of the wall setting, it must be remembered that an effective uniform thinning of the wall boundary layer is accomplished only by proper convergence of the walls. Consequently, the more converged wall settings are generally superior for transonic testing from the viewpoint of consistent wall cross-flow characteristics and proper shock-wave cancellation characteristics (see Chapters 8 and 11).

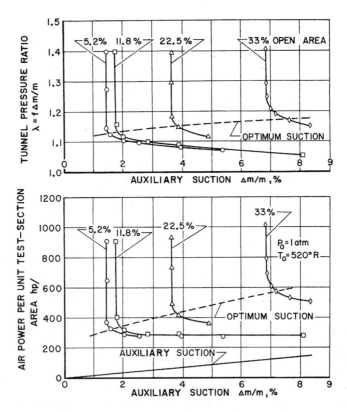

FIG. 14.10. *Influence of open-area ratio of perforated walls on tunnel pressure ratio and power requirements at $M = 1.0$ (diffuser flaps closed, $\frac{1}{16}$ in. holes, parallel walls) (AEDC 1 ft transonic model tunnel*[4]*)*.

4. INFLUENCE OF WALL GEOMETRY (PERFORATED WALLS)

a. Variation of open-area ratio

According to the results reported in Chapter 12, the open-area ratio of a perforated wall is one of the most significant parameters since it influences both the boundary layer thickness and, as a consequence, the suction and power required to operate the wind tunnel. In experiments conducted in the AEDC transonic model tunnel perforated walls with open-area ratios ranging between 5.2 and 33 per cent (Ref. 4) were used. Some significant

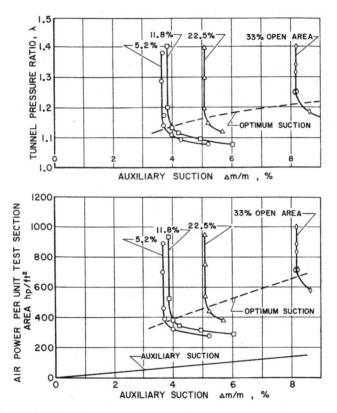

Fig. 14.11. *Influence of open-area ratio of perforated walls on tunnel pressure ratio and power requirements at $M = 1.20$ (diffuser flaps closed, $\frac{1}{16}$ in. holes, parallel walls) (AEDC 1 ft transonic wind tunnel[4]).*

results on establishment of Mach number one flow in the test section are presented in Fig. 14.10. In these experiments the walls were set parallel, and the diffuser flaps were closed. An increase in the open-area ratio greatly increased the auxiliary suction requirements and the necessary tunnel pressure ratio. The points of optimum suction, corresponding to the operating conditions at which uniform Mach one flow was established over the entire test section, are indicated in the graph. When the tunnel pressure ratio results were converted into power requirements (Fig. 14.10), it was evident that the open-area ratios of test sections in transonic wind tunnels must be kept as small as possible while remaining compatible with the requirements for interference-free flow.

The same results can also be shown for flow establishment at Mach number 1.20 (Fig. 14.11). At this Mach number, both tunnel pressure ratio and suction requirements increased rapidly with increasing open-area ratio. It should be noted that in this case a sonic nozzle was used to establish the desired Mach number flow; as a result auxiliary suction was needed to to satisfy both the requirement for boundary layer removal and the additional requirement for establishment of supersonic flow.

b. Variation of hole size in perforated walls

In previous sections it has been demonstrated that wall characteristics depend not only upon the geometric open-area ratio of the wall but also upon the hole size itself, that is, on the ratio of hole diameter to the displacement thickness of the boundary layer. Because of this influence of hole size, the requirements for auxiliary suction and tunnel pressure ratio are also influenced by the relative hole size, as shown in Fig. 14.12 (Ref. 4). At both

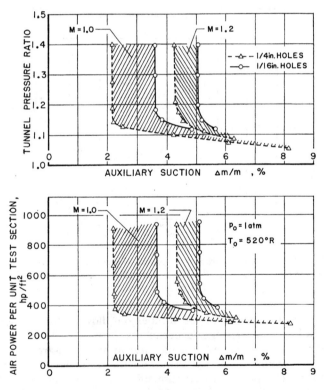

FIG. 14.12. *Influence of hole size of perforated walls on tunnel pressure ratio and power requirements at $M = 1.0$ and 1.2 (Diffuser flaps closed, 22.5 per cent open walls, parallel walls) (AEDC 1 ft transonic model tunnel[4])*.

the Mach number 1.0 and 1.2 conditions, suction and tunnel pressure ratio requirements were considerably reduced in the case of the walls with the larger holes. Consequently, when power requirements are considered, as large a hole diameter should be selected as is compatible with the intensity of the small supersonic waves produced by the individual holes (see Chapters 8 and 13).

5. INFLUENCE OF MODEL SIZE (PERFORATED TEST SECTION)

Models of rotational symmetry with 1, 3 and 5 per cent blockage were installed in the same perforated test sections discussed previously (AEDC transonic model tunnel experiments). Since these models produce disturbances which cause part of the test-section flow to cross the perforated walls and enter the plenum chamber, it is apparent that both auxiliary suction and tunnel power requirements must increase because of the thicker boundary layer approaching the diffuser. Some representative data showing the influence of a model in a perforated test section equipped with a $22\frac{1}{2}$ per cent open-area ratio ($\frac{1}{4}$ in. holes) are presented in Fig. 14.13 for Mach number one.

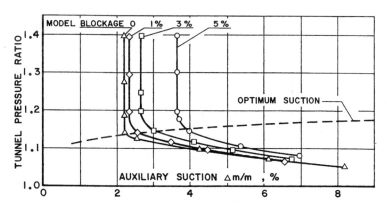

Fig. 14.13. *Influence of model size on pressure ratio and suction requirements of perforated wind tunnel at $M = 1.0$ (22.5 per cent open walls, $\frac{1}{4}$ in. holes, parallel walls) (AEDC 1 ft transonic model tunnel[4]).*

The expected effects are clearly evident. However, note that a 5 per cent blockage model increased the minimum auxiliary suction requirement by only approximately 1.5 per cent of the test-section mass flow, m_T. Since at Mach number one, the amount of air which must enter the plenum chamber corresponds to the blockage of the model, this increase of only 1.5 per cent in auxiliary suction (instead of the 5 per cent which corresponds to the percentage of model blockage) is an indication of the fact that the wall boundary layer has been accelerated to velocities having greater flow density. This fact accounts for part of the observed difference. Moreover, part of the air which enters the plenum chamber in the area of the maximum cross section of the model re-enters the test section in the downstream portion and relieves the requirement for mass flow removal. However, this re-entry causes the wall boundary layer in the downstream region of the test section to thicken and, consequently, the tunnel pressure ratio to increase in spite of the increased plenum chamber mass flow removal (see Fig. 14.13).

6. COMPARISON OF PERFORATED TEST SECTIONS EQUIPPED WITH SONIC AND SUPERSONIC NOZZLES

As discussed previously, supersonic Mach numbers can be established in partially open test sections either by using a sonic nozzle in conjunction with auxiliary suction or by using a conventional supersonic nozzle preceding the partially open test section. At low supersonic Mach numbers no great difference is expected in the power requirements of the two methods of operation since only small amounts of mass flow removal are required to establish the desired Mach number. At higher Mach numbers, for instance, at Mach numbers in excess of 1.20, the mass flow removals required for flow establishment with a sonic nozzle grow rapidly and cause a proportional increase in the auxiliary power required of the auxiliary suction compressor. In the latter case, the power increase is due mainly to the fact that the sonic throat of the wind tunnel must accommodate a noticeably larger airflow

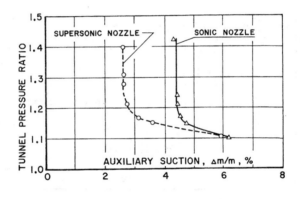

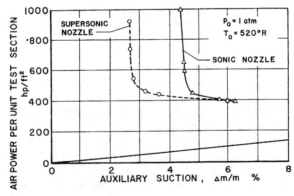

Fig. 14.14. *Tunnel pressure ratio and power requirements for perforated test sections with sonic and supersonic nozzles at $M = 1.20$ (diffuser flaps closed, 6 per cent open inclined holes, parallel walls) (AEDC 1 ft transonic model tunnel[1]).*

than the test section and that the mass flow loses its entire kinetic energy after entry into the plenum chamber (see Chapter 13).

Some typical test results obtained in the AEDC transonic model tunnel at Mach number 1.20 with perforated walls (inclined holes with a 6 per cent open-area ratio) are presented in Fig. 14.14 for configurations having either a sonic nozzle or a flexible supersonic nozzle. At this Mach number, tunnel pressure ratio and total power input per unit test-section area were not too different for the two modes of operation. The main difference was the greatly increased requirement for mass flow removal in the case of the test section having a sonic nozzle. The choice between the two configurations will thus depend largely upon the specific layout of each individual wind tunnel. If Mach number must be increased without changing the power or the compressor configuration of the main drive system, it is usually advantageous to install an auxiliary suction system or to increase its capacity. On the other hand, if a tunnel is equipped with a supersonic nozzle and the main drive power of the tunnel can be increased, optimum conditions will usually be established by an increase in the power of the main compressor and a moderate change in the auxiliary suction.

As noted previously, results cited for sonic and supersonic nozzle operation are valid only for comparatively low supersonic Mach numbers, as can be seen from the results obtained for test-section Mach numbers between 0.8 and 1.46 (see Fig. 14.15). In this figure the conditions for optimum operation are again indicated and a drastic difference is evident between the trends of the mass flow removal curves for sonic and supersonic nozzles. The difference in the auxiliary suction is reflected in the trend of the total power required for operation with sonic or supersonic nozzles (see Fig. 14.15).

In summary, it may be stated that for Mach numbers up to 1.2, operation with a sonic nozzle is not particularly detrimental from the standpoint of total power requirements if provisions are made for moderate amounts of mass flow removal. At Mach numbers in excess of 1.2, the power requirements of wind tunnels having sonic nozzles increase rapidly, as do the extremely large requirements for auxiliary suction. Consequently, in this high supersonic Mach number range, significant savings in total power requirements can be achieved when supersonic nozzles are used preceding the test section.

7. POWER REQUIREMENTS OF WIND TUNNELS WITH SLOTTED TEST SECTIONS

a. Tunnel pressure ratio as a function of plenum chamber mass flow removal

Relationships similar to those noted for perforated test sections also hold true for slotted test sections. It is possible to establish the desired Mach number in a slotted test section using different combinations of main drive pressure boost and mass flow removal from the plenum chamber. It is also possible, as in perforated test sections, to accomplish the mass flow removal from the plenum chamber by means of re-entry flaps, that is, by diffuser suction using the main flow as an ejector pump or by means of a separate auxiliary compressor. Both of these methods of operation were explored briefly in tests conducted in the AEDC transonic model tunnel.[5]

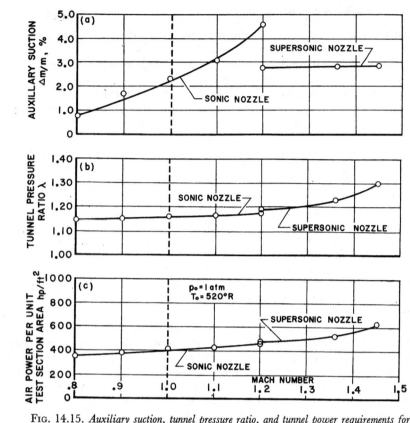

FIG. 14.15. *Auxiliary suction, tunnel pressure ratio, and tunnel power requirements for perforated test sections with sonic and supersonic nozzles at various Mach numbers for optimum operating conditions (6 per cent open, inclined holes, parallel walls, diffuser flaps closed) (AEDC 1 ft transonic model tunnel[1]).*

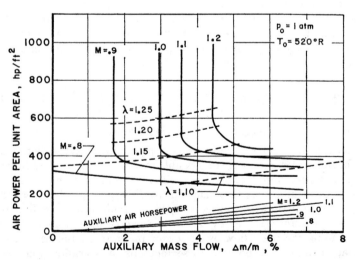

FIG. 14.16. *Power requirements and auxiliary mass flow of slotted test section for various Mach numbers (sixteen slots, 11 per cent open, diffuser flaps closed) (AEDC 1 ft transonic model tunnel[5]).*

Some results showing the relationship between power requirements and auxiliary mass flow removal are presented in Fig. 14.16 for a typical slotted test section having sixteen slots, 11 per cent open-area ratio, and a length of the slotted test section three times the tunnel height. As in the case of the perforated test sections, a minimum amount of auxiliary mass flow removal is required for each Mach number above 0.8 when the diffuser flaps are closed. Increase of the mass flow removal beyond this minimum value results in a rapid decrease in the total power requirements until the minimum power conditions are finally reached at large auxiliary mass flow removal.

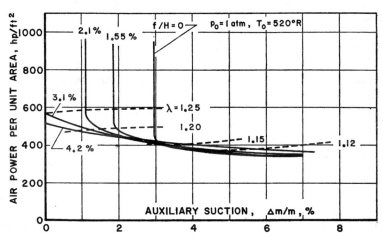

Fig. 14.17. *Influence of diffuser flap opening on power requirements of slotted test section at $M = 1.0$ (sixteen slots, 11 per cent open, parallel walls) (AEDC 1 ft transonic model tunnel[5]).*

This latter condition corresponds approximately to an equal power input from both the main drive and the auxiliary compressor. However, it should be noted that this test section was not equipped with a supersonic nozzle preceding the test section and that supersonic flow was established by mass flow removal from the test section. It is remarkable that with a reasonable amount of mass flow removal, the Mach number range between 0.9 and 1.2 can be covered using the same pressure boost from the main drive compressor, that is, with a pressure ratio of 1.15. This fact represents a large advantage in the design of main drive compressors. However, in such cases, considerably larger mass flows must be handled by the auxiliary compressors.

If no auxiliary compressor is available for mass flow removal from the plenum chamber, openings in the diffuser (i.e. diffuser inlets with re-entry flaps) must be provided to avoid choking or to establish a supersonic Mach number. Results obtained in the AEDC model tunnel using re-entry flaps similar to those described previously for the tests in perforated test sections (see Fig. 14.3) are presented in Fig. 14.17 for Mach number one. Mach number one flow was established in the test section without auxiliary mass flow removal using flap openings equal to 3.2 and 4.2 per cent of the tunnel height. However, with these flap openings, it was necessary to provide a tunnel pressure ratio of 1.24 compared with the ratio of 1.15 required when minimum auxiliary mass flow removal was provided and the diffuser flaps were kept closed. Also, the total power requirements were increased approximately 25 per cent when diffuser suction, instead of minimum auxiliary suction, was provided in conjunction with the closed flaps.

In summary, in slotted test sections, as in perforated test sections, considerable savings in total power can be realized when adequate mass flow removal from the plenum chamber is provided by auxiliary compressors. Diffuser re-entry flaps are advantageous only when equipment limitations make it impossible to establish the desired Mach number flow by any other means.

b. Influence of re-entry flap configuration

On numerous previous occasions it has been mentioned that diffuser losses can become extremely large when air from the plenum chamber is sucked into the upstream portion of the diffuser where it thickens the boundary layer and thus greatly reduces the available pressure recovery (see Section 2a of this chapter). Consequently, all wind tunnels which operate without the use of auxiliary suction compressors have been extensively investigated in order to arrange the re-entry of air into the diffuser as efficiently as possible. Experiments conducted in the NACA Langley 8 ft high-speed wind tunnel[6] are cited as a typical example of such a study.

The 8 ft Langley wind tunnel is equipped with twelve longitudinal slots having a maximum total open-area ratio of approximately 11 per cent. Initially the tunnel had re-entry inlets (shown schematically in Fig. 14.18a).

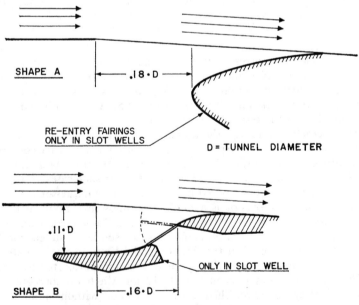

Fig. 14.18. *Various re-entry slot shapes of NACA 8 ft wind tunnel.*[6]

The nose of the re-entry fairings was located at 0.18 per cent of the tunnel diameter behind the test-section end. Behind the slots, the diffuser flow was practically unguided over a considerable length of the diffuser. After extensive and detailed flow studies, another re-entry shape was developed which provided better guidance for the re-entry flow and improved the diffuser efficiency (see Fig. 14.18b). Though the flow in the test section proper was not greatly influenced by the change in shape of the re-entry nose, an essential reduction in the power requirements was obtained (see Fig. 14.19).

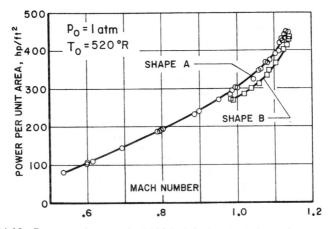

Fig. 14.19. *Power requirements for NACA 8 ft slotted wind tunnel with different re-entry slot shapes.*[6]

When the power data from these tests are compared with the data obtained from other test sections, it should be remembered that the length of the slotted test section in the NACA experiments was approximately 1.25 the test-section diameter. The length of the perforated test section of the AEDC model tunnel (Figs. 14.16 and 14.17) was three times the height of the test section.

When transonic or supersonic flow is to be established in a test section by diffuser suction, the re-entry air must not only flow smoothly into the diffuser but sufficient open area must also be provided at the minimum cross section of the diffuser to avoid choking of the flow. When diffuser flaps are used this minimum cross section is usually located near the downstream end of the re-entry flaps. Consequently, the higher the Mach number desired in the test section, the further downstream in the diffuser are the re-entry flaps located. A large number of systematic experiments have been carried out by the Boeing Co. to develop suitable transonic wind tunnel re-entry flaps. The slot shape and re-entry flaps used for the Langley test series just cited were designed to establish a maximum Mach number of 1.14 in the test section. Since Mach numbers of at least 1.25 were desired in the Boeing wind tunnel, arrangements other than the rigid two-position configurations of the NACA re-entry shapes were explored. Diffuser re-entry flaps which could be rotated around their downstream end were therefore provided.[2,7] With these remotely controllable flaps, the test-section Mach number could be changed either by controlling the power input to the main tunnel compressor or by adjusting the setting of the re-entry flaps.

Some typical results obtained with the new re-entry flaps are shown in Fig. 14.20. In these tests, the opening of the flaps was kept constant (equal

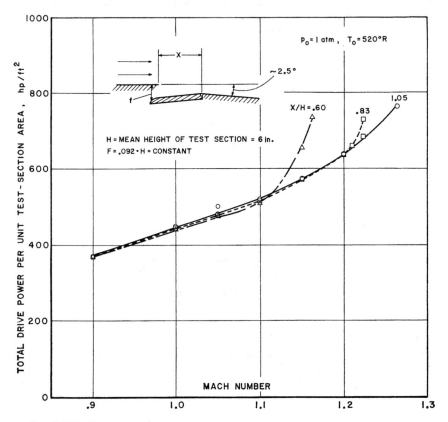

FIG. 14.20. *Compressor power requirements of slotted wind tunnel with different diffuser flap configurations (flaps open at four walls, sixteen slots, 11.1 per cent open, no auxiliary suction) (Boeing (4.8 × 7.2 in.) × 9 in. wind tunnel[7]).*

to 0.092 times the wind tunnel height), and only the location of the downstream hinge point of the flaps was varied. By shifting the hinge point further downstream into the diffuser better guidance of the flow into the diffuser was accomplished and a larger minimum diffuser area was provided at the flap inlet. As expected, the downstream location of the hinge points provided the larger Mach numbers desired, that is, choking in the diffuser was effectively delayed. At the extreme downstream position (station $x =$ approximately one times the mean height of the wind tunnel), a maximum Mach number somewhat in excess of 1.26 was established. Mach number range was limited only by the available power supply of the drive system. On the other hand, in the speed region where choking of the diffuser is not a problem (at Mach numbers below 1.10 for the flaps investigated), the power requirements for operation with all three Boeing flaps differed only slightly.

It must again be emphasized that the power requirement data presented in Fig. 14.20 for the Boeing investigation were obtained in a wind tunnel which had a slotted test section approximately 1.5 times the mean height of the tunnel. Therefore, before comparisons can be made, the effect of this length-to-height parameter should be evaluated. To this end a study of a typical loss distribution in the Boeing model wind tunnel[8] is cited here. As the loss survey of the model tunnel indicates, the return duct losses, that is, the losses between the diffuser downstream end and the inlet to the test section, amount to only 18 per cent of the total losses (Fig. 14.21). The

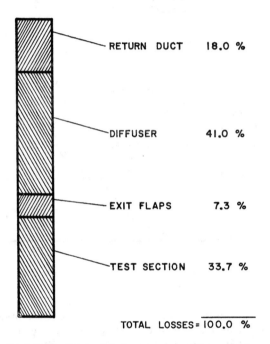

Fig. 14.21. *Loss distribution in slotted test section with re-entry flaps at* $M = 1.25$ *(flaps open at four walls, sixteen slots, 11.6 per cent open, no auxiliary suction)* *(Boeing (4.8 × 7.2 in.) × 9 in. model tunnel*[8]*)*.

remainder were produced in the test section, the exit flap area, and the diffuser, and amounted to 82 per cent of the total tunnel losses. Thus an increase in test-section length affects the direct test-section losses (34 per cent in the cited example), causes growth of the boundary layer and increases the requirement for re-entry mass flow into the plenum chamber, and increases the exit flap losses and the diffuser losses. Obviously comparisons of various wind tunnels can be made only when due consideration is given the length–height ratio of the test sections.

8. FULL-SCALE WIND TUNNELS

a. Comparison between full-scale and model tunnels

The power requirement data presented earlier for the Langley 8 ft slotted wind tunnel of the NACA (see Figs. 14.18 and 14.19) are re-plotted in Fig. 14.22 with other data obtained from the NACA 16 ft slotted transonic wind tunnel[9] and a slotted model tunnel 1 ft in diameter.[10] These

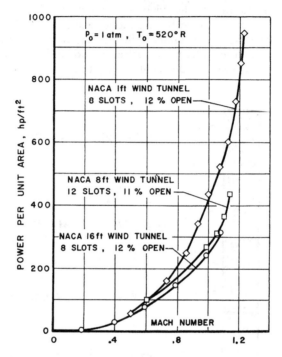

Fig. 14.22. *Power requirements for various NACA slotted wind tunnels without auxiliary suction.*[9,10,6]

wind tunnels have similar test sections with maximum open-area ratios ranging between 11 and 12 per cent. They differ somewhat in the shape of the diffuser re-entry openings but extensive investigations have been made to determine the optimum re-entry shapes for both the 8 and 16 ft tunnels. A comparison of these two full-scale tunnels indicates a slight advantage for the larger wind tunnel, which is probably caused by the difference in scale. Also the power curves for the model wind tunnel deviate considerably from the curves for the full-scale tunnels. Even though the simple diffuser openings of the model tunnel were not optimized and the test-section slots had a plain rectangular shape, it is believed that a considerable scale effect is indicated by the noted deviation of the model power curves. However, the general trend of the power requirements and the significance of devices designed to alleviate choking are correctly reflected by the small-scale model data.

Another comparison between the power requirements for model and full-scale tunnels was possible for the AEDC transonic wind tunnel[11,1] and the AEDC transonic model tunnel. Both wind tunnels were equipped with inclined-hole perforated walls. The full-scale wind tunnel power curves are presented in Fig. 14.23 for both minimum and maximum auxiliary suction. At the time these tests were made the auxiliary suction equipment in the full-scale wind tunnel had a maximum capacity of 5 per cent removal at Mach number 1.0. During the full-scale tests the supersonic nozzle preceding the test section was set at the proper value, and the re-entry flaps were

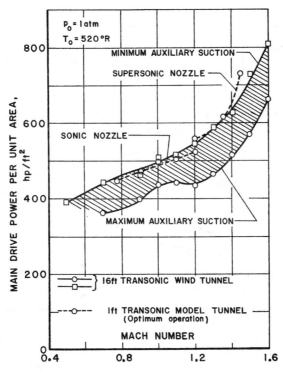

Fig. 14.23. *Power requirements for full-scale 16 ft and model 1 ft transonic wind tunnels of AEDC (inclined holes, 6 per cent open, parallel walls).*[11,1]

closed. Data from similar tests in the AEDC 1 ft transonic model tunnel were selected near the optimum operating point, that is, at auxiliary suction values somewhat larger than the minimum values required. Then, in order to facilitate comparison, the model tunnel data were converted into shaft horsepower data by assuming a compressor efficiency of $\eta = 0.90$. Also the losses in the return duct of the model tunnel were introduced additively as an estimated 20 per cent of the total losses at Mach one and were assumed to remain constant through the Mach number range. The comparison of the model tunnel and full-scale tunnel data indicates that the trend of the power curves is correctly predicted by the model tunnel results. However, the numerical values of the losses in the model tunnel are larger than those of the full-scale tunnel.

b. Full-scale wind tunnels

The main shaft horsepower requirements of several full-scale wind tunnels with partially open test sections are presented in Fig. 14.24.* The 8 ft and 16 ft NACA Langley wind tunnels were selected as typical "slotted test-section" wind tunnels; the AEDC 16 ft propulsion wind tunnel and the British ARA 9 × 8 ft wind tunnel are shown as examples of perforated test-section tunnels. Data from the slotted wind tunnels and the British perforated wind tunnel indicate that the power requirements of these tunnels have

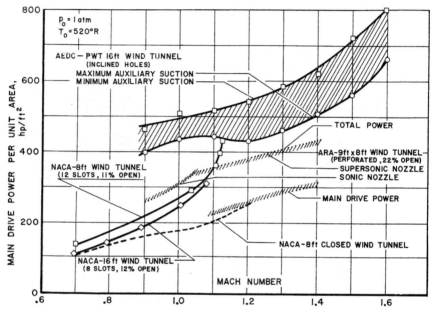

Fig. 14.24. *Main drive power requirements for various full-scale wind tunnels,*
AEDC 16 ft transonic wind tunnel: $L = 2/8 \times H$,[11]
NACA 8 ft wind tunnel: $L = 1.1 \times D$,[6]
NACA 16 ft wind tunnel: $L = 1.4 \times D$,[9]
ARA 9 × 8 ft wind tunnel: $L = 1.4 \times H$,[14]
NACA 8 ft closed wind tunnel: $L = 1.25 \times D$,[6]
(L = test-section length; H, D = test-section height or diameter).

approximately the same magnitude. Therefore, since the open-area ratio of the walls of these tunnels was selected to provide interference-free characteristics and the test sections had approximately the same relative length (1.1 to 1.4 times the height of the tunnel), it may be concluded that perforated and slotted wind tunnels do not exhibit significant differences in their power requirements. However, the data presented on the same graph for a closed wind tunnel having the same overall geometry as the 8 ft NACA wind

---

* The test-section length, L, indicated in Fig. 14.24 is the axial extent of the test section exhibiting satisfactory uniform velocity distribution.

tunnel do show that partially open transonic wind tunnels require approximately 50 per cent more power than closed wind tunnels.

The power requirements of the AEDC 16 ft propulsion wind tunnel are approximately twice as great as those of the Langley Field and British tunnels (Fig. 14.24). A large part of this difference is due to the fact that the AEDC wind tunnel has a test-section length *approximately* 2.8 *times the height* of the tunnel compared with lengths approximately 1.1 *to* 1.4 *times the height* for the other tunnels. This difference is believed to account for the greater portion of the power increase. In addition, the AEDC wind tunnel has special installations to permit full-scale propulsion testing such as a large scavenging scoop in the diffuser and a large cooler. When proper allowances are made for these unusual features of the AEDC 16 ft wind tunnel, the previous conclusions that no significant differences exist in the power requirements of perforated and slotted wind tunnels seems to be borne out.

REFERENCES

[1] GARDENIER, H. E. "Flow Establishment and Power Requirements in Slotted and Perforated Wind Tunnels." Arnold Engineering Development Center, July, 1956.

[2] OSBORNE, JAMES I. and ZECKY, H. "Progress Report on Development of Transonic Test Section for the Boeing Wind Tunnel, Vol. I." Boeing Airplane Co., D-11955, 1951.

[3] CHEW, W. L. "Total Head Profiles Near the Plenum and Test-Section Surfaces of Perforated Walls." AEDC-TN-55-59, December, 1955.

[4] CHEW, W. L. "Wind Tunnel Investigations of Transonic Test Sections—Phase II: Comparison of Results of Tests on Five Perforated-Wall Test Sections in Conjunction with a Sonic Nozzle." AEDC-TR-54-52, March, 1955.

[5] ALLEN, E. C. "Wind Tunnel Investigations of Transonic Test Sections—Phase III: Tests of a Slotted-Wall Test Section in Conjunction with a Sonic Nozzle." AEDC-TR-54-53, February, 1955.

[6] WRIGHT, R. H. and RITCHIE, V. S. "Characteristics of a Transonic Test Section with Various Slot Shapes in the Langley 8 ft High-Speed Tunnel." NACA RM L51H10, October 18, 1951.

[7] OSBORNE, J. I. and ZECKY, H. "Progress Report on Development of a Transonic Test Section for the Boeing Wind Tunnel (BWT 188), Vol. II: June 1951 to September 1951." Boeing Airplane Co., D-11955, February, 1952.

[8] OSBORNE, J. I. "Distribution of Losses in Model Transonic Wind Tunnel at a Mach Number of 1.25." Boeing Airplane Co., D-11934, June 19, 1951.

[9] WARD, W. G., WHITCOMB, C. F. and PEARSON, M. D. "Air Flow and Power Characteristics of the Langley 16 ft Transonic Tunnel with Slotted Test Sections." NACA RM L52E01, July, 1952.

[10] WRIGHT, R. H. and WARD, V. G. "NACA Transonic Wind Tunnel Sections." NACA RM L8J06, October 25, 1948.

[11] DICK, R. S. "Calibration of the PWT 16 ft Transonic Circuit With a Full Test Cart Having Inclined-Hole Perforated Walls." AEDC-TN-58-97, January, 1959.

[12] GOETHERT, B. H. "Flow Establishment and Wall Interference in Transonic Wind Tunnels." AEDC-TR-54-44, June, 1954.

[13] CHEW, W. L. "Wind Tunnel Investigations of Transonic Test Sections—Phase I: Tests of a 22.5 per cent Open-Area Perforated Wall Test Section in Conjunction with a Sonic Nozzle." AEDC-TR-53-10, October, 1953.

[14] HAINES, A. B. and JONES, J. C. M. "The Centre-Line Mach Number Distributions and Auxiliary Suction Requirement for the A.R.A. 9 ft × 8 ft Transonic Wind Tunnel." Aircraft Research Association Ltd. (G.B.), ARA–R–2, April, 1958.

BIBLIOGRAPHY

CHEW, W. L. "Experimental and Theoretical Studies on Three-dimensional Wave Reflection in Transonic Test Sections—Part III: Characteristics of Perforated Test-Section Walls with Differential Resistance to Cross Flow." AEDC TN–55–44, March, 1956.

CZARNECKI, A. W. "Data Report—Boeing Wind Tunnel Test No. 120—Test No. 3 of the Slotted Test Sections in the 1/20 Scale Model Wind Tunnel with a Two-Stage Compressor." Boeing Airplane Co., D–9626, March 30, 1950.

DICK, R. S. "Calibration of the PWT 16 ft Transonic Circuit With an Aerodynamic Test Cart Having 20 per cent Open Perforated Walls and Without Plenum Auxiliary Suction." AEDC TN–58–24, June, 1958.

DICK, R. S. "Calibration of the 16 ft Transonic Circuit of the Propulsion Wind Tunnel with an Aerodynamic Test Cart Having 6 per cent Open Inclined-Hole Walls." AEDC TN–58–90, November, 1958.

ESTABROOKS, B. B. and MILILLO, J. R. "Aerodynamic Performance of the AEDC-PWT Transonic Circuit Compressor." AEDC–TR–57–15, October, 1957.

GARDENIER, H. E. "Description of the Transonic Model Tunnel—Arnold Engineering Development Center." AEDC–TR–54–70, May, 1955.

NICHOLS, J. H. "An Investigation of the PWT Transonic Circuit Full Test Cart with Inclined-Hole Perforated Walls and without Plenum Auxiliary Suction." AEDC–TN–58–86, October, 1958.

RITCHIE, W. S. and Pearson, A. O. "National Advisory Committee for Aeronautics Calibration of the Slotted Test Section of the Langley 8 ft Transonic Tunnel and Preliminary Experimental Investigation of Boundary Reflected Disturbances." NACA RM L51K14, July 7, 1952.

TAYLOR, H. D. "Small-Scale Studies of a Porous Wall Transonic Test Section in Combination with a Laval Nozzle, Concluding Report." United Aircraft Corp., R–95434–19, April 4, 1952.

WARD, V. G., WHITCOMB, C. F., and PEARSON, M. D. "An NACA Transonic Test Section with Tapered Slots Tested at Mach Numbers to 1.26." NACA RM L50B14, March, 1950.

CHAPTER 15

LIST OF TRANSONIC WIND TUNNELS AND THEIR MAIN CHARACTERISTIC DATA

THE list of transonic wind tunnels presented in this section has been accumulated on the basis of information contained in the published literature. Due to this fact, all those wind tunnels for which such information was not available are omitted from this summary list.

Abbreviations used

AEDC	Arnold Engineering Development Center
ARA	Aircraft Research Association
DTMB	David Taylor Model Basin
MIT	Massachusetts Institute of Technology
NACA	National Advisory Committee for Aeronautics (now National Aeronautics and Space Administration—NASA)
NAMTC	Naval Air Missile Test Center
NLL	National Luftvaart Laboratorium
NPL	National Physical Laboratory
ONERA	Office National d'Études et de Recherches Aéronautiques
RAE	Royal Aircraft Establishment
WADC	Wright Air Development Center

Transonic Wind Tunnels

Tunnel designation	Location	Owner-operator	Test-section dimension	Test-section type	Mach number range	Pressurized	Intermittent or continuous	Horsepower a. Main b. Auxiliary	Special capabilities
Belgium TCEA 51	Rhode-St. Genèse	TCEA	40 × 40 cm	Slotted (2 walls)	0–1.2	No	Continuous	a. 615 kW (diffuser suction)	
France S.MA ONERA S.5 Sud-Aviation 40 cm	Modane Châtillon Suresnes	ONERA ONERA Sud-Aviation	8 m 20 × 30 cm 40 × 40 cm	Slotted Perforated	0.7–1.28 0.7–1.65	No No Yes	Continuous Continuous Intermittent	a. 110,000 a. 870 kW a. 120 compressor air inject drive	
Netherlands N.L.L. High-speed Tunnel 9	Amsterdam	NLL	2.0 × 1.6 m	Slotted (2 walls)	Subsonic to 1.3	Yes	Continuous	a. 20,000	
N.L.L. Tunnel 8	Amsterdam	NLL	0.42 × 0.55 m	Slotted (2 walls)	0–1.0	No	Continuous	a. 1200	Model of Tunnel No. 9
U.K. ARA Co-op	Bedford	Aircraft Research Assoctn.	9 × 8 ft	Perforated	0–1.4	No	Continuous	a. 25,000 b. 12,000	Four walls perforated with flexible nozzle
Short Bros. & Harland 2.5 ft	Belfast	Short Bros. & Harland	2.5 × 2.5 ft		0–1.2				

LIST OF TRANSONIC WIND TUNNELS

Transonic Wind Tunnels (contd.)

Tunnel designation	Location	Owner-operator	Test-section dimension	Test-section type	Mach number range	Pressurized	Intermittent or continuous	Horsepower a. Main b. Auxiliary	Special capabilities
U.K. (contd.)									
A. V. Roe 20 in.	Manchester	A. V. Roe	20 × 20 in.		0.4–1.6				
English Electric 18 in.	Warton	English Electric	18 × 18 in.	Slotted	0–1.6				
English Electric 12 in.	Warton	English Electric	12 × 12 in.	Slotted	0–1.1				
De Havilland 2 ft	Hatfield	De Havilland	2 × 2 ft		1.0–1.6				
RAE 8 ft	Farnborough	RAE	8 × 6 ft	Slotted (4 walls)	0.8–1.25	Yes	Continuous	a. 12,000 b. 8,000	
RAE 2 ft	Farnborough	RAE	2 × 1.5 ft	Slotted (4 walls)	0–1.4	No	Continuous	a. 8,000 b. diffuser suction	Provision for perforated walls
RAE 3 ft	Bedford	RAE	3 × 3 ft	Slotted (4 walls)	0.9–2.0	Yes	Continuous	a. 16,000 b. diffuser suction	
NPL 25 in.	Teddington	NPL	25 × 20 in.	Slotted	0–1.2	No	Intermittent	Injector	Solid liners for Mach 1.0–1.8
NPL 36 in.	Teddington	NPL	36 × 14 in.	Slotted	0–1.2	No	Intermittent	Injector	Solid liners for Mach 1.0–1.8
NPL 18 in.	Teddington	NPL	18 × 14 in.	Slotted	0–1.2	Yes 3 atm	Intermittent	Injector	Solid liners for Mach 1.0–1.8
NPL 20 in.	Teddington	NPL	20 × 8 in.	Slotted	0–1.2	No	Continuous	Injector	Solid liners for Mach 1.0–1.6

Transonic Wind Tunnels (contd.)

Tunnel designation	Location	Owner-operator	Test-section dimension	Test-section type	Mach number range	Pressurized	Intermittent or continuous	Horsepower a. Main b. Auxiliary	Special capabilities
U.S.A. United Aircraft 17×17 in. transonic	E. Hartford, Connecticut	UAC Research Dept.	17×17 in.	Perforated (slotted)	0.6–1.5	Yes	Intermittent	a. 1,250	
NACA 8×6 SS	Cleveland, Ohio	NACA	8×6 ft	Perforated	0.6–2.1	No	Continuous	a. 87,000 b. 8,850	Propulsion testing
NACA 8 ft transonic pressure	Langley Field, Virginia	NACA	7.1×7.1 ft	Slotted	0.6–1.3	Yes	Continuous	a. 22,500 b. 8,000	
NACA 16 ft transonic	Langley Field, Virginia	NACA	15.5×15.5 ft	Slotted	0.2–1.1*	No	Continuous	a. 60,000 b. 35,000	
NACA 19 ft	Langley Field, Virginia	NACA	16×16 ft	Slotted	0–1.15	Yes	Continuous	a. 20,000 b. 8,000†	Flutter and aeroelasticity (available 1958)
NACA Ames Unitary	Moffett Field, California	NACA	11×11 ft	Slotted	0.75–1.5	Yes	Continuous	a. 180,000 b. 19,350	
NACA 14 ft transonic	Moffett Field, California	NACA	13.5×13.5 ft	Slotted, perforated	0.6–1.2	No	Continuous	a. 110,000 b. 2,000	
NACA 6×6 ft	Moffett Field, California	NACA	6×6 ft		0.6–2.4	Yes	Continuous	a. 60,000 b. 5,950	
NACA Flutter	Langley Field, Virginia	NACA	2×4 ft		0–1.2	Yes	Continuous	a. 1,500	Flutter

* $M_{max} = 1.1$ obtained *without* auxiliary suction. † The auxiliary suction is presently *not* available.

LIST OF TRANSONIC WIND TUNNELS

Transonic Wind Tunnels (contd.)

Tunnel designation	Location	Owner-operator	Test-section dimension	Test-section type	Mach number range	Pressurized	Intermittent or continuous	Horsepower a. Main b. Auxiliary	Special capabilities
U.S.A. (contd.)									
NACA 9×12 in. SS	Langley Field, Virginia	NACA	9×12 in.		0.7–1.9	Yes	Intermittent	Blowdown	
NACA 26 in. transonic	Langley Field, Virginia	NACA	26 in. Octagon		0.7–1.4	Yes	Intermittent	Blowdown	Flutter
NACA 2×2 ft	Moffett Field, California	NACA	24×24 in.		0.0–1.4	Yes	Continuous	a. 4,000	
NACA 1×3.5 ft	Moffett Field, California	NACA	12×35 in.		0.3–1.05	No	Continuous	a. 2,000	
WADC 10 ft transonic	WPAFB, Dayton, Ohio	USAF	10 ft	Slotted, perforated	0–1.24	Yes	Continuous	a. 40,000 b. 13,000	
NAMTC 16 ft aerodynamic test division	Pt. Mugu, California	U.S. Navy, University of South California	16×16 in.	Slotted	0–1.6	Yes	Continuous	a. 15,500 b. 760	
Brown University, transonic	Providence, Rhode Island	Brown University	9×9 in.	Slotted, perforated	0.25–2.0	No	Continuous	a. 550 b. 40	
Boeing, transonic	Seattle, Washington	Boeing Airplane Co.	8×12 ft	Slotted	0–1.2	No	Continuous	a. 54,000 b. 700	

Transonic Wind Tunnels (contd.)

Tunnel designation	Location	Owner-operator	Test-section dimension	Test-section type	Mach number range	Pressurized	Intermittent or continuous	Horsepower a. Main b. Auxiliary	Special capabilities
U.S.A. (contd.)									
Cornell Aero Lab. 8 ft transonic	Buffalo, New York	USAF–CAL	8 × 8 ft	Perforated	0–1.3	Yes	Continuous	a. 22,500 b. 5,800	High-speed dynamometer for supersonic propeller testing
N. American trisonic	Los Angeles, California	North American	7 × 7 ft	Perforated	0.2–3.5	Yes	Intermittent	a. 10,000	
MIT, transonic blowdown	Cambridge, Mass.	MIT	22 in. Octagon		0.5–1.3	Yes	Intermittent	a. 150 b. 6	
N. American 16 in. SS	Los Angeles, California	North American	15.75 × 15.75 in.		0.4–3.75	No	Intermittent	a. 400 b. 14	
Ohio State University, Tunnel 1	Columbus, Ohio	Ohio State University	12 × 12 in.	Perforated	0.6–3.0	Yes	Intermittent	a. 800 b. 47	High Reynolds number
Rosemount, 12 × 16 in. transonic	Rosemount Res. Lab., Minneapolis, Minnesota	University of Minnesota	12 × 16 in.	Porous wall	0–1.3	No	Continuous	a. 3,800 b. 1,000	
AEDC 12 in. transonic	Tullahoma, Tennessee	USAF–ARO, Inc.	12 × 12 in.	Perforated, slotted	0.5–1.5	Yes	Continuous	a. 2,000	
AEDC-PWT (transonic leg)	Tullahoma, Tennessee	USAF–ARO, Inc.	16 × 16 ft	Perforated	0.5–1.6	Yes	Continuous	a. 216,000 b. 180,000	Propulsion testing

LIST OF TRANSONIC WIND TUNNELS

Transonic Wind Tunnels (concluded)

Tunnel designation	Location	Owner-operator	Test-section dimension	Test-section type	Mach number range	Pressurized	Intermittent or continuous	Horsepower a. Main b. Auxiliary	Special capabilities
U.S.A. (contd.) DTMB 18 in.	Carderock, Maryland	U.S. Navy	18 × 18 in.		0–1.2	No	Intermittent	a. 450	
DTMB 9.5 in.	Carderock, Maryland	U.S. Navy	9.5 × 9.5 in.		0–1.2	No	Intermittent	b. 450	
DTMB 7 × 10 ft transonic	Carderock, Maryland	U.S. Navy	7 × 10 ft	Slotted	0.2–1.2	Yes	Continuous	a. 24,000 b. 5,400	
Southern California Cooperative Wind Tunnel, transonic	Pasadena, California	Calif. Inst. of Technology (operator), Douglas, Convair, North American, McDonnell (owners)	7.5 × 9.75 ft	Slotted	0–1.30	Yes 4 atm	Continuous	a. 40,000	
Southern California Cooperative Wind Tunnel, subsonic	Pasadena, California		8.5 × 11.25 ft	Solid	0–1.75 except near Mach 1.0	Yes 4 atm	Continuous	a. 40,000	

389

CHAPTER 16

CONCLUDING REMARKS

THE preceding review of transonic wind tunnel testing techniques indicates that during the last fifteen years remarkable progress has been made in providing reliable and economical transonic test facilities. Numerous complex and detailed problems have been subjected to extensive and thorough investigation, both in the theoretical as well as experimental fields. As a result of these efforts, the reliability and the test limits of modern transonic wind tunnels have been enough improved to provide the aerodynamic data required for the design of airplanes and other vehicles.

The preceding review also indicates that in many areas compromise solutions in wind tunnel design have been accepted in order to arrive at timely, economical, and feasible systems. For example, the slotted or perforated walls of modern wind tunnels satisfy the requirements of a complex model only as far as the mean boundary conditions are concerned. Undoubtedly, further research is required in order to provide greater knowledge of the detailed aerodynamic laws involved. Furthermore, even in the present state of the art when the relative size of a model in a transonic wind tunnel is comparable with that of models in regular subsonic wind tunnels, more information is desired to determine whether, in tests of complex models, the relative wind tunnel size can be reduced further than is the practice today.

It is also apparent that most transonic testing techniques have been evolved by experimental means. Nonetheless, despite the fact that wind tunnels are *experimental* tools, it is still desirable to develop the supporting theory. However, a formidable obstacle to such development exists due to the well-known difficulties produced by the mixed-type transonic flow, and due to the interaction between the potential flow and the boundary layer.

The transonic wind tunnel has been in the forefront of test facilities during the last several years while the first research or military aircraft were being developed to fly safely through the transonic range. This period of airplane development is past. The forefront of the effort lies now in the high supersonic and hypersonic fields as far as military airplane development is concerned. However, even for advanced supersonic aircraft and space vehicles transonic aerodynamic problems remain to be solved during the launch and re-entry phases. Thus, even in the military field, the continuous availability of transonic wind tunnels is required. This is particularly true because no relief is expected from a satisfactory transonic theory in the foreseeable future.

As in previous periods of airplane development, commercial aircraft will gradually move into the speed range which has been successfully explored and left behind by military aircraft. Since economical operation is considerably more important for commercial than for military aircraft, transonic

CONCLUDING REMARKS

wind tunnels will experience another period of extraordinary significance. This second period will end only after the commercial airplane has left the transonic speed range behind and moved towards supersonic, or possibly even hypersonic, velocities. However, because these aircraft, too, must fly through the transonic speed range during their acceleration or deceleration phase the transonic wind tunnel will always remain as a necessary supporting test facility.

INDEX

AEDC (see Arnold Engineering Development Center)
Aerodynamic Research Institute, Goettingen (Germany), 7
AGARD calibration Model B, 211, 222
AGARD calibration Model C, 218, 222–226
Aircraft Research Association (Gt. Britain), 9 × 8 ft. tunnel, 128, 130, 306–310, 324, 380
Airfoils, two-dimensional,
 flow calculations in,
  closed tunnel, 51–53
  free flight, 51–53
 flow disturbances in slotted tunnels, 274
 lift coefficient effect of, 119
 pressure distribution of,
  in closed tunnel, 53, 111, 112
  in slotted tunnel, 114–117
 test results from,
  closed tunnel, 109–113
  slotted tunnel, 113, 116–119
  slotted tunnel compared with theory for $M = 1$, 113–116
 vortex formation, 59
Airplane models (see also AGARD calibration models B and C, and Wing-body configurations),
 in flow non-uniformity, 200–202
 in perforated tunnels, 223, 226–230
 in wind tunnel and free flight, comparison, 229–233
Airplanes,
 Bell X-1, 24, 25
 comparison of free flight and tunnel data, 229–234
 DC-2, 4
 drag, 2
 He-70, 6
 Heinkel 178, 9
 in free flight,
  sonic flow, 37–39
  subsonic flow, 32–37
  supersonic flow, 39
 Messerschmitt 262, 9
 North American F-110A, 27, 29
 P-51D, 15
 Wright Brothers', 3
Arnold Engineeering Development Center,
 1 ft transonic model tunnel, 70, 123, 157, 178, 184, 185, 190, 191, 193–195, 197, 203, 217, 256, 265, 269, 270, 272, 275, 305, 311, 312, 322, 323,

Arnold Engineering Development Center
—*continued*
 325, 327, 331, 346, 347, 349, 351–354, 357–361, 363–373, 375, 379
 16 ft. transonic wind tunnel, 27, 28, 123, 190, 195, 217, 311–312, 326–330, 379, 380

Bodies of revolution (see Cone-cylinder, Cone-ogive, and Cone-ogive-cylinder models),
Boeing Airplane Company, (4·8 × 7·2 in.) × 9 in. wind tunnel, 376, 377
Boundary layer growth,
 in perforated tunnels, 334–352
 in slotted tunnels, 352, 355
Brown University,
 4 × 8 in. wind tunnel, 337, 339
 7-1/2 × 9 in. wind tunnel, 109

Chalais-Meudon (France),
 open-circuit wind tunnel, 5
Choking, wind tunnel, 10, 11, 23
 boundary layer effects on, 18, 19
 definition of, 75
 in closed tunnel, 48–52, 90, 91, 318
 in partially open test sections, 319, 320–322
 in slotted tunnels, 115, 211
 prevention of by re-entry flaps in diffuser, 373
Compressibility effect, 7, 54, 55
Compression waves, 25, 26, 43, 136–138, 144, 155–157, 167, 169, 178–181, 301, 302
Cone-cylinder models,
 flow disturbances,
  calculations, 167–170
  in perforated tunnel, 170–174
  in perforated-slotted tunnel, 176–178
  in slotted tunnels, 174–176
 test results,
  in perforated tunnels, comparison, 123–125
  in perforated tunnels, 178–182, 184
  in perforated tunnel with inclined holes, 184, 190, 191, 197
  in slotted tunnels, 184, 185
   with inclined-hole perforated cover plates, 187, 189
   with perforated cover plates, 187, 188
  in supersonic flow, 167, 168

INDEX

Cone-ogive models,
 NACA RM-10, 170, 71, 177
Cone-ogive-cylinder models, pressure distribution in inclined-hole perforated tunnels, 195, 196
Cornell Aeronautical Laboratory, Inc., 147
 1 × 1 ft. tunnel, 227–230
 3 × 4 ft. tunnel, 227–233
 7 ft. tunnel, 46, 47
 12 ft. tunnel, 228, 229
Cross-flow characteristics,
 in perforated tunnels, 171–174, 183, 240, 248–262, 266–270, 275–279
 in perforated tunnels (inclined hole), 183, 238, 240, 263–270, 273, 279
 in porous tunnels, 237
 in slotted tunnels (longitudinal) 174–178, 270–275
 in slotted tunnels (transverse), 240–248

Diffusers, wind tunnel,
 suction, 362–364
 with re-entry flaps, 371, 374–377
Drag, aerodynamic, 2, 4
Drop tests, 11, 12
DVL, Berlin (Germany),
 2·7 m diam. high-speed wind tunnel, 8, 9
 drop tests, 11

"educated hole", 254–256
Expansion waves,
 cancellation, 25, 26, 140, 155–158, 178
 reflection, 43, 51, 136–140, 144, 149, 150, 167–169, 173, 179–181, 186
Flow establishment,
 general, 282, 283
 in perforated tunnels, 283, 285–296, 300–312
 in slotted tunnels, 282, 296–300, 312–318, 323, 324

Institut Aéronautique, Saint-Cyr (France),
 closed wind tunnel with movable inserts, 17

Kutta condition, 241, 247

Missiles, Sergeant, 28

NACA RM-10 model (see Cone-ogive models)
National Advisory Committee for Aeronautics, 7, 9, 14, 23, 61, 281, 351
 Ames 16 ft tunnel, 17, 24
 Langley Field,
  annular transonic tunnel, 13, 14
  3 × 3 in. tunnel, 350
  6·25 × 4·50 in. tunnel, 209, 210
  12 in. tunnel, 378

National Advisory Committee for Aeronautics—*continued*
 Langley Field—*continued*
  7 × 10 ft tunnel, 16, 209, 211
  8 ft tunnel, 211, 213–215, 231, 232, 316, 317, 374, 375, 378, 380
 Lewis 8 × 6 ft tunnel, 27
  16 ft tunnel, 120, 121, 123, 213–215, 296, 314–316, 352, 378, 380
  30 × 60 ft tunnel, 4, 5
National Physical Laboratory (Gt. Britain),
 7-1/2 × 3 in. tunnel, 116
 20 × 8 in. tunnel, 116
Nozzles, wind tunnel,
 contouring, 4
 sonic,
  power requirements with, 309, 312, 323
  wave patterns in, 303
 supersonic, 305
  flow establishment with, 309, 312, 323
  power requirements with 370–372

Ottobrun (Germany),
 7 ft high-speed wind tunnel, 46

Perforated test section walls (straight hole),
 AGARD model B test results, comparison with interference-free data, 217, 219–222
 AGARD model C test results, comparison with interference-free data, 218, 219, 223–227
 airplane model aerodynamic characteristics, comparison with flight test results, 230–232
 comparison with interference-free data, 222, 227–229
 boundary layer growth,
  displacement thickness, 348–352
  influence of open-air ratio, 349
  influence of wall setting, 351
  mass-flow removal requirements, comparison between theory and experiment, 350, 351
  experimental, 343–346
  theoretical, 337–339, 342–345
  refined theory, 339–343
  theoretical calculations, 334–337
  velocity profiles, 346, 347
 calibration, 329–331
 characteristics,
  basic, 86
  in isentropic and viscous flows, 88
  of thick walls, 238–239
  of thin walls, 238–240
  pressure drop through cross flow, 86–88
 cone-cylinder, test results, 178–182
 cross-flow characteristics,
  at M = 0·90, 183
  comparison with inclined-hole wall, 266

INDEX

Perforated test section walls, (straight hole) —*continued*
  cross flow characteristics—*continued*
    flow disturbances produced by holes, 275–279
    influence of boundary layer thickness, 267–270
    influence of hole size, 260, 261
    influence of Mach number, 251–255, 261, 262
    of various types of perforations, 249–251, 254
    pressure drop calculations, 86, 240, 250
    test results, 248–256
    with "educated holes", 254–256
  disadvantage in subsonic flow, 25, 89, 91
  early development, 25–29
  expansion wave cancellation, 157
  flow behind lifting wing, 107, 108
  flow distortion, 88–91
  flow disturbances, one wall, due to model displacement, 98–100
  flow disturbances, two walls, calculation, 100, 101
    due to model displacement, 101–103
  flow establishment, subsonic, Mach number distribution, 285–296
    mass flow removal requirements, 320, 325
    test section venting, 283
  flow establishment, supersonic,
    effects of mass flow removal, 301, 303
    theory, 300–303
    with sonic nozzle, 303–308
    with supersonic nozzle, 309–312
  fuselage test results, comparison with closed tunnel data, 123, 124
  model disturbances compared with cross-flow characteristics, 171–174
    cone-cylinder model, 167–174
    cone-ogive-cylinder model, 171, 173
    NACA RM-10 model, 170, 171, 172
    three-dimensional models, general, 174
    wedge-flate-plate model, 171, 172
  performance, 25–30
  power requirements,
    full scale tunnels, 379, 380
    influence of auxiliary plenum chamber suction, 356, 357
    influence of hole size, 368, 369
    influence of model size, 369
    influence of open-area ratio, 366–368
    influence of wall setting, 364–366
    model tunnels, 379
    with auxiliary plenum chamber and/or diffuser suction, 362–364
    with sonic nozzles, 370, 371
    with supersonic nozzles, 370, 371
    without auxiliary plenum chamber suction, 358–362

Perforated test section walls (straight hole) —*continued*
  shock wave cancellation theory, 142–144
  shock wave reflections, 207, 209
    boundary layer influence, 153–155
    partial, 155, 156
    test results, qualitative, 147–149
    test results, quantitative, 149–153
  velocity corrections, circular tunnel, calculation, 103, 104
    model displacement and lift, 104–106
  velocity corrections, one wall, 92–97
    Mach number influence, 97, 98
    wing without lift, 98–100
  wing-body test results, 128–130
    comparison with closed tunnel data, 125–128
    comparison with interference-free data, 204, 207
Perforated test section walls (inclined hole),
  calibration, 326–331
  characteristics, 181–183
  cone-cylinder test results, 123–125, 181–184
    at various Mach numbers, 190, 191
    with various blockage ratios, 192–197
    with various open-area ratios, 191, 192
  cone-ogive-cylinder test results, 195, 196
  cross-flow characteristics,
    at M = 0·90, 183
    comparison with straight hole wall, 262
    flow disturbances produced by holes, 278, 279
    influence of boundary layer thickness, 267–270
    influence of hole diam vs. plate thickness, 265, 267
    influence of hole inclination angle, 263–265
    influence of Mach number, 269
    pressure drop of thick walls, 238–240
    pressure drop of thin walls, 238, 239
  definition of, 269
  expansion wave cancellation, 157, 158
  model disturbances,
    compared with cross-flow characteristics, 171–174
    cone-cylinder model, 172, 173
    NACA RM-10 model, 173, 174
    three-dimensional models, general, 174
    wedge-flat-plate model, 171, 172
  power requirements,
    full-scale tunnels, 378–381
    model tunnels, 378, 379
  wall interference, wave reflection, 157, 158
  wing-body test results, 125–130
Porous test section walls,
  cross-flow characteristics, 237
  performance, 25
  pressure drop in cross flow, 88

INDEX

Power requirements, wind tunnel, 11
 AEDC 16 ft transonic tunnel, 27, 379
 air bleed in scheme for, 21
 for perforated wall tunnels, 356–372, 378, 379
 for slotted wall tunnels, 371–381
 for Wieselberger-type tunnels, 57, 58
 in full-scale tunnels, 378–381
 in model tunnels, 378, 379
Propeller dynamometer installation, 47

Rotating rigs, 12, 14

Saint-Cyr (France) (see Institut Aéronautique, Saint-Cyr)
Screens, turbulence, 4, 19
Secondary flow fields, 185–187
Shock wave cancellation, 25
 in inclined hole perforated tunnels, 157, 158
 in oblique supersonic flow, 161
 theory for perforated tunnels, 142–144
 theory for slotted tunnels, 140, 141
Shock wave reflection, 17, 26, 43
 cancellation in perforated tunnels, compared with theory, 147–153
 cancellation in slotted tunnels, compared with theory, 144–147
 effect of boundary layer in perforated tunnels, 153–156
 effect of open-area ratio, 151, 152
 effect of outflow and inflow, partially open tunnels, 139, 140
 in closed tunnel, 49–51, 136
 in open-jet tunnels, 136
 in partially open tunnels, 136–139
Shock waves,
 delayed movement at sonic speed, 134
 fluctuations, 59
 stabilization with sonic throat, 60
Slotted test section walls (longitudinal)
 airfoil test results, 113–115
  comparison with closed tunnel data, 116–120
  comparison with open-jet tunnel data, 118–120
  comparison with theory, $M = 1$, 113, 115, 116
 basic characteristics, 237
 blockage correction for 2-dimensional airfoil, 117, 118
 bodies of revolution test results, compared with free-fall data, 120–123
 boundary layer growth, displacement thickness, 354, 355
  velocity profiles, 352–354
 cone-cylinder test results, 184–186

Slotted test section walls (longitudinal)
 —*continued*
 cross-flow characteristics,
  flow curvature effect, 273–275
  in oblique flow, many slots, 271, 273
  in oblique flow, one slot, 270–272
  pressure drop, one slot, 70, 71
 early development, 21–24
 effectiveness,
  with perforated plates, 68, 70, 71
  with protruding slats, 68
  without lift, Mach number influence, 74
 flow behind lifting wing, 106, 107
 flow establishment, subsonic,
  Mach number distribution, 282, 296–300
  test section venting, 296–300
 flow establishment, supersonic
  influence of slot shape, 314, 315
  influence of wall divergence, 316
  mass flow removal requirements, 324, 325
  with perforated cover plates, 316–318
  with sonic nozzle, 312–318
  with straight taper slots, 312–314
  with supersonic nozzle, 312, 323, 324
 model disturbances, 3-dimensional, 174, 175
 oblique supersonic flow, 159, 160
  momentum considerations, 161
  shock wave cancellation, 161
 power requirements, 58
  full-scale tunnels, 378, 380
  model tunnels, 378
  with mass flow removal, 371–373
  with re-entry flaps, 371, 374–377
 shock wave cancellation theory, 140–142
 shock wave reflection, test results, 144–147
 velocity corrections,
  as function of open-area ratio, 61, 67
  open-area ratio as function of slot number, 66
  open-area ratio as function of slot spacing, 67
  slots with constant width, 71, 72
  slots with varying width, 72–74
 wall interference,
  influence of slot width, 65–67
  pressure build-up, 62–65, 72, 73
  with few slots, 57, 59, 61
  with many slots, 61–63
 wing test results,
  comparison with interference-free data, 209–213
 wing-body test results,
  comparison with interference-free data, 203–206, 213–216
  comparison with perforated tunnel data, 207, 209
  with inclined hole perforated plates, 187, 189

INDEX

Slotted test section walls, (longitudinal)—*cont.*
  with lifting wing, circular tunnel,
    flow inclination corrections, 76–83
    lift influence, 75, 76
    Mach number influence, 84
  with movable inserts, 17
  with perforated plates,
    characteristics, 68–71
    cone-cylinder test results, 187, 188
    flow establishment with, 316, 318
    model disturbances, 175–178
  with protruding slats, 68, 69, 175
Slotted test section walls (transverse),
  cross-flow pressure drop, 247–249
  oblique supersonic flow, 162, 163
    momentum considerations, 163, 164
    shock wave cancellation, 163
    shock wave reflection, partial, 164, 165
  one transverse slot,
    pressure drop, thick wall, 247
    pressure drop, thin wall, 240–243, 246, 247
  several transverse slots, 243–245

Test techniques, transonic,
  drop tests, 11
  free-flight tests, 11, 12
  rotating rigs, 12–14
  transonic bump, 15–17
  wing-flow method, 14, 15
Trefftz theorem, 76, 107
Test section walls, transonic (see also Perforated, Porous, and Slotted test section walls),
  basic requirements, 236
  flow establishment problems, 282, 283
  flow non-uniformities in, 200–202
Test sections, transonic (see Perforated, Porous, and Slotted test section walls and Wind tunnels, closed, open-jet, and partially open).
Transonic bump, 15–17

United Aircraft Corporation,
  flow through "educated holes", 255–256
  shock wave cancellation studies, 144, 147, 149
  two-stream wind tunnel, 19–21

WADC (see Wright Air Development Center)
Wall interference,
  corrections,
    at sonic speed, 130–134
    in closed tunnels, 44–46, 48, 49, 51, 54, 55
    in open-jet tunnels, 41–44, 49–51, 53, 54
    in partially open tunnels, 41, 54, 56
    in test sections, subsonic flow, 53–56

Wall interference—*continued*
  expansion wave cancellation, 157, 158
  flow disturbances,
    three-dimensional models, 167–198
    two-dimensional models, 167–169, 171, 172
  shock wave cancellation,
    in perforated tunnels, 142–144, 147–153, 155, 156
    in slotted tunnels, 140–142, 144–147, 163
  shock wave reflection,
    in closed test sections, 136
    in open-jet test sections, 136
    in partially open test sections, 136–139
    in slotted tunnels, 155–161
Wedge-flat-plate models, flow disturbances,
  calculations, 167, 168
  in perforated tunnel, 171, 172
  in shock wave cancellation studies, 144–150
Whirling-arm method (see Rotating rigs)
Wieselberger-type wind tunnel, 57, 58, 61
Wind tunnels (see also Perforated, Porous, and Slotted test section walls)
  closed,
    air bleed-in effects, 21, 22
    airfoil test results, 109–113
      comparison with open-jet tunnel data, 118–120
      comparison with slotted tunnel data, 116–120
    boundary layer effects, 18–21
    choking, 18, 19, 48–52, 75, 90
    flow establishment, mass flow removal requirements, boundary layer growth, 318
    fuselage test results, comparison with perforated tunnel data, 123, 124
    model size effects, 10
    power requirements, 58
    pressure wave control, 61
    shock wave reflections, 136
    two-stream tunnel, 19–21
    wall interference, subsonic flow, contoured walls, 46
      Mach number influence, 46, 48, 49
      straight walls, 44, 45
      velocity corrections, 45, 54, 55
    wall interference, supersonic flow, Mach number influence, 49, 51
      wave reflections, 49–51
    wing-body test results, comparison with perforated tunnel data, 125–128
    with contoured walls, 46
    with movable sidewalls, 17, 178
  full-scale, power requirements, 380, 381
  high-speed,
    closed, 7–11
    compressibility effect in, 7
    open jet, 11

INDEX

Wind tunnels—*continued*
  low-speed,
    after 1930, 2–4
    before 1930, 2
  model tunnels, power requirements, 378, 379
  open-jet,
    airfoil test results,
      comparison with closed tunnel data, 118–120
      comparison with slotted tunnel data, 118–120
    power requirements, 58
    shock wave reflections, 136
    wall interference,
      Mach number influence, 42, 44
      subsonic flow, 41, 42
      velocity corrections, 41, 53, 54
      wave reflections, 43, 49–51
  partially open (see also Perforated, Porous, and Slotted test section walls),
    boundary conditions,
      with cone-cylinder model, 167–170
      with NACA RM-10 model, 170, 171
      with wedge-flat-plate model, 167, 168
    calibration, 325–331
      subsonic flow, 326
      supersonic flow, 327
    cross-flow characteristics, flow requirements, 236
    flow establishment, mass flow removal requirements, 304
      boundary layer growth, 319
      comparison of slotted and perforated test sections, 324, 325

Wind tunnels,—flow establishment—*cont.*
      influence of convergence and divergence, 322
      influence of hole size, 320, 322, 323
      influence of open-area ratio, 322
      with sonic nozzle, 303–305
    shock wave reflections, 136–139
      influence of outflow and inflow, 139, 140
    wall interference, velocity corrections, 41, 54, 56
    with supersonic nozzles, 323, 324
  sonic, 130–134
  transonic, list of, 383–389
Wing-body configurations,
  aerodynamic characteristics in slotted tunnels, 213–217
  force measurements in,
    perforated tunnels, 128–130, 204, 207–209
    slotted tunnels, 203, 204
  test results, perforated tunnel, 129–134
Wing-flow technique, 14, 15
Wings, lifting,
  downwash corrections,
    perforated wall, 107, 108
    slotted wall, 76, 106, 107
  flow inclination, corrections for, 76–83, 105
Wings, semi-span, aerodynamic characteristics in slotted tunnel, 209–213
Wright Air Development Center, 150, 156
  6 in. supersonic wind tunnel, 145, 149
  10 ft transonic wind tunnel, 17, 18, 21, 22, 69, 203, 204, 317, 318
Wright Brothers, airplane, 3
  wind tunnel, 2, 3